W9-BEB-111

A Handbook of
Critical Approaches
to Literature

A Handbook of Critical Approaches to Literature

THIRD EDITION

WILFRED L. GUERIN

Louisiana State University in Shreveport

EARLE LABOR
LEE MORGAN

Centenary College of Louisiana

JEANNE C. REESMAN

University of Texas at San Antonio

JOHN R. WILLINGHAM

The University of Kansas

New York   Oxford
OXFORD UNIVERSITY PRESS
1992

Oxford University Press

Oxford   New York   Toronto
Delhi   Bombay   Calcutta   Madras   Karachi
Kuala Lumpur   Singapore   Hong Kong   Tokyo
Nairobi   Dar es Salaam   Cape Town
Melbourne   Auckland

and associated companies in
Berlin   Ibadan

Copyright © 1992 by Oxford University Press, Inc.

Copyright © 1966, 1979 by Wilfred L. Guerin, Earle Labor,
Lee Morgan, and John R. Willingham.

Published by Oxford University Press, Inc.
200 Madison Avenue, New York, New York 10016

Oxford is a registered trademark of Oxford University Press

All rights reserved. No part of this publication may be reproduced,
stored in a retrieval system, or transmitted, in any form or by any means,
electronic, mechanical, photocopying, recording, or otherwise,
without the prior permission of Oxford University Press.

Library of Congress Cataloging-in-Publication Data
A Handbook of critical approaches to literature / Wilfred L. Guerin
. . . [et al.] — 3rd ed., rev. and enl.
p. cm.   Includes bibliographical references and index.
ISBN 0-19-506948-X
1. Criticism.   I. Guerin, Wilfred L.
PN81.G8   1992
801'.95—dc20   91-37865

2 4 6 8 9 7 5 3

Printed in the United States of America
on acid-free paper

TO OUR FIRST CRITICS

Jeannine Thing Campbell
Carmel Cali Guerin
Rachel Higgs Morgan
Sylvia Kirkpatrick Steger
Grace Hurst Willingham

Preface

This book, now in its third edition, has been from the first the product of our shared conviction that the richness of great literature merits correspondingly rich responses—responses that may be reasoned as well as felt. Corollary to this conviction is our belief that such responses come best when the reader appreciates a great work from as many perspectives as it legitimately opens itself to. Nothing, of course, replaces the reader's initial *felt* responses: the sound of poetry on both the outer and the inner ear; the visions of fiction in the mind's eye; the kinesthetic assault of "total theater." But human responses seldom remain dead-level: they reverberate through multiple planes of sensibility, impelled toward articulation—in short, toward criticism. To answer the inevitable classroom questions, "Why can't we simply *enjoy* this poem [story, novel, play]? Why must we spoil the fun by criticizing?" we would rejoin, "The greatest enjoyment of literary art is never simple!" Furthermore, we should recall, in T. S. Eliot's words, "that criticism is as inevitable as breathing, and that we should be none the worse for articulating what passes in our minds when we read a book and feel an emotion about it."

Eliot's reminder was instrumental in the genesis of the first edition of *A Handbook of Critical Approaches* in the early 1960s, when the four original coauthors were colleagues in the English Department at Centenary College of Louisiana. At that time we had become sensitive to the problems of teaching literary analysis to young college students in the absence of a comprehensive yet elementary guide to some of the major critical approaches to works of literature. No work of that sort existed at the time, yet students clearly could have profited from a more formalized and contemporary introduction to the serious study of literature than they generally had received in lower levels of education. We found that most lower– and many upper–division students were entering and emerging from courses in literature still unenlightened about the most rewarding critical techniques that a keen reader could apply to good imaginative writing. Even students whose exposure to literature had been extensive often possessed only a narrow and fragmented concept of such interpretive approaches. Consequently, one of our first aims—then and now—has been to help establish a healthy balance in the student's critical outlook. We still fervently believe that any college or university student—or, for that matter, any advanced high school student—should have at hand the main lines of the most useful approaches to literary criticism.

With these assumptions in mind, we marked off our areas of concern and laid claim to fill the need we sensed. We have been gratified with the success of that claim, indicated by the acceptance of the book by our professional colleagues and by thousands of students throughout the land and abroad. (The book has now been published in Spanish, Portuguese, Japanese, and Chinese, and in English in Korea.) However, there has also been an acceptance we did not anticipate. Our original concern was to offer critical approaches to students in the early years of college work, but we have found that in instance after instance the book is being used at upper–division levels and in graduate classes. Even so, this extended use has not precluded the book's acceptance by numerous high school teachers as well.

We hope that in this third edition we have preserved that versatility, and we have worked strenuously to improve upon it. Since the publication of the second edition in the late seventies, we have witnessed a veritable explosion of critical theories, along with a radical expansion and revision of the literary canon. These extraordinary developments have prompted corresponding revisions in our handbook: two new chapters have been added on "Feminist Approaches" and "Structuralism and Poststructuralism." (Our earlier chapter on "The Exponential Approach" has been condensed and combined with "The Formalistic Approach" to make room for these other approaches.) Significant revisions and additions have been made in our chapter on "Additional Approaches." There, our section on Marxist criticism has been radically expanded to include more recent Neo-Marxist theories; and we have added substantial new sections on Dialogics, Hermeneutics, the New Historicism, and (coming full circle to our opening Prologue on "The Precritical Response") Reader-Response Criticism. In this seventh chapter we have tried to survey thirteen approaches to literature in such a way not only that the least sophisticated reader can become aware of approaches other than the six which form the core of our handbook but also, conversely, that the more knowledgeable student of literature can find both orientation and bibliographical guides to seeking out much more information about these several approaches. Perhaps our most important change has been the addition of a new coauthor for this third edition, who brings to our collective experience a fresh vision of contemporary criticism.

Despite these changes, our aim in this new edition is still much the same as it was in the first and second editions: to provide a basic introduction to the major critical-interpretive perspectives that a reader beginning a serious study may bring to bear on literature. This book describes and demonstrates the critical tools that have come to be regarded as indispensable for the sensitive reader; these tools are what we call "approaches." Furthermore, because this is a *handbook* of critical approaches, we have tried to make it suggestive rather than exhaustive. We make no claim to being definitive; on the contrary, the book's value lies, in part, in opening the student's eyes to the *possibilities* in literature and criticism. Today we read much about heuristics, the process of discovery. This sense of discovery was important in the previous editions, and it continues to be important here.

But heuristics can be guided, and for that reason we have selected six main approaches to literary criticism, all of which we consider viable not only for the critical expert but also for the critical neophyte. These approaches constitute the first six chapters of our handbook. Each of them begins with an introduction to and definition of a particular interpretive approach, usually followed by a detailed practical application of that approach to the same four major works, one each from the following genres: novel, short story, poetry, and drama (two British and two American). Each chapter also has many other literary works cited as occasional illustrations, thereby effectively extending the book's application beyond the four works treated more extensively, while at the same time permitting the student to apply the various critical approaches to the works thus briefly mentioned. There is no rigid sequence from chapter to chapter, and the four major works are not all treated with the same degree of detail in each chapter, since not all works lend themselves equally well to a given approach. Consequently, one important aspect of our treatment of critical reading should be the student's recognition of the need to select the most suitable approach for a given literary work.

Following a discussion of the "common reader's" precritical response, chapter 1 considers the cluster of perspectives generally accepted as traditional—the biographical, the historical, the moralistic, and the philosophical—with some attention to textual matters, to genre, and to the time-honored technique of interpretive summary. Chapter 1 weighs both the advantages and the limitations of such approaches. Chapter 2 deals with the formalistic approach, which has come to be especially associated with the New Criticism. Chapter 3 presents a treatment of psychology in literature, concentrating on Freudian criticism. Chapter 4 moves into the realm of Jungian archetypal patterns and cultural myths as they are manifested in literature. Chapter 5 serves as an extensive introduction to what has become perhaps the most complex of our major approaches, namely, the feminist approach, which is extraordinarily rich in its sociocultural and political as well as its aesthetic implications, comprising such widely varying attitudes as those of Marxists, Neo-Freudians, and deconstructionists as these are sometimes amalgamated with the reformation of sexist attitudes in literary criticism. Chapter 6 elucidates what has come to be regarded as the most esoteric and controversial of contemporary critical approaches: structuralism and poststructuralism (particularly deconstruction); we hope that we bring more light (with less heat) to these challenging theories. Since chapter 7 surveys thirteen different approaches, we have not attempted to treat the four major works chosen for detailed analyses in the six preceding chapters.

Those four works were chosen because of their rich potential for interpretation, because they will make the beginning student aware of the joys of reading at increasingly higher levels of ability, and because they are generally important in introductory courses in literature. Two of them—*Adventures of Huckleberry Finn* and *Hamlet*—are easily available in paperback, if not in the student's literature anthology. The other two—"To His Coy Mistress" and "Young Goodman Brown"— are included in this book. Regardless of the availability of these four works, we hope that this book will serve primarily as a model or guidebook for the interpretation of *other* literary works. In short, this handbook is intended as a complementary text, best used in conjunction with an anthology or a set of paperbacks.

Bibliographical references are provided throughout the various chapters as well as at the end of the handbook. These include the most helpful books and articles related to the respective interpretive approaches, most of which are generally available in academic libraries.

This handbook may be read from cover to cover as a continuous unit, of course, but it has been organized for both flexibility and adaptability. For example, although it is primarily organized by "approaches" rather than genres, at the beginning of the course the instructor may assign the introductory section of each chapter, later assigning the section of each of the six main chapters that deals with a certain genre. Thus, the instructor who decides to begin with the short story may assign the introductory section of each of the chapters and then the six discussions of "Young Goodman Brown." Another possible use of this book is to have students read several literary works early in the term and discuss them in class without immediate recourse to this handbook. Then they might read it, or pertinent sections of it, and bring their resulting new insights to bear on the literature read earlier, as well as on subsequent readings. This double exposure has the advantage of creating a sense of discovery that awaits the perceptive reader.

Finally, our debt to the canon of literary criticism and scholarship is obvious, and we acknowledge it with gratitude. Our bibliographies suggest the breadth ond depth of critical scholarship and teaching upon which this volume is based. We once again acknowledge the many helpful suggestions that were made for the first edition. For these we thank Laurence Perrine, William B. Allmon, James A. Gowen, Donald F. Warders, Arthur Schwartz, Richard Coanda, and James Wilcox. Assistance in the preparation of the first edition was given by Kathleen Owens, Czarena Stuart, Irene Winterrowd, Yvonne B. Willingham, Mildred B. Smith, Melinda M. Carpenter, Alyce Palpant, and Jeanette DeLine. To these names we added the following, who helped prepare the second edition: Betty Labor, Ruby George, and the library staffs of Centenary College of Louisiana and Louisiana State University in Shreveport; Dean Mary McBride and Professor Robert C. Leitz III, of Louisiana State University in Shreveport; former colleague Stephen J. Mayer; and the students of Earle Labor's literary criticism seminar at the University of Aarhus in Denmark. For this third edition we wish to thank especially Ellen Brown, Bernard Duyfhuizen, Michael L. Hall, Jeff Hendricks, Kyle Labor, Phillip Leininger, Bettye Leslie, Teresa Mangum, Barry Nass, Steven Shelburne, Frederick C. Stern, and Keith G. Thomas. Finally, we are most deeply indebted to Patty Roberts, without whose untiring efforts in helping to prepare our manuscript this edition might never have come forth.

W.L.G.
E.L.
L.M.
J.C.R.
J.R.W.

Contents

A Handbook of
Critical Approaches
to Literature

The Precritical Approach

No less a professional critic than Dr. Samuel Johnson in the *Life of Gray* made the precritical response the ultimate criterion in determining literary merit:

> I rejoice to concur with the common reader; for by the common sense of readers, uncorrupted by literary prejudices, after all the refinements of subtility and the dogmatism of learning, must be finally decided all claim to poetical honours.

Now, Dr. Johnson's common reader in eighteenth-century England probably bears little resemblance to what would have to be defined as a common reader in contemporary, industrial, urban, democratic, mass-educated America. But if we make some allowance for changes in taste and the literary curriculum from the eighteenth to the twentieth century, we can perhaps agree that for the most part we are talking about relatively similar types of readers.

It was inevitable that literature should one day become a part of the academic curriculum. Anything that could so move and interest large numbers of people, including the most cultivated and enlightened, and that had such obvious and pronounced didactic uses was in the judgment of academicians bound to be worthy of intellectual analysis. Such a view may well have motivated educators to make literature an academic subject: it "taught" something; it was a source of "knowledge." The study of modern literature as a subject in school, however, is relatively recent. For centuries in western Europe, only the literature of classical antiquity was thought to have sufficient merit for systematic study. In any event, once literature was established in the curriculum, it was subjected to the formal discipline of criticism, which ultimately consisted of taking it apart (and putting it back together again) to see how and why as well as what it was and meant.

A popular opinion has it that because literary "technicians" have so rigorously pursued their studies, many "common readers" shy away from the rich and pleasurable insights that balanced, intelligent literary criticism can lead to. Whatever the reason, many students not innately hostile to literature may well have come almost to despise it, certainly to dread it in school, and to think of it in the same category as quantum physics, Erse philology, macroeconomic theory, or—worse yet—English grammar.

Some professional critics have apparently sympathized with this negative view of the effects of criticism and have espoused subjective and appreciative critical criteria that bear scrutiny in such a discussion as this. Susan Sontag, for example, in "Against Interpretation" mounted a frontal attack on most kinds of current criticism, which, she maintained, actually usurp the place of a work of art (*Evergreen Review*, 1964; rpt. in *Against Interpretation and Other Essays* [New York: Dell, 1969]: 13–23). In her free-swinging assault Sontag is at once defending a precritical response somewhat similar to the one elaborated in this prologue and asserting that critical analysis is the desecration of an art form. She sees art as the uninhibited creative spirit in action, energetic and sensual. She sees criticism—at least, most of it—as a dry-as-dust intellectual operation, the intent of which is to control and manage art and the method of which is to reduce the work of art to content and then to interpret that. Her approach is highly provocative and stimulating. Yet despite some last-minute disclaimers that she is not condemning all critical commentary and some advice to critics to pay more attention to form, it is difficult to escape the conclusion that in her opinion interpretation impoverishes art and that its practice for the last half century by most academic and professional critics has been unquestionably harmful. She concludes with the pronouncement that "in place of a hermeneutics we need an erotics of art."

Such a view would seem to place her in general agreement with Leslie Fiedler, who, addressing a national convention of the College English Association in Philadelphia, advocated "ecstatics" as a response to literature. Professor Fiedler would make the gut reaction the be-all and end-all of art. The traditionally accepted standards and classics are in his view elitist, academic opinions and productions that have been forced on the reading public, who demonstrably prefer sentimental literature, horror stories, and pornography—all of the Pop variety. Such popular writings produce almost exclusively emotional effects—particularly feelings of pathos, terror, and sexual titillation. They cause readers, says Fiedler, "to go out of control, out of [their] heads." He continues by pointing out that

> we do have a traditional name for the effect sought, and at its most successful achieved, by Pop; the temporary release from the limits of rationality, the boundaries of the ego, the burden of consciousness; the moment of privileged insanity[;] that traditional name is, of course, "Ekstasis," which Longinus spoke of in the last centuries of the Classic Era, not in terms of Popular Art or High Art, which in fact cannot be distinguished in terms of this concept; but of *all* art at its irrational best—or, to use his other favorite word, most "sublime."

That political principles underlie Fiedler's position is clear in his closing remarks:

> Once we have made *ekstasis* rather than instruction or delight the center of critical evaluation, we will be freed from the necessity of ranking mass-produced and mass-distributed books in a hierarchal order viable only in a class-structured society, delivered from the indignity of having to condescend publicly to works we privately relish and relieved of the task of trying to define categories like "high" and "low," "majority" and "minority" which were from the beginning delusive and unreal. ("Is There a Majority Literature?" *The CEA Critic* 36 [May 1974]: 3–8)

Both Sontag's and Fiedler's points of view are instructive for readers interested in familiarizing themselves with the variety of critical responses to a literary work. Whether one subscribes to them in their entirety or in part or disagrees with them categorically, they are invigorating polemics that can spark further intellectual exchange on the issue in the classroom, in the learned journals, and in magazines and newspapers.

Subjective, less rational responses to literature in the classroom have not gone unchallenged. Among the earliest spirited rebuttals were J. Mitchell Morse's "Are English Teachers Obsolete?" (*The CEA Critic* 36 [May 1974]: 9–18); Ann Berthoff's "Recalling Another Freudian Model—A Consumer Caveat" (*The CEA Critic* 35 [May 1953]: 12–14); and Eva Touster's "Tradition and the Academic Talent" (*The CEA Critic* 33 [May 1971]: 14–17). And John Ciardi in the second edition of *How Does a Poem Mean?* (Boston: Houghton, 1975) emphatically condemns appreciation and free association in discussing poetry in the classroom, calling the one "not useful," the other "permissive and pointless," and both together "dull" (xix–xxi).

Perhaps as a result of this controversy, a dilemma has arisen in the classroom for some teachers of literature, namely, whether to discuss material in an essentially subjective manner—the extreme of which could be relativistic and nonrational—or whether to employ the tools of logical and intellectual analysis. We believe that these options do not necessarily constitute a dilemma.

There is unquestionably a kind of literary analysis that is like using an elephant gun to shoot a gnat. It is practiced by riders of all kinds of scholarly hobbyhorses and may manifest itself in such ways as ascertaining the number of feminine rhymes in *The Rape of the Lock* or instances of trochees in book 4 of *Paradise Lost* or the truth about Hamlet's weight problem. The early pages of Charles Dickens's *Hard Times* illustrate the imagination-stifling effect of one such technique. Thomas Gradgrind, patron of a grammar school in an English industrial town, is listening to a class recite. He calls on one of the pupils, "girl number twenty," for the definition of a horse. "Girl number twenty" (in Gradgrind's world there is no personal identity) cannot produce the expected rote answer. A better-conditioned classmate can: "'Quadruped. Graminivorous. Forty teeth, namely twenty-four grinders, four eye-teeth and twelve incisive. Sheds coat in the spring; in marshy countries sheds hoofs, too. Hoofs hard, but requiring to be shod with iron. Age known by marks in mouth.' 'Now girl number twenty,' said Mr. Gradgrind, 'You know what a horse is.'" It hardly needs pointing out that such a definition would not do justice to the likes of Bucephalus, Pegasus, Black Beauty, Traveller, or Flicka. But absorption with extraneous, irrelevant, or even too practical considerations that detract from aesthetic perception seems to be an occupational disease of many literary critics. This appears to be a problem, however, rather than a dilemma, and its solution is among the several aims of this book.

Our purpose in this chapter is to show that the precritical response is not only desirable but indeed essential in the fullest appreciation of literature. In doing so, we do not mean to suggest that analysis or expertise detracts from aesthetic sensitivity any more than we mean to suggest that a precritical response is an unworthy one. It is a truism to say that our senses can sometimes mislead us, hence the need to analyze literature that is being studied as well as read for pure pleasure.

We maintain that knowledge, even of a specialized kind, is not in and of itself a deterrent to the enjoyment of literature. On the contrary, this book is predicated on the assumption that such knowledge and the intelligent application of several interpretive techniques can enhance the pleasure that the common reader can derive from a piece of literature.

Let us illustrate with an analogy. A college student in an American university decides to take in a movie on a Friday evening as a reward for a week of grinding study. She rounds up a group of friends similarly disposed, and they head for a nearby mall, the site of a huge theater where eight films are being shown simultaneously in different auditoriums. The sheer joy of weekend freedom and the anticipation of an attractive choice of films afford an ecstasy denied to many. Even that pleasure is heightened by the sight of hordes of other students laughing and clowning about their release from labs and libraries into the wonderful world of cinema. America's future business and professional leaders are stocking up on mouth-watering tubs of hot buttered popcorn and mammoth cups of soft drinks before disappearing into dark caverns full of luxuriously upholstered reclining theater seats, there to thrill vicariously to torrid love scenes, gory detective brutality, wild and crazy comedy, complex psychological drama, and amazing tales of the future. Everything combines to immerse them in a pool of sensation.

Not far away from these avid fans, a smaller, somewhat less flamboyant group of students are making their way into one of the auditoriums, but they lack none of the other group's excitement and anticipation. They are members of one of the university's film classes, and they are accompanied by their professor. They are thoroughly informed on the history of moviemaking; they know both classic and contemporary films; they understand the technical operations of the camera and its myriad effects; they are familiar with acting styles, past and present. On the level of sense experience, they are receiving the same impressions as the other group of students. But because of their special knowledge, they *comprehend* what they are witnessing. Their knowledge does not dim their pleasure; it does not nullify any precritical, amateur response. It may intensify it; it certainly complements it. For there is no real opposition of responses here. These more knowledgeable movie-goers do not say to themselves at one point, "Now we're *feeling*," and at another, "Now we're *knowing*." By this stage the knowing is almost as instinctive as the feeling.

What the academic needs to keep always in mind is that the precritical response is not an inferior response to literature. (After all, we may be sure that Shakespeare did not write *Hamlet* so that scholarly critical approaches to it could be formulated.) Rather, the precritical response employing primarily the senses and the emotions is an indispensable one if pleasure or delight is the aim of art. Without it the critic might as well be merely proofreading for factual accuracy or correct mechanical form. It may be said to underlie or even drive the critical response.

Setting ✓

The students' precritical response to a movie parallels the common reader's precritical response to literature. The glitzy Southern California scene of *Lethal Weapon* corresponds to the **setting** of the work of literature (the antebellum South of *Huckleberry Finn*; Puritan Massachusetts in "Young Goodman Brown"; Cavalier England in "To His Coy Mistress"; eleventh-century Denmark in *Hamlet*).

Precritical responses to setting in the works to be dealt with in this handbook are likely to be numerous and freewheeling. One reader of *Huckleberry Finn* will respond to the nostalgia of an earlier, rural America, to the lazy tempo and idyllic mood of Huck and Jim's raft trip down the Mississippi. Still another will delight in the description of the aristocratic Grangerfords' bourgeois parlor furnishings or the frontier primitivism of Arkansas river villages and the one-horse plantation of the Phelpses. The Gothic texture of the New England forest in "Young Goodman Brown" will sober some readers, as will the dark and brooding castle of Hamlet. The actual setting of "To His Coy Mistress" must be inferred (a formal garden? the spacious grounds of a nobleman's estate? some Petit Trianon type of apartment?), but romantically connotative settings such as the "Indian Ganges" and the "tide of Humber" are alluded to, as are macabre or mind-boggling places like "marble vaults" and "deserts of vast eternity."

Plot ✓

The students' uncomplicated view of an individual film itself equals the reader's precritical response to the **conflict (plot)** involving **protagonist** and **antagonist** (Hamlet versus his uncle; Mel Gibson versus hoods, killers, and drug dealers). Readers who delight in action will thrill to the steps in Hamlet's revenge, even when it lights on the innocent, and will feel the keen irony that prevents him from knowing his Ophelia to be true and guiltless and from enjoying the fruit of his righteous judgment. Such time-honored plot ingredients as the escape, the chase, the capture, the release—sensationally spiced with lynching, tar-and-feathering, swindling, feuding, murder, and treachery—may form the staple of interest for precritical readers of *Huckleberry Finn*. Such readers will also be rooting for the white boy and his black slave friend to elude their pursuers and attain their respective freedoms. Enigma and bewilderment may well be the principal precritical response elicited by the plot of "Young Goodman Brown": is Brown's conflict an imaginary one, or is he really battling the Devil in this theological *Turn of the Screw?* Or in "To His Coy Mistress," will the young Cavalier prevail with his Coy Mistress to make love before they are crushed in the maw of Time?

Character ╱

The young moviegoers assess, after a fashion, the roles of the actors. Although these are frequently cultural stereotypes, they bear some analogy to the common reader's common-sense **character** analysis of literary figures (the self-effacing, sacrificial nature of Sidney Carton in *A Tale of Two Cities*, the matter-of-fact courage and resourcefulness of Robinson Crusoe, the noble but frustrated humanity of John Savage in *Brave New World*). Precritical reactions to the characters in "To His Coy Mistress" will no doubt vary with the degree to which the reader subscribes to situation ethics or adheres to a clearly articulated moral code. Strict constructionists will judge the male aggressor a wolf and the woman a tease at best. Libertines will envy the speaker his line. Feminists will deplore the male-chauvinist exploitation that is being attempted. Although characters more complex than hot-shot L.A. detectives appear in *Huckleberry Finn*, a precritical perusal of the book will probably divide them into good (those basically sympathetic to Huck and Jim) and bad (those not). Similarly, the dramatis personae of *Hamlet* will be judged according to whether they line up on the side of the tormented Hamlet or on that of his diabolically determined uncle. In more complex character analysis, the simplistic grouping into good and bad will not be adequate; it may in fact necessitate an appreciation of ambiguity. From this viewpoint, Gertrude and Polonius and Rosencrantz and Guildenstern appear more weak and venal than absolutely vicious. Just as ambiguity was prominent in the plot of "Young Goodman Brown," so does it figure largely in a reader's precritical evaluation of character. Brown may appear to be a victim of trauma, an essentially shallow man suddenly made to seem profoundly disturbed.

Structure ✓

The students' awareness of the major complications and developments of a film plot and the importance of each to the outcome is akin to the reader's or viewer's unconscious sense of **plot structure,** the relatedness of actions, the gradual buildup in suspense from a situation full of potential to a climax and a resolution (as in Macbeth's rise to be king of Scotland through foul and bloody means and the poetic justice of his defeat and death by one he had wronged). A precritical response to the structure of "To His Coy Mistress" could certainly involve the recognition of the heightening intensity, stanza by stanza, of the lover's suit—from the proper and conventional complimentary forms of verbal courting to more serious arguments about the brevity of life and finally, to the bold and undisguised affirmation that sexual joy is the central goal of the lover's life. The common reader can discern the plot development in *Hamlet* step by step, from mystery, indecision, and torment to knowledge, resolute action, and catharsis. He or she may be fascinated by the stratagems that Hamlet and Claudius employ against each other and amused by the CIA-like techniques of Polonius to ferret out Hamlet's secret. Spellbinding horror and, later, cathartic pathos are possible emotions engendered by the climax and denouement of this revenge tragedy. The episodic plot of *Huckleberry Finn* is somehow coherent even though precritical readers must confront in rapid order thrill, suspense, danger, brutality, outrage, absurdity, laughter, tears, anger, and poetic justice as they respond to Huck and Jim's attempt to elude capture; the side-splitting charlatanism as well as the sinister and criminal behavior of the King and the Duke; wrecked steamboats; tent revivals; feuding, shooting in the street, and thwarted lynching; and finally the mixed triumph of the heroes. The structural stages in "Young Goodman Brown" may result in ambivalent reactions by the reader: on the one hand, plain recognition of the destructive effects of the events of the plot on Brown; on the other, bewilderment as to whether the events really took place or were all fantasy.

Style

The acting technique in the movie (realistic, stylized) has its counterpart in the verbal **style** of a literary work (the spare, understated prose of Hemingway; the sophisticated wit of *The Importance of Being Earnest*; the compressed, highly allusive idiom of poets like Eliot and Yeats; the crude, earthy plain talk of Norman Mailer's *The Naked and the Dead*, John Updike's *Couples*, and Kurt Vonnegut's *Slaughterhouse-Five*). The precritical reader feels it in the Pike County dialect of *Huckleberry Finn*, its vocabulary and rhythms seemingly ringing true in every line; in the urbane diction, learned allusion, and polished couplets of "To His Coy Mistress"; in the magnificent blank verse of *Hamlet*, alternately formal and plain, yet somehow organic and right for both dramatic action and philosophical soliloquy; in the solemn, cadenced phraseology of "Young Goodman Brown," echoing what one imagines Puritan discourse to have been like, both in and out of the pulpit, its lightest touches still somehow ponderous.

Atmosphere

Defined as the mood or feeling that permeates an environment, **atmosphere** is a further common ingredient in the two parts of our analogy. Several factors combine to create it: at the movie, the weather, the physical setting, the tension, the acting itself; in a literary work, such similar factors as the eerie locales and stormy weather in Mary Shelley's *Frankenstein* and Emily Brontë's *Wuthering Heights*, the panic of the green troops in Stephen Crane's *Red Badge of Courage*, the suspense and terror in Edgar Allan Poe's "Tell-Tale Heart," the indifference and listlessness of the characters in William Faulkner's "That Evening Sun."

The four works that we are emphasizing for precritical responses afford interesting possibilities. "To His Coy Mistress," which on the surface seems to have fewer overt atmosphere-producing elements, in fact has a fairly pronounced atmosphere (or atmospheres, since there are shifts). The atmosphere results from the diction and the tone the speaker employs. The formal honorific "Lady" and its implied politeness create, if not a drawing-room atmosphere, a stylized one where there is romantic badinage, where gallants wax hyperbolic in a formulary way, and where fair maidens drop their eyes demurely or, if hyperbole becomes too warm, tap male wrists with a delicate fan. It is a mannered, controlled, ritualistic atmosphere. But in the second stanza, compliments give way to professorial lectures as the aggressive male grows impatient with coyness carried too far, hence a dispiriting philosophical discussion about the brevity of life and the nothingness of afterlife. Finally, in the third stanza, the atmosphere becomes electric and potentially physical as the diction becomes explicitly erotic.

In *Huckleberry Finn*, on a very obvious plane, setting contributes to atmosphere. The Mississippi River, sleepy villages, small towns, one-horse plantations, Victorian parlors: all combine to present an essentially "normal" nineteenth-century-Americana kind of security along with zest for life. Diction, character, and costume, however, also function to add subtle features to the atmosphere: the casual use of expressions like "nigger" and "harelip" (most of our nineteenth-century ancestors did not share our aversion to using racial epithets or to making fun of physical deformity); the toleration and acceptance of violence, cruelty, and inhumanity observable in conversation and exposition; the radical inconsistency of basically decent, religious people breaking up slave families while evincing genuine affection for them and concern for their welfare. The amalgam of their shocking and sometimes contradictory attitudes and actions results in an utterly convincing atmosphere.

Both setting and plot make for a gloomy, foreboding atmosphere in *Hamlet* and "Young Goodman Brown." The Shake-spearean drama opens with sentries walking guard duty at midnight on the battlements of a medieval castle where a ghost has recently appeared. It is bitter cold and almost un-naturally quiet. Though later the scene changes many times, this atmosphere persists, augmented by the machinations of the principals, by dramatic confrontations, by reveries on death, by insane ravings, and finally by wholesale slaughter. In only slightly less melodramatic form, Hawthorne's story takes the reader to a witches' sabbath deep in the forests of seventeenth-century Massachusetts, where a cacophony of horrid sounds makes up the auditory background for a scene of devilish human countenances and eerie, distorted images of trees, stones, clouds. The protagonist's ambiguous role in the evil ceremony, which ruins his life, adds to the dark atmosphere pervading the story.

Theme

The often rich and varied underlying idea of the action is the **theme.** In a movie, theme may be no more than "Arrest those lawbreakers!" "Convict those criminals!" "Overcome those obstacles to true love!" In a literary work, theme may be as obvious as the message in *Uncle Tom's Cabin* that "Slavery is cruel and morally degrading and must go" or the implicit point of *Robin Hood* that "Some rich folks deserve to be taken from, and some poor folks need to be given to." These scarcely compare with such profound thematic implications as those in *Adam Bede, The Scarlet Letter,* or "The Love Song of J. Alfred Prufrock." As theme is a complex aspect of literature, one that requires very intentional thinking to dis-cern, it is not likely to elicit the precritical response that the more palpable features do. This is not to say that it will not be felt. Twain's criticisms of slavery, hypocrisy, chicanery, vio-lence, philistine aesthetic taste, and other assorted evils will move both the casual reader and the scholar. So will Mar-vell's speaker's spunky defiance of all-conquering Time. The poignancy of young Hamlet's having to deal with so many of life's insolubles at once and alone is certainly one of the play's major themes, and is one available at the precritical level of response. There are others. Despite complexity and ambiguity, the precritical reader will sense the meaning of faith and the effects of evil in "Young Goodman Brown" as two of the more urgent themes in the story.

None of these feelings, whether at a movie or in private reading, is contingent upon a technical knowledge of motion pictures or a graduate degree in the humanities. Without either, people may appreciate and respond precritically to both Oscar-award-winning films and the cold setting of Jack London's "To Build a Fire," to the sequence of events that causes Oedipus to blind himself, or to the phantasmagoric atmosphere of horror pervading Poe's "Masque of the Red Death."

In short, regardless of the extent to which close scrutiny and technical knowledge aid in literary analysis, there is no substitute for an initial personal, appreciative response to the basic ingredients of literature: setting, plot, character, structure, style, atmosphere, and theme. The reader who manages to proceed without that response sacrifices the spontaneous joy of seeing any art object whole, the wondrous sum of myriad parts.

▪ 1 ▪

Traditional Approaches

I. NATURE AND SCOPE OF THE TRADITIONAL APPROACHES

Some years ago, a story was making the rounds in academic circles and was received in good humor by all the enlightened teachers of literature. A professor of English in a prestigious American university, so the story goes, entered the classroom one day and announced that the poem under consideration for that hour was to be Andrew Marvell's "To His Coy Mistress." He then proceeded for the next fifty minutes to discuss Marvell's politics, religion, and career. He described Marvell's character, mentioned that he was respected by friend and foe alike, and speculated on whether he was married. At this point the bell rang, signaling the end of the class. The professor closed his sheaf of notes, looked up, smiling, and concluded, "Damn' fine poem, men. Damn' fine."

The story was told to ridicule the type of criticism that dominated the study of literature until the 1930s and that is still employed in some classrooms even today. In this approach the work of art frequently appeared to be of secondary importance, something that merely illustrated background. Such an approach often (many would say inevitably) led to the study of literature as essentially biography, history, or other branch of learning, rather than as art.

16

Well on in the twentieth century, however, a new type of literary analysis emerged in which the literary work per se (that is, as a separate entity divorced from extrinsic considerations) became the dominant concern of scholars. The New Critics, as the proponents of this position were called, insisted that scholars concentrate on the work itself, on the text, examining it as art. This method revolutionized the study of literature. It frequently divided critics and teachers into opposing factions: those of the older school, for whom literature provided primarily an opportunity for exercising what they perceived to be the really relevant scholarly and cultural disciplines (for example, history, linguistics, and biography) and the New Critics, who maintained that literature had an intrinsic worth, that it was not just one of the means of transmitting biography and history. Now that the controversy has lessened, the rationale of the New Criticism seems to have put into clearer focus what a poem or play or piece of fiction is trying to do; it has unquestionably corrected many wrongheaded interpretations resulting from an unwise use of the older method. To this extent it has expanded our perception and appreciation of literary art.

Nevertheless, in their zeal to avoid the danger of interpreting a literary work solely as biography and history—the end result of the traditional method, they thought—many twentieth-century followers of New Criticism have been guilty of what may well be a more serious mistake, that of ignoring any information not in the work itself, however helpful or necessary it might be. Fortunately, the most astute critics have espoused a more eclectic approach and have fused a variety of techniques. They have certainly insisted on treating literature as literature, but they have not ruled out the possibility of further aesthetic illumination from traditional quarters. Oscar Cargill, in the introduction to his *Toward a Pluralistic Criticism* (Carbondale: Southern Illinois UP, 1965), endorsed the eclectic approach unequivocally:

> I have always held that any method which could produce the meaning of a work of literature was a legitimate method. . . . I came to the conclusion that . . . the critic's task was . . . to procure a viable meaning appropriate to the critic's time and place. Practically, this meant employing not any one method in

interpreting a work of art but every method which might prove
efficient. (xii–xiv)

In any event, while we may grant the basic position that
literature is primarily art, it must be affirmed also that art
does not exist in a vacuum. It is a creation by someone at
some time in history, and it is intended to speak to other
human beings about some idea or issue that has human relevance. Any work of art for that matter will always be more
meaningful to knowledgeable people than to uninformed
ones. Its greatness comes from the fact that when the wisest,
most cultivated, most sensitive minds bring all their information, experience, and feeling to contemplate it, they are
moved and impressed by its beauty, by its unique kind of
knowledge, and even by its nonaesthetic values. It is surely
dangerous to assume that a work of art must always be judged
or looked at or taught as if it were disembodied from all
experience except the strictly aesthetic. Many literary classics are admittedly autobiographical, propagandistic, or topical (that is, related to contemporary events). These concerns
are, in fact, central to one of the most recent theoretical
approaches—the new historicism (see chapter 7).

Thus, although we have not yet elaborated these critical
methods, let us be aware from the outset that in succeeding
chapters we will be dealing with some widely divergent interpretive approaches to literature and that, regardless of
what newer modes of analysis may be in the ascendant, the
traditional methods retain their validity.

A. Textual Scholarship: A Prerequisite
to Criticism

Before we embark upon any interpretive ventures, we should
look to that branch of literary studies known as textual criticism. In the words of James Thorpe, author of one of the best
modern books on the subject, *Principles of Textual Criticism*
(San Marino, CA: Huntington, 1972), textual criticism has as
its ideal the establishment of an *authentic* text, or the "text
which the author intended" (50). This aim is not so easy to
achieve as one might think, however, and it is a problem not

only with older works, where it might be more expected, but also in contemporary literature. There are countless ways in which a literary text may be corrupted from what the author intended. The author's own manuscript may contain omissions and errors in spelling and mechanics; these mistakes may be preserved by the text copyists, be they scribes or printers, who may add a few of their own. Or, as has often happened, copyists or editors may take it upon themselves to improve, censor, or correct what the author wrote. If the author or someone who knows what the author intended does not catch these errors during proofreading, they can be published, disseminated, and perpetuated. (Nor does it help matters when authors themselves cannot decide what the final form of their work is to be but actually release for publication several different versions or, as is frequently the case, delegate broad editorial powers to others along the line.) So many additional mishaps can befall a manuscript in the course of producing multiple copies for the public that, to quote Thorpe again, the "ordinary history of the transmission of a text, without the intervention of author or editor, is one of progressive degeneration" (51). Shocking as such an assertion may sound, it is nevertheless true and can be documented.

We frequently assume that the text before us has come down unchanged from its original form. More often than not, the reverse is the case; what we see is the result of painstaking collation of textual variants, interpretation, and emendation or conjecture. Because it is pointless to study inaccurate versions of anything, from economic theories to works of literature, except with a view to ascertaining the true (that is, the authorial) version, our debt to textual criticism is wellnigh incalculable. For example, the student who uses the eight-volume Chicago edition of *The Canterbury Tales*, a collation of scores of medieval manuscripts, should certainly appreciate the efforts of precomputer scholars. Similarly, the studies of W. W. Greg, A. W. Pollard, Fredson Bowers, Charlton Hinman, Stanley Wells, Garry Taylor, and a host of others have gone far toward the establishment of a satisfactory Shakespearean text. This type of scholarship should create in the student a healthy respect for textual criticism and expert editing, and well it might, for as Thorpe has aptly phrased it, "where there is no editing the texts perish" (54).

Textual criticism plays an especially important role in studying the genesis and development of a piece of literature. Thus it has enabled us to see how Ezra Pound's editorial surgery transformed T. S. Eliot's *The Waste Land* from a clumsy and diffuse poem to a modern classic. (The poem still presents textual problems, however, because Eliot himself authorized versions containing substantive differences.) Other famous textual cases include Dickens's two endings for *Great Expectations:* after seeing the first "unhappy" ending in proof, Dickens wrote another and authorized only it. Later editors have published the first version as having more aesthetic integrity, but Dickens never authorized it. Thomas Hardy made so many substantive character and plot alterations in the four versions of *The Return of the Native,* all of which he authorized for publication between 1878 and 1912, that James Thorpe understandably asks, "Which is the real *Return of the Native?*" (34). Moreover, textual criticism is, contrary to what ill-informed people may think, anything but an essentially mechanical operation. Although its practitioners are very much concerned with such matters as spelling, punctuation, capitalization, italicization, and paragraphing (accidentals, as they are called in textual criticism) in the establishment of an authentic text, they deal with much more than close proofreading. They must be highly skilled in linguistics, literary history, literary criticism, and bibliography, to mention only the most obvious areas.

However, though textual critics must and do make aesthetic judgments, not only in accidentals but also in substantives (actual verbal readings), they do so in order to establish by means as scientific as possible an authentic text for the literary critic, who may then proceed to interpret and evaluate. Textual criticism is therefore treated in this book not as a traditional interpretive approach to literature but as an indispensable tool for further meaningful analysis. This relationship between textual and strictly interpretive criticism may be expressed in a surgical metaphor: textual critics are the first in a team of critics who prepare the literary corpus for further dissection. Nevertheless, we should not push any analogy between textual criticism and science too far. Textual critics are not and should not be considered scientists. They have no predetermined or inviolable laws that they can use to come out with an authentic text. Perhaps it would be more accurate to concede that textual critics are scientists of sorts; they simply are not exact scientists (that is, ones dealing in an exact science). They are, more precisely, a combination of scientist and artist. As A. E. Housman says, textual criticism is the "science of discovering error in texts and the art of removing it" ("The Application of Thought to Textual Criticism," in Ronald Gottesman and Scott Bennett, eds., *Art and Error: Modern Textual Editing* [Bloomington: Indiana UP, 1970]: 2).

Thorpe, however, is highly critical of any scientific claims for textual criticism. Indeed, one of the main points of his book is the failure of textual studies to measure up to their alleged scientific status. Somewhat resignedly he concludes:

> It would be cheerful to be able to report that a mastery of sound principles, an application of effective methods, and an exercise of conscientious care will enable the textual critic to reach the ideal which is incorporated in the first principle of his craft. But it would not be true. In textual criticism, the best that one can do is to cut the losses, to reduce the amount of error, to improve or clarify the state of textual affairs, to approach the ideal. After all has been done that can be done, however, the results of textual criticism still are necessarily imperfect. (55)

Whether one agrees with Thorpe or with those who view textual criticism as less tentative and more scientific, all critics can agree on one thing: it is far more preferable to have a version of a literary work that textual criticism can make available to us than to have one that has not been subjected to the rigorous methodology of that discipline.

B. Types of Traditional Approaches

We present two types of traditional critical approaches to literature, the historical-biographical and the moral-philosophical; each will be defined, discussed, and subsequently applied to each of the four works selected for emphasis. Early in the discussions of each work we will also treat textual matters, summarizing the narrative line of the literary work and making some nontechnical observations about genre. (The generic approach will receive additional attention in chapter 7.) These steps may be thought of as preliminaries to traditional literary analysis and can certainly be useful in other interpretive approaches as well.

1. Historical-Biographical

Although the historical-biographical approach has been evolving over many years, its basic tenets are perhaps most clearly articulated in the writings of the nineteenth-century French critic H. A. Taine, whose phrase *race, milieu, et moment,* elaborated in his *History of English Literature,* bespeaks a hereditary and environmental determinism. Put simply, this approach sees a literary work chiefly, if not exclusively, as a reflection of its author's life and times or the life and times of the characters in the work.

At the risk of laboring the obvious, we will mention the historical implications of William Langland's *Piers Plowman,* which is, in addition to being a magnificent allegory, a scorching attack on the corruption in every aspect of fourteenth-century English life—social, political, and religious. So timely, in fact, were the poet's phrases that they became rallying cries in the Peasants' Revolt. John Milton's sonnet "On the Late Massacre in Piedmont" illustrates the topical quality that great literature may and often does possess. This poem commemorates the slaughter in 1655 of the Waldenses, members of a Protestant sect living in the valleys of northern Italy. A knowledge of this background clarifies at least one rather factual reference and two allusions in the poem. Several of Milton's other sonnets also reflect events in his life or times. Two such are "On His Blindness," best understood when one realizes that the poet became totally blind when he was forty-four, and "On His Deceased Wife," a tribute to his second wife, Katherine Woodcock. Milton was already blind when he married her, a fact that explains the line, "Her face was veiled." In fact, Milton affords us an excellent example of an author whose works reflect particular episodes in his life. *Samson Agonistes* and *The Doctrine and Discipline of Divorce* may be cited as two of the more obvious instances.

A historical novel is likely to be more meaningful when either its milieu or that of its author is understood. James Fenimore Cooper's *Last of the Mohicans,* Sir Walter Scott's *Ivanhoe,* Charles Dickens's *Tale of Two Cities,* and John Steinbeck's *Grapes of Wrath* are certainly better understood by readers familiar, respectively, with the French and Indian War (and the American frontier experience generally), Anglo-Norman Britain, the French Revolution, and the American Depression. And, of course, there is a very real sense in which these books are *about* these great historical matters, so that the author is interested in the characters only to the extent that they are molded by these events.

What has just been said applies even more to ideological or propagandist novels. Harriet Beecher Stowe's *Uncle Tom's Cabin,* Frank Norris's *The Octopus,* and Upton Sinclair's *The Jungle* ring truer (or falser as the case may be) to those who know about the antebellum South, railroad expansion in the late nineteenth century, and scandals in the American meat-packing industry in the early twentieth century. Sinclair Lewis's satires take on added bite and fun for those who have lived in or observed the cultural aridity of *Main Street,* who have been treated by shallow and materialistic physicians like some of those in *Arrowsmith,* who have sat through the sermons and watched the shenanigans of religious charlatans like Elmer Gantry, or who have dealt with and been in service clubs with all-too-typical American businessmen like Babbitt. Novels may lend themselves somewhat more readily than lyric poems to this particular interpretive approach; they usually treat a broader range of experience than poems do and thus are affected more by extrinsic factors.

It is a mistake, however, to think that poets do not concern themselves with social themes or that good poetry cannot be written about such themes. Actually, poets have from earliest times been the historians, the interpreters of contemporary culture, and the prophets of their people. Take, for example, a poet as mystical and esoteric as William Blake. Many of his best poems can be read meaningfully only in terms of Blake's

England. His "London" is an outcry against the oppression of man by society: he lashes out against child labor in his day and the church's indifference to it, against the government's indifference to the indigent soldier who has served his country faithfully, and against the horrible and unnatural consequences of a social code that represses sexuality. His "Preface" to *Milton* is at once a denunciation of the "dark Satanic Mills" of the Industrial Revolution and a joyous battle cry of determination to build "Jerusalem/In England's green and pleasant Land." It has been arranged as an anthem for church choirs, is widely used in a hymn setting, and was sung in London in the 1945 election by the victorious Labour party. The impact of the Sacco and Vanzetti case upon young poets of the 1920s or of the opposition to the war in Vietnam upon almost every important American poet in the 1960s resulted in numerous literary works on these subjects. Obviously, then, even some lyric poems are susceptible to historical-biographical analysis.

Political and religious verse satires like John Dryden's in the seventeenth century and personal satires like Alexander Pope's in the eighteenth century have as one of their primary purposes the ridiculing of contemporary situations and persons. Dryden propounds his own Anglican faith and debunks the faith of both Dissenters and Papists in *Religio Laici.* Later, when he had renounced Anglicanism and embraced Roman Catholicism, he again defended his position, and in *The Hind and the Panther* he attacked those who differed. His *Absalom and Achitophel* is a verse allegory using the biblical story of Absalom's rebellion against his father, King David, to satirize the Whig attempt to replace Charles II with his illegitimate son, the Duke of Monmouth. Pope's *Dunciad* is certainly a satire against all sorts of literary stupidity and inferiority, but it is also directed against particular literary people who had the bad fortune to offend Pope. All these works may be understood and appreciated without extensive historical or biographical background. Most readers, however, would probably agree with T. S. Eliot that "No poet, no artist of any art, has his complete meaning alone" (from "Tradition and the Individual Talent") and with Richard D. Altick that "almost every literary work is attended by a host of outside circumstances which, once we expose and explore them, suffuse it with additional meaning" (*The Art of Literary Research*, rev. ed. [New York: Norton, 1975]: 5).

The triumph of such verse satires as those of Dryden and Pope is that they possess considerable merit as poems, merit that is only enhanced by their topicality. That it should ever have been necessary to defend them because they were topical or "unpoetic" is attributable to what Ronald S. Crane calls, in *A Collection of English Poems, 1660–1800* (New York: Harper, 1932), the tyranny of certain Romantic and Victorian "presuppositions about the nature of poetry" and the "inhibitions of taste which they have tended to encourage." He mentions among such presuppositions the notions that

> true poetry is always a direct outpouring of personal feeling; that its values are determined by the nature of the emotion which it expresses, the standards being naturally set by the preferences of the most admired poets in the nineteenth-century tradition; that its distinctive effort is "to bring unthinkable thoughts and unsayable sayings within the range of human minds and ears"; that the essence of its art is not statement but suggestion. (v)

In short, even topical poetry can be worthwhile when not limited by presuppositions that make poetry a precious, exclusively personal, even esoteric thing.

2. Moral-Philosophical

The moral-philosophical approach is as old as classical Greek and Roman critics. Plato, for example, emphasized moralism and utilitarianism; Horace stressed that literature should be *dulce et utile* (delightful and instructive). Among its most famous exemplars are the commentators of the age of neoclassicism in English literature (1660–1800), particularly Samuel Johnson. The basic position of such critics is that the larger function of literature is to teach morality and to probe philosophical issues. They would interpret literature within a context of the philosophical thought of a period or group. From their point of view Jean-Paul Sartre and Albert Camus can be read profitably only if one understands existentialism. Similarly, Pope's *Essay on Man* may be grasped only if one understands the meaning and the role of reason in eighteenth-century thought. Such teaching may also be religiously oriented. Henry Fielding's *Tom Jones*, for example, illustrates the moral superiority of a hot-blooded young man like Tom, whose sexual indulgences are decidedly atoned for by his humanitarianism, tenderheartedness, and instinctive honor (innate as opposed to acquired through training). Serving as foils to Tom are the real sinners in the novel—the vicious and the hypocritical. Hawthorne's *Scarlet Letter* is likewise seen essentially as a study of the effects of secret sin on a human soul—that is, sin unconfessed before both God and man, as the sin of Arthur Dimmesdale with Hester Prynne, or, even more, the sin of Roger Chillingworth. Robert Frost's "Stopping by Woods on a Snowy Evening" suggests that duty and responsibility take precedence over beauty and pleasure.

A related attitude is that of Matthew Arnold, the Victorian critic, who insisted that a great literary work must possess "high seriousness." (Because he felt that Chaucer lacked it, Arnold refused to rank him among the very greatest English poets.) In each instance critics working from a moral bent are not unaware of form, figurative language, and other purely aesthetic considerations, but they consider them to be secondary. The important thing is the moral or philosophical teaching. On its highest plane this is not superficially didactic, though it may at first seem so. In the larger sense, all great literature teaches. The critic who employs the moral-philosophical approach insists on ascertaining and stating *what* is taught. If the work is in any degree significant or intelligible, this meaning will be there.

It seems reasonable, then, to employ historical-biographical or moral-philosophical analyses among other methods (such as textual study and recognition of genre) in getting at the total meaning of a literary work when the work seems to call for them. Such approaches are less likely to err on the side of overinterpretation than are more esoteric methods. And overinterpretation is a particularly grievous critical error. A reader who stays more or less on the surface of a piece of literature has at least understood part of what it is about, whereas a reader who extracts interpretations that are neither supportable nor reasonable may miss a very basic or even key meaning. Obviously, a dull, pedestrian, uniformly literal approach to literary analysis is the antithesis of the informed, imaginative, and creative approach that this book advocates. But it must be remembered that, brilliant and ingenious criticism notwithstanding, words in context, though they may mean many things, cannot mean just anything at all. Daring, inventive readings of metaphorical language must have defensible rationales if they are to be truly insightful and convincing.

The enemies of the traditional approach to literary analysis have argued that it has tended to be somewhat deficient in imagination and has neglected the newer sciences, such as psychology and anthropology and that it has been too content with a commonsense interpretation of material. But it has nevertheless performed one valuable service: in avoiding cultism and faddism, it has preserved scholarly discipline and balance in literary criticism. We do not mean that we favor traditional criticism over predominantly aesthetic interpretive approaches. We do suggest, however, that any knowledge or insight (with special reference to scholarly disciplines like history, philosophy, theology, sociology, art, and music) that can help to explain or clarify a literary work ought to be given the fullest possible chance to do so. Indeed, in some sense these approaches represent a necessary first step that precedes most other approaches.

Readers who intend to employ the traditional approaches to a literary work will almost certainly employ them simultaneously. That is, they will bring to bear on a poem, for instance, all the information and insights these respective disciplines can give in seeing just what the poem means and does.

II. THE TRADITIONAL APPROACHES IN PRACTICE

A. Traditional Approaches to Marvell's "To His Coy Mistress"

1. *The Text of the Poem*

Some words on textual problems in Andrew Marvell's "To His Coy Mistress" will set the stage for our consideration of the poem. One of these problems is the last word in this couplet:

> Now therefore, while the youthful hue
> Sits on thy skin like morning dew.

Instead of "dew," the first edition of the poem had "glew," which we now know is a dialectal variant of "glow," although it was earlier thought to be another spelling of "glue," a senseless reading in the context. "Lew" (dialectal "warmth") was also suggested as a possibility. But when someone conjectured "dew," probably in the eighteenth century, it was apparently so happy an emendation that virtually all textbooks have long printed it without any explanation. The first edition of this handbook followed those textbooks. But two recent texts restore the earliest reading. Both *The Anchor Anthology of Seventeenth-Century Verse* (ed. Louis L. Martz [Garden City, NY: Doubleday, 1969], vol. 1) and *Andrew Marvell, Complete Poetry* (ed. George de F. Lord [New York: Random (Modern), 1968]) print "glew" (meaning "glow") as making more sense in the context and being quite sound linguistically. Two other words in the poem that must be explained are "transpires" and "instant" in this couplet:

> And while thy willing soul transpires
> At every pore with instant fires.

In each case, the word is much nearer to its Latin original than to its twentieth-century meaning. "Transpires" thus means literally "breathes forth," and "instant" means "now present" and "urgent." Admittedly, this sort of linguistic information borders on the technical, but an appreciation of the meaning of the words is imperative for a full understanding of the poem.

2. *The Genre of the Poem*
Most critics are careful to ascertain what literary type or genre they are dealing with, whether a poem (and if so, what particular kind), a drama, a novel, or a short story. This first step—the question "What are we dealing with?"—is highly necessary, because different literary genres are judged according to different standards. We do not expect, for example, the sweep and grandeur of an epic in a love lyric, nor do we expect the extent of detail in a short story that we find in a novel. From a technical and formal standpoint, we do expect certain features in particular genres, features so integral as to define and characterize the type (for example, rhythm, rhyme, narrative devices such as a point-of-view character, and dramatic devices such as the soliloquy). The lyric, the genre to which "Coy Mistress" belongs, is a fairly brief poem characterized primarily by emotion, imagination, and subjectivity.

Having ascertained the genre and established the text, the employer of traditional methods of interpretation next determines what the poem says on the level of statement or, as John Crowe Ransom has expressed it, its "paraphrasable content." The reader discovers that this poem is a proposition, that is, an offer of sexual intercourse. At first it contains, however, little of the coarseness or crudity usually implied in the word *proposition*. On the contrary, though impassioned, it is graceful, sophisticated, even philosophical. The speaker, a courtier, has evidently urged an unsuccessful suit on a lady. Finding her reluctant, he is, as the poem opens, making use of his most eloquent line. But it is a line that reveals him to be no common lover. It is couched in the form of an argument in three distinct parts, going something like this: (1) If we had all the time in the world, I could have no objection to even an indefinite postponement of your acceptance of my suit. (2) But the fact is we do not have much time at all; and once this phase of existence (that is, life) is gone, all our chances for love are gone. (3) Therefore the only conclusion that can logically follow is that we should love one another now, while we are young and passionate, and thus seize what pleasures we can in a world where time is all too short. After all, we know nothing about any future life and have only the grimmest observations of the effects of death.

This is, as a matter of fact, a specious argument, viewed from the rigorous standpoint of formal logic. The fallacy is called denying the antecedent, in this case the first part of the conditional statement beginning with "if." The argument goes like this: If we have all the time and space in the world, your coyness is innocent (not criminal). We do not have all the time and space in the world. Therefore, your coyness is not innocent. Both premises are true, and the conclusion is still false. The lady's coyness may not be innocent for other reasons besides the lovers' not having all the time and space in the world. The male arguer undoubtedly does not care whether his argument is valid or not as long as it achieves his purpose. As Pope so well expressed it in *The Rape of the Lock:*

> For when success a Lover's toil attends,
> Few ask, if fraud or force attained his ends. (2.33–34)

3. Historical-Biographical Considerations

We know several facts about Marvell and his times that may help to explain this framework of logical argument as well as the tone and learned allusions that pervade the poem. First, Marvell was an educated man (Cambridge B.A., 1639), the son of an Anglican priest with Puritan leanings. Because both he and his father had received a classical education, the poet was undoubtedly steeped in classical modes of thought and literature. Moreover, the emphasis on classical logic and polemics in his education was probably kept strong in his mind by his political actions. (He was a Puritan, a Parliamentarian, an admirer of Oliver Cromwell, a writer of political satires, and an assistant to John Milton, who was Latin secretary to the government.) That it should occur, therefore, to Marvell to have the speaker plead his suit logically should surprise no one.

There is, however, nothing pedantic or heavy-handed in this disputatious technique. Rather, it is playful and urbane, as are the allusions to Greek mythology, courtly love, and the Bible. When the speaker begins his argument, he establishes himself in a particular tradition of love poetry, that of courtly love. No one would mistake this poem for love in the manner of "O my love's like a red, red rose" or "Sonnet from the Portuguese." It is based on the elevation of the beloved to the status of a virtually unattainable object, one to be idolized, almost like a goddess. This status notwithstanding, she is capable of cruelty, and in the first couplet the speaker accuses her of a crime, the crime of withholding her love from him. Moreover, because she is like a goddess, she is also capricious and whimsical, and the worshiper must humor her by following the conventions of courtly love. He will complain (of her cruelty and his subsequent pain and misery) by the River Humber. He will serve her through praise, adoration, and faithful devotion from the fourth millennium B.C. (the alleged time of Noah's flood) to the conversion of the Jews to Christianity, an event prophesied to take place just before the end of the world. Doubtless, this bit of humor is calculated to make the lady smile and to put her off her guard against the ulterior motive of the speaker.

However pronounced courtly love may be in the opening portion of the poem (the first part of the argument), by the time the speaker has reached the conclusion, he has stripped the woman of all pretense of modesty or divinity by his accusation that her "willing soul" literally exudes or breathes forth ("transpires") urgent ("instant") passion and by his direct allusion to kinesthetic ecstasy: "sport us," "roll all our strength," "tear our pleasures with rough strife/Thorough the iron gates of life" (the virginal body). All of this is consistent with a speaker who might have been schooled as Marvell himself was.

Many allusions in the poem that have to do with the passage of time show Marvell's religious and classical background. Two have been mentioned: the Flood and the conversion of the Jews. But there are others that continue to impress the reader with the urgency of the speaker's plea. "Time's winged chariot" is the traditional metaphor for the vehicle in which the sun, moon, night, and time are represented as pursuing their course. At this point, the speaker is still in the humorous vein, and the image is, despite its serious import, a pleasing one. The humor grows increasingly sardonic, however, and the images become in the second stanza downright repulsive. The allusions in the last stanza (the conclusion to the argument or case) do not suggest playfulness or a Cavalier attitude at all. Time's "slow-chapped [slow-jawed] power" alludes to the cannibalism of Kronos, chief of the gods, who, to prevent ever being overthrown by his own children, devoured all of them as they were born except Zeus. Zeus was hidden, later grew up, and ultimately became chief of the gods himself. The last couplet,

> Thus, though we cannot make our sun
> Stand still, yet we will make him run,

suggests several possible sources, both biblical and classical. Joshua commanded the sun to stand still so that he could win a battle against the Amorites (Josh. 10:13). Phaeton took the place of his father, the sun, in a winged chariot and had a wild ride across the sky, culminating in his death (Ovid, *Metamorphoses*). Zeus bade the sun to stand still in order to lengthen his night of love with Alcmene, the last mortal woman he embraced. In this example it is, of course, easy to see the appropriateness of the figures to the theme of the poem. Marvell's speaker is saying to his mistress that they are human, hence mortal. They do not have the ear of God as Joshua had, so God will not intervene miraculously and stop time. Nor do they possess the power of the pagan deities of old. They must instead cause time to pass quickly by doing what is pleasurable.

In addition to Marvell's classical and biblical background, further influences on the poem are erotic literature and Metaphysical poetry. Erotic poetry is, broadly speaking, simply love poetry, but it must emphasize the sensual. In "Coy Mistress" this emphasis is evident in the speaker's suit through the references to his mistress's breasts and "the rest" of her charms and in the image of the lovers rolled up into "one ball." The poem is Metaphysical in its similarities to other seventeenth-century poems that deal with the psychology of love and religion and—to enforce their meaning—employ bizarre, grotesque, shocking, and often obscure figures (the Metaphysical conceit). Such lines as "My vegetable love should grow," the warning that worms may violate the mistress's virginity and that corpses do not make love, the likening of the lovers to "amorous birds of prey," and the allusion to Time's devouring his offspring ("slow-chapped") all help identify the poem as a product of the seventeenth-century revolt against the saccharine conventions of Elizabethan love poetry. As for its relation to vers de société, "To His Coy Mistress" partakes more of the tone than the subject matter of such poetry, manifesting for the most part wit, gaiety, charm, polish, sophistication, and ease of expression—all of these despite some rough Metaphysical imagery.

4. Moral-Philosophical Considerations

An examination of what "Coy Mistress" propounds morally and philosophically reveals the common theme of *carpe diem*, "seize the day," an attitude of "eat and drink, for tomorrow we shall die." Many of Marvell's contemporaries treated this idea (for example, Robert Herrick in "To the Virgins, To Make Much of Time" and Edmund Waller in "Go, Lovely Rose"). This type of poetry naturally exhibits certain fundamental moral attitudes toward the main issue this poem treats—sex. These attitudes reflect an essentially pagan view. They depict sexual intercourse as strictly dalliance ("Now let us sport us while we may"), as solely a means of deriving physical sensations. Although not a Cavalier poet, Marvell is here letting his speaker express a more Cavalier (as opposed to Puritan) idea.

One more aspect of the historical background of the composition of the poem may be helpful in understanding its paradoxically hedonistic and pessimistic stance. The seventeenth century, it should be remembered, was not only a period of intense religious and political struggle, but a period of revolutionary scientific and philosophical thought. It was the century when Francis Bacon's inductive method was establishing itself as the most reliable way of arriving at scientific truth; it was the century when the Copernican theory tended to minimize the uniqueness and importance of the earth, hence of man, in the universe; it was the century when Thomas Hobbes's materialism and degrading view of human nature tended to outrage the orthodox or reflective Christian. Given this kind of intellectual milieu, readers may easily see how the poem might be interpreted as the impassioned utterance of a man who has lost anything resembling a religious or philosophical view of life (excluding, of course, pessimism). The paradox of the poem consists in the question of whether the speaker is honestly reflecting his view of life—pessimism—and advocating sensuality as the only way to make the best of a bad situation or whether he is simply something of a cad—stereotypically male, conceited, and superior, employing eloquence, argument, and soaringly passionate poetry merely as a line, a devious means to a sensual end. If the former is the case, there is something poignant in the way the man must choose the most exquisite pleasure he knows, sensuality, as a way of spitting in the face of his grand tormentor and victorious foe, Time. If the latter, then his disturbing images of the female body directed at his lady only turn upon him to reveal his fears and expose his lust. A feminist reading, as in chapter 5, sees the rhetoric of the poem very differently than does a traditional reading.

B. Traditional Approaches to *Hamlet*

1. *The Text of the Play*
Few literary works have received the amount and degree of textual study that Shakespeare's *Hamlet* has. There are some obvious reasons for this. To begin with, even the earliest crude printings, shot through with the grossest errors, revealed a story and a mind that excited and challenged viewers, producers, readers, critics, and scholars—so much so that the scholars decided to do everything possible to ascertain what Shakespeare actually wrote. The other reasons are all related to this one. Shakespearean editors ever since have realized the importance of establishing an accurate text if students and audiences are to discover the meaning of *Hamlet*.

It is difficult at this remove in time for the college student embarking on a serious reading of *Hamlet* to realize that the beautiful anthology or the handy paperback before him, each edited by an eminent authority, contains the product of nearly four hundred years of scholarly study of four different versions of *Hamlet* and that it still includes some moot and debatable readings. Besides questionable readings, there are a number of words whose meanings have changed over the years but that must be understood in their Elizabethan senses if the play is to be properly interpreted. To be sure, modern editors explain the most difficult words, but occasionally they let some slip by or fail to note that reputable scholars differ. Obviously, it is not possible here to point out all the variants of a given passsage or to give the seventeenth-century meaning of every puzzling construction, but the student can catch at least a glimpse of the multiplicity and the richness of interpretations by examining some of the more famous ones.

One of the best-known examples of such textual problems occurs in act I, scene ii: "O that this too too solid flesh would melt." This is perhaps the most common rendering of this line. The word "solid" appears in the first folio edition (1623) of Shakespeare's complete works. Yet the second quarto edition (1604–5), probably printed from Shakespeare's own manuscript, has "sallied," a legitimate sixteenth-century form of "sully" (to dirty, or make foul). These words pose two rather different interpretations of the line: if one reads "solid," the line seems to mean that Hamlet regrets the corporeality of the flesh and longs for bodily dissolution in order to escape the pain and confusion of fleshly existence. If, on the other hand, one reads "sullied," the line apparently reveals Hamlet's horror and revulsion upon contemplating the impurity of life and, by extension, his own involvement in it through the incest of his mother. J. Dover Wilson, in *What Happens in "Hamlet"* (London: Cambridge UP, 1935), sees "sullied flesh" as the clue to many significant passages in the play (for example, to Hamlet's imaginations "foul as Vulcan's stithy"); to his preoccupation with sexuality, particularly with the sexual nature of his mother's crime; and to his strange conduct toward Ophelia and Polonius. This view becomes even more credible when one considers Hamlet's seemingly incomprehensible remark to Polonius in act II, scene ii, where he calls the old man a "fishmonger" (Elizabethan slang for "pimp"); implies that Ophelia is a prostitute by referring in the same speech to "carrion" (Elizabethan "flesh" in the carnal sense); and warns Polonius not to let her "walk i' the sun" (that is, get too close to the "son" of Denmark, the heir apparent, him of the "sullied flesh" and "foul" imaginations). Wilson explains Hamlet's ambiguous remark as obscene because Hamlet is angry that Polonius would stoop to "loose" his daughter to him (as stockmen "loose" cows and mares to bulls and stallions to be bred) in order to wheedle from him the secret of his behavior, and he is angry and disgusted that his beloved would consent to be used in this way. Hence his later obscenities to her, as in act III, scene i, when he tells her repeatedly to go to a "nunnery" (Elizabethan slang for "brothel").

One final example must suffice to illustrate the importance of textual accuracy in interpreting this piece of literature. In the second quarto the speeches of the officiant at Ophelia's funeral are headed "Doct." This is probably "Doctor of Divinity," the term that one editor of *Hamlet*, Cyrus Hoy, inserts in the stage directions (Norton Critical Edition [New York: Norton, 1963]). The "Doctor of Divinity" reading was one reason for J. Dover Wilson's asserting positively that Ophelia's funeral was a Protestant service, contrary to the way directors often stage it. Indeed, the point seems to be relevant, because it affects one's interpretation of the play. Although Shakespeare used anachronisms whenever they suited his purpose, a careless disregard of facts and logic was not typical of him. For example, both Hamlet and Horatio are students at Wittenberg. That this university was founded several hundred years after the death of the historical Hamlet is beside the point. What does seem important is that Wittenberg was the university of Martin Luther and a strong center of Protestantism. It is not unreasonable to assume, then, that Shakespeare wanted his audience to think of Denmark as a Protestant country (it was so in his day)—indeed that he wanted the entire drama to be viewed in contemporary perspective, a point that will be elaborated later in this chapter.

2. A Summary of the Play

The main lines of the plot of *Hamlet* are clear. Hamlet, Prince of Denmark and heir presumptive to the Danish throne, is grief-stricken and plunged into melancholy by the recent death of his father and the "o'erhasty" remarriage of his mother to her late husband's brother, who has succeeded to the throne. The ghost of the prince's father appears to him and reveals that he was murdered by his brother, who now occupies the throne and whom he describes as "incestuous" and "adulterate." Enjoining young Hamlet not to harm his mother, the ghost exhorts him to take revenge on the murderer. In order to ascertain beyond question the guilt of his uncle and subsequently to plot his revenge, Hamlet feigns madness. His sweetheart Ophelia and his former schoolfellows Rosencrantz and Guildenstern attempt to discover from him the secret of his "antic behavior" (Ophelia because her father, Polonius, has ordered her to do so, Rosencrantz and Guildenstern because the king has ordered them to do so). All are unsuccessful.

Before actually initiating his revenge, Hamlet wants to be sure it will hit the guilty person. To this end, he arranges for a company of traveling players to present a drama in the castle that will depict the murder of his father as the ghost has described it. When the king sees the crime reenacted, he cries out and rushes from the assembly. This action Hamlet takes to be positive proof of his uncle's guilt, and from this moment he awaits only the right opportunity to kill him. After the play, Hamlet visits his mother's apartment, where he mistakes Polonius for the King and kills him. The killing of Polonius drives Ophelia mad and also convinces the king that Hamlet is dangerous and should be gotten out of the way. He therefore sends Hamlet to England, accompanied by Rosencrantz and Guildenstern, ostensibly to collect tribute, but in reality to be murdered. However, Hamlet eludes this trap by substituting the names of his erstwhile schoolfellows on his own death warrant and escaping through the help of pirates. He reaches Denmark in time for the funeral of Ophelia, who has apparently drowned herself. Laertes, her brother, has returned from Paris vowing vengeance on Hamlet for the death of his father. The king helps Laertes by arranging a fencing match between the two young men and seeing to it that Laertes's weapon is naked and poisoned. To make doubly sure that Hamlet will not escape, the king also poisons a bowl of wine from which Hamlet will be sure to drink. During the match, Laertes wounds Hamlet, the rapiers change hands, and Hamlet wounds Laertes; the Queen drinks the poisoned wine; and Laertes confesses his part in the treachery to Hamlet, who then stabs the king to death. All the principals are thus dead, and young Fortinbras of Norway becomes king of Denmark.

3. Historical-Biographical Considerations

It will doubtless surprise many students to know that *Hamlet* is considered by some commentators to be topical and autobiographical in certain places. In view of Queen Elizabeth's advanced age and poor health—hence the precarious state of the succession to the British crown—Shakespeare's decision to mount a production of *Hamlet*, with its usurped throne and internally disordered state, comes as no surprise. (Edward Hubler has argued that *Hamlet* was probably written in 1600 [in Sylvan Barnet, gen. ed., *The Complete Signet Classic Shakespeare* (New York: Harcourt, 1972): 912, n.2].) There is some ground for thinking that Ophelia's famous characterization of Hamlet may be intended to suggest the Earl of Essex, formerly Elizabeth's favorite, who had incurred her severe displeasure and been tried for treason and executed:

> The courtier's, soldier's, scholar's, eye, tongue, sword
> The expectancy and rose of the fair state,
> The glass of fashion and the mould of form,
> The observed of all observers. . . . (III.i)

Also, something of Essex may be seen in Claudius's observation on Hamlet's madness and his popularity with the masses:

> How dangerous it is that this man goes loose!
> Yet must we not put the strong law on him:
> He's loved of the distracted multitude,
> Who like not in their judgment but their eyes;
> And where 'tis so, the offender's scourge is weighed,
> But never the offence. (IV.viii)

Yet another contemporary historical figure, the Lord Treasurer, Burghley, has been seen by some in the character of Polonius. Shakespeare may have heard his patron, the young Henry Wriothesley, Earl of Southampton, express contempt for Elizabeth's old Lord Treasurer; indeed, this was the way many of the gallants of Southampton's generation felt. Burghley possessed most of the shortcomings Shakespeare gave to Polonius; he was boring, meddling, and given to wise old adages and truisms. (He left a famous set of pious yet shrewd precepts for his son, Robert Cecil.) Moreover, he had an elaborate spy system that kept him informed about both friend and foe. One is reminded of Polonius's assigning Reynaldo to spy on Laertes in Paris (II.i). This side of Burghley's character was so well known that it might have been dangerous for Shakespeare to portray it on stage while the old man was alive (because Burghley had died in 1598, Shakespeare could with safety do so in this general way).

Other topical references include Shakespeare's opinion (II.ii) about the revival of the private theater, which would employ children and which would constitute a rival for the adult companies of the public theater, for which Shakespeare wrote. It is also reasonable to assume that Hamlet's instructions to the players (III.ii) contain Shakespeare's criticisms of contemporary acting, just as Polonius's description of the players' repertoire and abilities (II.ii) is Shakespeare's satire on dull people who profess preferences for rigidly classified genres. Scholars have also pointed out Shakespeare's treatment of other stock characters of the day: Osric, the Elizabethan dandy; Rosencrantz and Guildenstern, the bootlicking courtiers; Laertes and Fortinbras, the men of action; Horatio, the "true Roman" friend; and Ophelia, the courtly love heroine.

In looking at *Hamlet* the historical critic might be expected to ask, "What do we need to know about eleventh-century Danish court life or about Elizabethan England to understand this play?" Similar questions are more or less relevant to the traditional interpretive approach to any literary work, but they are particularly germane to analysis of *Hamlet*. For one thing, most twentieth-century American students, largely unacquainted with the conventions, let alone the subtleties, of monarchical succession, wonder (unless they are aided by notes) why Hamlet does not automatically succeed to the throne after the death of his father. He is not just the oldest son; he is the only son. Such students need to know that in Hamlet's day the Danish throne was an elective one. The royal council, composed of the most powerful nobles in the land, named the next king. The custom of the throne's descending to the oldest son of the late monarch had not yet crystallized into law.

As true as this may be in fact, however, Dover Wilson maintains that it is not necessary to know it for understanding *Hamlet*, because Shakespeare intended his audiences to think of the entire situation—characters, customs, and plot— as English, which he apparently did in most of his plays, even though they were set in other countries. Wilson's theory is based upon the assumption that an Elizabethan audience could have but little interest in the peculiarities of Danish government, whereas the problems of royal succession, usurpation, and potential revolution in a contemporary English context would be of paramount concern. He thus asserts that Shakespeare's audience conceived Hamlet to be the lawful heir to his father and Claudius to be a usurper and the usurpation to be one of the main factors in the play, important to both Hamlet and Claudius. Whether one accepts Wilson's theory or not, it is certain that Hamlet thought of Claudius as a usurper, for he describes him to Gertrude as

> A cutpurse of the empire and the rule,
> That from a shelf the precious diadem stole
> And put it in his pocket! (III.iv)

and to Horatio as one

> . . . that hath killed my king and whored my mother,
> Popped in between th' election and my hopes. . . . (V.ii)

This last speech suggests strongly that Hamlet certainly expected to succeed his father by election if not by primogeniture.

Modern students are also likely to be confused by the charge of incest against the Queen. Although her second marriage to the brother of her deceased husband would not be considered incestuous today by many civil and religious codes, it was so considered in Shakespeare's day. Some dispensation or legal loophole must have accounted for the popular acceptance of Gertrude's marriage to Claudius. That Hamlet considered the union incestuous, however, cannot be emphasized too much, for it is this repugnant character of Gertrude's sin, perhaps more than any other factor, that plunges Hamlet into the melancholy of which he is victim.

And here it is necessary to know what "melancholy" was to Elizabethans and to what extent it is important in understanding the play. A. C. Bradley tells us that it meant to Elizabethans a condition of the mind characterized by nervous instability, rapid and extreme changes of feeling and mood, and the disposition to be for the time absorbed in a dominant feeling or mood, whether joyous or depressed. If Hamlet's actions and speeches are examined closely, they seem to indicate symptoms of this disease. He is by turns cynical, idealistic, hyperactive, lethargic, averse to evil, disgusted at his uncle's drunkenness and his mother's sensuality, and convinced that he is rotten with sin. To appreciate his apparent procrastination, his vacillating from action to contemplation, and the other superficially irreconcilable features in his conduct, readers need to realize that at least part of Hamlet's problem is that he is a victim of extreme melancholy. (For more detailed discussions of Hamlet's melancholy, see A. C. Bradley's *Shakespearean Tragedy* [London: Macmillan, 1914], J. Dover Wilson's *What Happens in "Hamlet,"* and Weston Babcock's *"Hamlet": A Tragedy of Errors* [Lafayette, IN: Purdue UP, 1961].)

One reason for the popularity of *Hamlet* with Elizabethan audiences was that it dealt with a theme they were familiar with and fascinated by—revenge. *Hamlet* is in the grand tradition of revenge tragedies and contains virtually every stock device observable in vastly inferior plays of this type. Thomas Kyd's *Spanish Tragedy* (ca. 1585) was the first successful English adaptation of the Latin tragedies of Seneca. The typical revenge tragedy began with a crime (or the recital of it); continued with an injunction by some agent (often a ghost) to the next of kin to avenge the crime; grew complicated by various impediments to the revenge, such as identifying the criminal and hitting upon the proper time, place, and mode of the revenge; and concluded with the death of the criminal, the avenger, and frequently all the principals in the drama.

One additional fact about revenge may be noted. When

Claudius asks Laertes to what lengths he would go to avenge his father's death, Laertes answers that he would "cut [Hamlet's] throat i' th' church" (IV.vii). It is probably no accident that Laertes is so specific about the method by which he would willingly kill Hamlet. In Shakespeare's day it was popularly believed that repentance had to be vocal to be effective. By cutting Hamlet's throat, presumably before he could confess his sins, Laertes would deprive Hamlet of this technical channel of grace. Thus Laertes would destroy both Hamlet's soul and his body and would risk his own soul, a horrifying illustration of the measure of his hatred. Claudius's rejoinder

> No place indeed should murder sanctuarize;
> Revenge should have no bounds

indicates the desperate state of the king's soul. He is condoning murder in a church, traditionally a haven of refuge, protection, and legal immunity for murderers.

Elizabethan audiences were well acquainted with these conventions. They thought there was an etiquette, almost a ritual, about revenge; they believed that it was in fact a fine art and that it required a consummate artist to execute it.

4. Moral-Philosophical Considerations

Any discussion of *Hamlet* should acknowledge the enormous body of excellent commentary that sees the play as valuable primarily for its moral and philosophical insights. Little more can be done here than to summarize the most famous of such interpretations. They naturally center on the character of Hamlet. Some explain Hamlet as an idealist temperamentally unsuited for life in a world peopled by fallible creatures. He is therefore shattered when he discovers that some humans are so ambitious for a crown that they are willing to murder for it and that others are so highly sexed that they will violate not only the laws of decorum (for example, by remarrying within a month of a spouse's death) but also the civil and ecclesiastical laws against incest. He is further crushed when he thinks that his fiancée and his former schoolfellows are tools of his murderous uncle. Other critics see Hamlet's plight as that of the essentially moral and virtuous intellectual man, certainly aware of the gentlemanly code that demands satisfaction for a wrong, but too much the student of philosophy and the Christian religion to believe in the morality or the logic of revenge. Related to this is the view of Hamlet as a kind of transitional figure, torn between the demands and the values of the Middle Ages and those of the modern world. The opposed theory maintains that Hamlet *is* a man of action, thwarted by such practical obstacles as how to kill a king surrounded by a bodyguard. Many modern critics emphasize what they term Hamlet's psychoneurotic state, a condition that obviously derives from the moral complexities with which he is faced.

Hamlet fulfills the technical requirements of the revenge play as well as the salient requirements of a classical tragedy; that is, it shows a person of heroic proportions going down to defeat under circumstances too powerful for him to cope with. For most readers and audiences the question of Hamlet's tragic flaw will remain a moot one. But this will not keep them from recognizing the play as one of the most searching artistic treatments of the problems and conflicts that form so large a part of the human condition.

C. Traditional Approaches to *Adventures of Huckleberry Finn*

There are few works of literature that lend themselves to so many interpretive analyses as *Huckleberry Finn*. Bernard De Voto has written that the novel contains "God's plenty"; in that verdict lies the key to the traditional critical approach. The phrase "God's plenty" was also applied by Dryden to Chaucer's *Canterbury Tales;* so we should remember those attributes of Chaucer's art that elicited such praise—narrative and descriptive power, keen knowledge of human nature, high comedy, biting satire, and lofty morality. All of these are also in *Huckleberry Finn*.

1. Dialect and Textual Matters
To Twain's good ear and appreciation of the dramatic value of dialect we owe not only authentic and subtle shadings of class, race, and personality, but also, as Lionel Trilling has

said, a "classic prose" that moves with "simplicity, directness, lucidity, and grace" ("Introduction," The [sic] Adventures of Huckleberry Finn [New York: Holt, 1948]: xvii).
T. S. Eliot called this an "innovation, a new discovery in the
English language," an entire book written in the natural
prose rhythms of conversation ("Introduction," The [sic] Adventures of Huckleberry Finn [London: Cresset, 1950] in Adventures of Huckleberry Finn, ed. Sculley Bradley, Richmond Croom Beatty, E. Hudson Long, and Thomas Cooley
[New York: Norton, 1962]: 323). This linguistic innovation is
certainly one of the features to which Ernest Hemingway
referred when he said that "all modern American literature
comes from one book by Mark Twain called Huckleberry
Finn" (Green Hills of Africa [New York: Scribner's, 1935]:
22). If we agree with Hemingway, therefore, we can think of
Twain as the "father of modern American literature."

Huckleberry Finn has an interesting textual history that
space will allow us only to touch on here. Writing a frontier
dialect, Twain was trying, with what success we have just
seen, to capture in both pronunciation and vocabulary the
spirit of the times from the lips of contemporary people.
Nevertheless, some of his editors (for example, Richard Watson Gilder of the Century Magazine, William Dean Howells,
and especially Twain's wife Livvie) bowdlerized and pret-
tified those passages they thought "too coarse or vulgar" for
Victorian ears, in certain cases with Twain's full consent. It is
a minor miracle that this censoring, though it has taken
something from the verisimiltude of the novel, seems not to
have harmed it materially. (Hamlin Hill and Walter Blair,
The Art of "Huckleberry Finn" [New York: Intext, 1962] is an
excellent succinct treatment of the textual history of this
novel.)

2. The Genre and the Plot of the Novel

Huckleberry Finn is a novel—that is, an extended prose narrative dealing with characters within the framework of a plot.
Such a work is usually fictitious, but both characters and
situations or events may be drawn from real life. It may emphasize action or adventure (for example, Treasure Island or
mystery stories); or it may concentrate on character delinea-

tion (that is, the way people grow or deteriorate or remain static in the happenings of life—*The Rise of Silas Lapham* or *Pride and Prejudice*); or it may illustrate a theme either aesthetically or propagandistically (*Wuthering Heights* or *Uncle Tom's Cabin*). It can, of course, do all three of these, as *Huckleberry Finn* does, a fact that accounts for the multiple levels of interpretation.

Huckleberry Finn is not only a novel; it is also a direct descendant of an important subgenre: the Spanish picaresque tale that arose in the sixteenth century as a reaction against the chivalric romance. In the latter type, pure and noble knights customarily rescued virtuous and beautiful heroines from enchanted castles guarded by fire-breathing dragons or wicked knights. In an attempt to debunk the artificiality and insipidity of such tales, Spanish writers of the day (notably the anonymous author of *Lazarillo de Tormes*) introduced into fiction as a central figure a kind of antihero, the picaro—a rogue or rascal of low birth who lived by his wits and his cunning rather than by exalted chivalric ideals. (Although not a pure picaro, Cervantes's Don Quixote is involved in a plot more rambling and episodic than unified and coherent.) Indeed, except for the fact that the picaro is *in* each of the multitude of adventures, all happening "on the road," the plot is negligible by modern standards. In these stories we simply move with this new type of hero from one wild and sensational experience to another, involving many pranks and much trenchant satire. Later treatments of the picaro have occasionally minimized and frequently eliminated his roguish or rascally traits. Dickens's picaros, for example, are usually model poor boys.

Many of the classics of world literature are much indebted to the picaresque tradition, among them René Le Sage's *Gil Blas*, Henry Fielding's *Tom Jones*, and Charles Dickens's *David Copperfield*, to mention only a few. *Huckleberry Finn* is an obvious example of the type. The protagonist is a thirteen- or fourteen-year-old boy living in the American antebellum South. He is a member by birth of the next-to-the-lowest stratum of Southern society, white trash—one who has a drunkard father who alternately abandons him and then returns to persecute him, but who has no mother, no roots, and no background or breeding in the conventionally accepted sense. He is the town bad boy who smokes, chews, plays hooky, and stays dirty, and whom two good ladies of St. Petersburg, Missouri, have elected to civilize.

The narrative moves onto "the road" when Huck, partly to escape the persecution of his drunken father and partly to evade the artificially imposed restrictions and demands of society, decides to accompany Jim, the slave of his benefactors, in his attempt to run for his freedom. The most immediate reason for Jim's deciding to run away is the fact that Miss Watson, his owner, has decided to sell him "down the river"—that is, into the Deep South, where instead of making a garden for nice old ladies or possibly being a house servant, he will surely become a field hand and work in the cane or cotton fields. These two, the teenaged urchin and the middle-aged slave, defy society, the law, and convention in a daring escape on a raft down the dangerous Mississippi River.

Continually in fear of being captured, Huck and Jim travel mostly at night. They board a steamboat that has run onto a snag in the river and has been abandoned; on it they find a gang of robbers and cutthroats, whom they manage to elude without detection. In a vacant house floating down the river they discover the body of a man shot in the back, who, Jim later reveals, is Huck's father. They become involved in a blood feud between two aristocratic pioneer families. They witness a cold-blooded murder and an attempted lynching on the streets of an Arkansas village. They acquire two disreputable traveling companions who force them to render menial service and to take part in burlesque Shakespearean performances, bogus revival meetings, and attempted swindles of orphans with newly inherited wealth. Finally, after some uneasy moments when Jim is captured, they learn that Jim has been freed by his owner, and Huck decides to head west—away from civilization.

3. Historical-Biographical Considerations

At the surface of narrative level, *Huckleberry Finn* is something of a thriller. The sensationalism may seem to make the story improbable, if not incredible, but we should consider its historical and cultural context. This was part of frontier

America in the 1840s and 1850s, a violent and bloody time. It was the era of Jim Bowie and his murderous knife, of gunslingers like Jack Slade, of Indian fighters like Davy Crockett and Sam Houston. Certainly there is a touch of the frontier, of the South or the West, in the roughness, the cruelty, the lawlessness, and even the humor of *Huckleberry Finn*. Indeed, Mark Twain was very much in the tradition of such humorists of the Southwest as Thomas Bangs Thorpe and such professional comedians as Artemus Ward and Josh Billings; in various writings he employed dialect for comedy, burlesque, the tall tale, bombast, the frontier brag. *Huckleberry Finn*, of course, far transcends the examples of early American humor.

Furthermore, we know from Mark Twain's autobiographical writings and from scholarly studies of him, principally those of Bernard De Voto, A. B. Paine, and Dixon Wecter, that the most sensational happenings and colorful characters in *Huckleberry Finn* are based on actual events and persons Twain saw in Hannibal, Missouri, where he grew up, and in other towns up and down the Mississippi. For example, the shooting of Old Boggs by Colonel Sherburn is drawn from the killing of one "Uncle Sam" Smarr by William Owsley on the streets of Hannibal on January 24, 1845. The attempted lynching of Sherburn is also an echo of something that Mark Twain saw as a boy, for he declared in later life that he once "saw a brave gentleman deride and insult a [lynch] mob and drive it away." During the summer of 1847 Benson Blankenship, older brother of the prototype Huck, secretly aided a runaway slave by taking food to him at his hideout on an island across the river from Hannibal. Benson did this for several weeks and resolutely refused to be enticed into betraying the man for the reward offered for his capture. This is undoubtedly the historical source of Huck's loyalty to Jim that finally resulted in his electing to "go to Hell" in defiance of law, society, and religion rather than turn in his friend.

A point about Jim's escape that needs clarification is his attempt to attain his freedom by heading *south*. Actually, Cairo, Illinois, free territory and Jim's destination, is farther south on the river than St. Petersburg, Missouri, from which he is escaping. Thus when the fugitives miss Cairo in the fog and dark, they have lost their only opportunity to free Jim by escaping southward. Still another point is that if it had been Jim's object simply to get to *any* free territory, he might as easily have crossed the river to Illinois right at St. Petersburg, his home. But this was not his aim. Although a free state, Illinois had a law requiring its citizens to return runaway slaves. Jim therefore wanted particularly to get to Cairo, Illinois, a junction of the underground railroad system where he could have been helped on his way north and east on the Ohio River by abolitionists.

The obscene performance of the "Royal Nonesuch" in Bricksville, Arkansas, where the King prances about the stage on all fours as the "cameleopard," naked except for rings of paint, was based on some of the bawdier male entertainments of the old Southwest. This particular type featured a mythical phallic beast called the "Gyascutus." There were variations, of course, in the manner of presentation, but the antics of the King illustrate a common version. (Both Mark Twain and his brother Orion Clemens recorded performances of this type, Orion in an 1852 newspaper account of a Hannibal showing, Mark in a notebook entry made in 1865 while he was in Nevada.)

The detailed description of the Grangerford house with its implied yet hilarious assessment of the nineteenth-century culture may be traced to a chapter from *Life on the Mississippi* entitled "The House Beautiful." Here may be observed the conformity to the vogue of sentimentalism, patriotism, and piousness in literature and painting and the general garishness in furniture and knickknacks.

One pronounced theme in *Huckleberry Finn* that has its origin in Twain's personality is his almost fanatical hatred of aristocrats. Indeed, aristocracy was one of his chief targets. A *Connecticut Yankee in King Arthur's Court* is less veiled than *Huckleberry Finn* in its attack on the concept. But it was not only British aristocracy that Twain condemned; elsewhere he made his most vitriolic denunciations of the American Southern aristocrat. Though more subtle, *Huckleberry Finn* nevertheless is the more searching criticism of aristocracy. For one thing, aristocracy is hypocritical. Aristocrats are not paragons of true gentleness, graciousness, courtliness, and selflessness. They are trigger-happy, inordinately proud, implacable bullies. But perhaps Twain's antipathy to aristocracy, expressed in virtually all his works, came from the obvious misery caused to all involved, perpetrators as well as victims. The most significant expression of this in *Huckleberry Finn* is, of course, in the notion of race superiority. Clinging as they did to this myth, aristocrats—as Alex Haley has portrayed them in *Roots*—could justify any kind of treatment of blacks. They could separate families, as in the case of Jim and the Wilks slaves; they could load them with chains, forget to feed them, hunt them like animals, curse and cuff them, exploit their labor, even think of them as subhuman, and then rationalize the whole sordid history by affirming that the slaves ought to be grateful for any contact with civilization and Christianity.

Moreover, not only aristocrats but every section of white society subscribed to this fiction; thus a degenerate wretch like pap Finn could shoulder a free Negro college professor off the sidewalk and later deliver an antigovernment, racist tirade to Huck replete with the party line of the Know-Nothings, a semisecret, reactionary political group that flourished for a brief period in the 1850s. (Its chief tenet was hostility to foreign-born Americans and the Roman Catholic Church. It derived its name from the answer its oath-bound members made to any question about it, "I know nothing about it.") We thus sense the contempt Twain felt for Know-Nothingism when we hear its chief doctrines mouthed by a reprobate like pap Finn. (Indeed, it may be more than coincidental that Twain never capitalizes the word "pap" when Huck is referring to his father.)

Closely related to this indictment of aristocracy and racism and their concomitant evils are Twain's strictures on romanticism, which he thought largely responsible for the harmful myths and cultural horrors that beset the American South of his day. In particular, he blamed the novels of Sir Walter Scott and their idealization of a feudal society. In real life this becomes on the adult level the blood feud of the Grangerfords and Shepherdsons and on the juvenile level the imaginative high jinks of Tom Sawyer and his "robber gang" and his "rescue" of Jim.

There are many other examples of historical and biographical influences on the novel. Years spent as a steamboat pilot
familiarized Mark Twain with every snag, sandbar, bend, or
other landmark on the Mississippi, as well as with the more
technical aspects of navigation—all of which add vivid authenticity to the novel. His vast knowledge of Negro superstitions was acquired from slaves in Hannibal, Missouri, and on
the farm of his beloved uncle, John Quarles, prototype of
Silas Phelps. Jim himself is modeled after Uncle Dan'l, a
slave on the Quarles place. These superstitions and examples
of folklore are not mere local color, devoid of rhyme or reason; but, as Daniel Hoffman has so clearly pointed out in
Form and Fable in American Fiction (New York: Oxford UP,
1961), they are "of signal importance in the thematic development of the book and in the growth toward maturity of its
principal characters" (321). Huck was in real life Tom Blankenship, a boyhood chum of Twain's who possessed most of
the traits Twain gave him as a fictional character. Although
young Blankenship's real-life father was ornery enough,
Twain modeled Huck's father on another Hannibal citizen,
Jimmy Finn, the town drunk.

Like *The Canterbury Tales*, where Dryden found "God's
plenty," *Huckleberry Finn* gives its readers a portrait gallery
of the times. Scarcely a class is omitted. The aristocracy is
represented by the Grangerfords, the Shepherdsons, and Colonel Sherburn. They are hardly Randolphs and Lees of tide-
water Virginia, and their homes reveal that. The Grangerford
parlor, for example, shows more of philistinism and puritanism than of genuine culture. These people are, nevertheless,
portrayed as recognizable specimens of the traditional aristocrat, possessed of dignity, courage, devotion to principle,
graciousness, desire to preserve ceremonious forms, and Calvinistic piety. Colonel Sherburn in particular illustrates another aspect of the traditional aristocrat—his contempt for
the common man, which is reflected in his cold-blooded
shooting of Old Boggs, his cavalier gesture of tossing the
pistol on the ground afterward, and his single-handedly facing down the lynch mob.

Towns of any size in *Huckleberry Finn* contain the indus-
trious, respectable, conforming bourgeoisie. In this class are

the Widow Douglas and her old-maid sister Miss Watson, the Peter Wilks family, and Judge Thatcher. The Phelpses too, although they own slaves and operate a "one-horse cotton plantation," belong to this middle class. Mrs. Judith Loftus, whose canniness undoes Huck when he is disguised as a girl, is, according to De Voto, the best-drawn pioneer wife in any of the contemporary records. The host of anonymous but vivid minor characters reflects and improves upon the many eyewitness accounts. These minor characters include the ferryboat owner, the boatmen who fear smallpox as they hunt Jim, the raftsmen heard from a distance joking in the stillness of the night. The Bible Belt poor white, whether whittling and chewing and drawling on the storefront benches of an Arkansas village or caught up in the fervor of a camp meeting or joining his betters in some sort of mob action, is described with undeniable authenticity.

Criminals like the robbers and cutthroats on the *Walter Scott* and those inimitable confidence men, the King and the Duke, play their part. Pap Finn is surely among the earliest instances of Faulkner's Snopes types—filthy, impoverished, ignorant, disreputable, bigoted, thieving, pitifully sure of only one thing, his superiority as a white man. Then we observe the slaves themselves, convincing because they include not just stereotyped minstrel characters or "moonlight and magnolia good darkies," but interesting human beings, laughable, strong, honorable, trifling, dignified, superstitious, illiterate, wise, loving, pathetic, loyal, victimized. Most make only brief appearances, yet we feel that we have known a group of engaging, complex, and gifted people.

4. Moral-Philosophical Considerations
Important as are its historical and biographical aspects, the chief impact of *Huckleberry Finn* derives from its morality. This is, indeed, the *meaning* of the novel. All other aspects are subservient to this one. Man's inhumanity to man (as Huck says, "Human beings *can* be awful cruel to one another") is the major theme of this work, and it is exemplified in both calm and impassioned denunciation and satire. Almost all the major events and most of the minor ones are variations on this theme. The cruelty may be manifested in

attempts to swindle young orphans out of their inheritance, to con village yokels with burlesque shows, to fleece religion-hungry frontier folk with camp meetings, or to tar and feather malefactors extralegally. Cruelty can and often does have even more serious consequences: for example, the brutal and senseless slaughter of the aristocratic Grangerfords and Shepherdsons and the murder of a harmless old windbag by another arrogant aristocrat.

The ray of hope that Mark Twain reveals is the relationship of Huck and Jim; Huck's ultimate salvation comes when of his own choice he rejects the values of the society of his time (he has all along had misgivings about them) and decides to treat Jim as a fellow human being. The irony is that Huck has made the right decision by scrapping the "right" reasons (that is, the logic of conventional theology) and by following his own conscience. He is probably too young to have intellectualized his decision and applied it to black people as a whole. Doubtless it applies only to Jim as an individual. But this is a tremendous advance for a boy of Huck's years. It is a lesson that is stubbornly resisted, reluctantly learned. But it is *the* lesson of *Huckleberry Finn*.

Huckleberry Finn is a living panorama of a country at a given time in history. It also provides insights, and it makes judgments that are no less valid in the larger sense today than they are about the period Mark Twain chronicled. This fidelity to life in character, action, speech, and setting; this personal testament; this encyclopedia of human nature; this most eloquent of all homilies—all of these are what cause this book to be not only a supreme artistic creation but also, in the words of Lionel Trilling, "one of the central documents of American culture" (6).

D. Traditional Approaches to "Young Goodman Brown"

"Young Goodman Brown," universally acclaimed as one of Hawthorne's best short stories, presents the student with not only several possible meanings but several rather ambiguous meanings. D. M. McKeithan, in an article entitled "'Young Goodman Brown': An Interpretation" (*Modern Language*

Notes 67 [1952]: 93), lists the suggestions that have been advanced as "the theme" of the story: "the reality of sin, the pervasiveness of evil, the secret sin and hypocrisy of all persons, the hypocrisy of Puritanism, the results of doubt or disbelief, the devastating effects of moral scepticism, . . . the demoralizing effects of the discovery that all men are sinners and hypocrites." Admittedly, these themes are not as diverse as they might at first appear. They are, with the possible exception of the one specifically mentioning Puritanism, quite closely related. But meaning is not restricted to theme, and there are other ambivalences in the story that make its meanings both rich and elusive. After taking into account some matters of text and genre, we shall look at "Young Goodman Brown" from our traditional perspectives.

1. The Text of the Story

Textually, "Young Goodman Brown," first published in 1835 in the *New England Magazine*, presents relatively few problems. Obsolete words in the story like "wot'st" (know), "Goody" (Goodwife, or Mrs.), "Goodman" (Mr.) are defined in most desk dictionaries, and none of the other words has undergone radical semantic change. Nevertheless, as we have seen, although a literary work may have been written in a day when printing had reached a high degree of accuracy, a perfect text is by no means a foregone conclusion. With Hawthorne, as with other authors, scholars are constantly working on more accurate texts.

For example, the first edition of this handbook used a version of "Young Goodman Brown" that contained at least two substantive variants. About three-fourths of the way through the story the phrase "unconcerted wilderness" appeared. David Levin, in an article entitled "Shadows of Doubt: Specter Evidence in Hawthorne's 'Young Goodman Brown'" (*American Literature* 34 [Nov. 1962]: 346, n. 8) points out that nineteen years after Hawthorne's death, a version of the story edited by George P. Lathrop printed "unconcerted" for the first time: every version before then, including Hawthorne's last revision, had had "unconverted." In that same paragraph the first edition of this handbook printed "figure" as opposed to "apparition," the word that Levin tells us occurred in the first published versions of the story. Obviously, significant interpretive differences could hinge on which words are employed in these contexts.

2. The Genre and the Plot of the Story

"Young Goodman Brown" is a short story; that is, it is a relatively brief narrative of prose fiction (ranging in length from five hundred to twenty thousand words) characterized by considerably more unity and compression in all its parts than the novel—in theme, plot, structure, character, setting, and mood. In the story we are considering, the situation is this: one evening near sunset sometime in the late seventeenth century, Goodman Brown, a young man who has been married only three months, prepares to leave his home in Salem, Massachusetts, and his pretty young bride, Faith, to go into the forest and spend the night on some mission that he will not disclose other than to say that it must be performed between sunset and sunrise. Although Faith has strong forebodings about his journey and pleads with him to postpone it, Brown is adamant and sets off. His business is evil by his own admission; he does not state what it is specifically, but it becomes apparent to the reader that it involves attending a witches' Sabbath in the forest, a remarkable action in view of the picture of Brown, drawn early in the story, as a professing Christian who admonishes his wife to pray and who intends to lead an exemplary life after this one night.

The rising action begins when Brown, having left the village, enters the dark, gloomy, and probably haunted forest. He has not gone far before he meets the Devil in the form of a middle-aged, respectable-looking man with whom Brown has made a bargain to accompany on his journey. Perhaps the full realization of who his companion is and what the night may hold in store for him now dawns on Brown, for he makes an effort to return to Salem. It is at best a feeble attempt, however, for, though the Devil does not try to detain him, Brown continues walking with him deeper into the forest.

As they go, the Devil shocks Goodman Brown by telling him that his (Brown's) ancestors were religious bigots, cruel exploiters, and practitioners of the black art—in short, full-fledged servants of the Devil. Further, the young man is told that the very pillars of New England society, church, and state are witches (creatures actually in league with the Devil), lechers, blasphemers, and collaborators with the Devil. Indeed, he sees his childhood Sunday School teacher, now a witch, and overhears the voices of his minister and a deacon of his church as they ride past conversing about the diabolical communion service to which both they and he are going.

Clinging to the notion that he may still save himself from this breakup of his world, Goodman Brown attempts to pray, but stops when a cloud suddenly darkens the sky. A babel of voices seems to issue from the cloud, many recognizable to Brown as belonging to godly persons, among them his wife. After the cloud has passed, a pink ribbon such as Faith wears in her cap flutters to the ground. Upon seeing it, Goodman Brown is plunged into despair and hastens toward the witches' assembly. Once there, he is confronted with a congregation made up of the wicked and those whom Brown had always assumed to be righteous. As he is led to the altar to be received into this fellowship of the lost, he is joined by Faith. The climax of the story comes just before they receive the sacrament of baptism: Brown cries to his wife to look heavenward and save herself. In the next moment he finds himself alone.

The dénouement (resolution, unraveling) of the plot comes quickly. Returning the next morning to Salem, Goodman Brown is a changed man. He now doubts that anyone is good—his wife, his neighbors, the officials of church and state—and he remains in this state of cynicism until he dies.

The supernaturalism and horror of "Young Goodman Brown" mark the story as one variant of the Gothic tale, a type of ghost story originating formally in late eighteenth-century England and characterized by spirit-haunted habitations, diabolical villains, secret doors and passageways, terrifying and mysterious sounds and happenings, and the like. Obviously, "Young Goodman Brown" bears some resemblance to these artificial creations, the aesthetic value of most of which is negligible. What is much more significant is that here is a variation of the Faust legend, the story of a man who makes a bargain with the Devil (frequently the sale of his soul) in exchange for some desirable thing. In this instance Goodman Brown did not go nearly so far in the original indenture, but it was not necessary from the Devil's point of view. One glimpse of evil unmasked was enough to wither the soul of Brown forever.

3. Historical-Biographical Considerations

So much for textual matters, paraphrasable content, and genre. What kind of historical or biographical information do we need in order to feel the full impact of this story, aesthetically and intellectually? Obviously, some knowledge of Puritan New England is necessary. We can place the story in time easily, because Hawthorne mentions that it takes place in the days of King William (that is, William III, who reigned from 1688 to 1702). Other evidences of the time of the story are the references to persecution of the Quakers by Brown's grandfather (the 1660s) and King Philip's War (primarily a massacre of Indians by colonists [1675–1676]), in which Brown's father participated. Specific locales like Salem, Boston, Connecticut, and Rhode Island are mentioned, as are terms used in Puritan church organization and government, such as ministers, elders, meetinghouses, communion tables, saints (in the Protestant sense of *any* Christian), selectmen, and lecture days.

But it is not enough for us to visualize a sort of first Thanksgiving picture of Pilgrims with steeple-crowned hats, Bibles, and blunderbusses. For one thing, we need to know something of Puritan religion and theology. This means at least a slight knowledge of Calvinism, a main source of Puritan religious doctrine. A theology as extensive and complex as Calvinism and one that has been the subject of so many misconceptions cannot be described adequately in a handbook of this type. But at the risk of perpetuating some of these misconceptions, let us mention three or four tenets of Calvinism that will illuminate to some degree the story of Goodman Brown. Calvinism stresses the sovereignty of God—in goodness, power, and knowledge. Correspondingly, it emphasizes the helplessness and sinfulness of human beings, who have been since the Fall of Adam, innately and totally depraved. Their only hope is in the grace of God, for God alone is powerful enough (sovereign enough) to save them. And the most notorious, if not the chief, doctrine is predestination, which includes the belief that God has, before their creation, selected certain people for eternal salvation, others for eternal damnation. Appearances are therefore misleading; an outwardly godly person might not be one of the elect. Thus it is paradoxical that Goodman Brown is so shocked to learn that there is evil among the apparently righteous, for this was one of the most strongly implied teachings of his church.

In making human beings conscious of their absolute reliance on God alone for salvation, Puritan clergymen dwelt long and hard on the pains of hell and the powerlessness of mere mortals to escape them. Brown mentions to the Devil that the voice of his pastor "would make me tremble both Sabbath day and lecture day." This was a typical reaction. In Calvinism, nobody could be sure of sinlessness. Introspection was mandatory. Christians had to search their hearts and minds constantly to purge themselves of sin. Goodman Brown is hardly expressing a Calvinistic concept when he speaks of clinging to his wife's skirts and following her to Heaven. Calvinists had to work out their own salvation in fear and trembling, and they were often in considerable doubt about the outcome. The conviction that sin was an ever-present reality that destroyed the unregenerate kept it before them all the time and made its existence an undoubted, well-nigh tangible fact. We must realize that aspects of the story like belief in witches and an incarnate Devil, which until the recent upsurge of interest in demonism and the occult world have struck modern readers as fantastic, were entirely credible to New Englanders of this period. Indeed, on one level, "Young Goodman Brown" may be read as an example of Satanism. Goody Cloyse and the Devil in the story even describe at length a concoction with which witches were popularly believed to have anointed themselves and a satanic worship attended by witches, devils, and lost souls.

It is a matter of historical record that a belief in witchcraft and the old pagan gods existed in Europe side by side with Christianity well into the modern era. The phenomenon has recurred in our own day, ballyhooed by the popular press as well as the electronic media. There was an analogous belief prevalent in Puritan New England. Clergymen, jurists, statesmen—educated people generally, as well as uneducated folk—were convinced that witches and witchcraft were realities. Cotton Mather, one of the most learned men of the period, attests eloquently to his own belief in these phenomena in *The Wonders of the Invisible World*, his account of the trials of several people executed for witchcraft. Some of the headings in the table of contents are instructive: "A True Narrative, collected by Deodat Lawson, related to Sundry Persons afflicted by Witchcraft, from the 19th of March to the 5th of April, 1692" and "The Second Case considered, viz. If one bewitched be cast down with the look or cast of the Eye of another Person, and after that recovered again by a Touch from the same Person, is not this an infallible Proof that the party accused and complained of is in Covenant with the Devil?"

Hawthorne's great-grandfather, John Hathorne (Nathaniel added the "w"), was one of the judges in the infamous Salem witch trials of 1692, during which many people were tortured, and nineteen hanged, and one crushed to death (a legal technicality was responsible for this special form of execution). Commentators have long pointed to "Young Goodman Brown," *The Scarlet Letter*, and many other Hawthorne stories to illustrate his obsession with the guilt of his Puritan forebears for their part in these crimes. In "The Custom House," his introduction to *The Scarlet Letter*, Hawthorne wrote of these ancestors who were persecutors of Quakers and witches and of his feeling that he was tainted by their crimes. The Devil testified that he helped young Goodman Brown's grandfather, a constable, lash a "Quaker woman . . . smartly through the streets of Salem," an episode undoubtedly related to Hawthorne's "Custom House" reference to his great-grandfather's "hard severity towards a woman of [the Quaker] sect."

Hawthorne's notebooks are also a source in interpreting his fiction. They certainly shed light on his preoccupation with the "unpardonable sin" and his particular definition of that sin. It is usually defined as blasphemy against the Holy Ghost, or continued conscious sin without repentance, or refusing to acknowledge the existence of God even though the Holy Spirit has actually proved it. The notebooks, however, and works of fiction like "Ethan Brand," "Young Goodman Brown," and *The Scarlet Letter* make it clear that for Hawthorne the Unpardonable Sin was to probe, intellectually and rationally, the human heart for depravity without tempering the search by a "human" or "democratic" sympathy. Specifically in the case of "Young Goodman Brown," Brown's obduracy of heart cuts him off from all, so that "his dying hour [is] gloom."

4. Moral-Philosophical Considerations

The terror and suspense in the Hawthorne story function as integral parts of the allegory that defines the story's theme. In allegory (a narrative containing a meaning beneath the surface one), there is usually a one-to-one relationship; that is, one idea or object in the narrative stands for only one idea or object allegorically. A story from the Old Testament illustrates this. The pharaoh of Egypt dreamed that seven fat cows were devoured by seven lean cows. Joseph interpreted this dream as meaning that seven years of plenty (good crops) would be followed by seven years of famine. "Young Goodman Brown" clearly functions on this level of allegory (while at times becoming richly symbolic). Brown is not just one Salem citizen of the late seventeenth century, but rather seems to typify humankind, to be in a sense Everyman, in that what he does and the reason he does it appear very familiar to most people, based on their knowledge of others and on honest appraisal of their own behavior.

For example, Goodman Brown, like most people, wants to experience evil—not perpetually, of course, for he is by and large a decent chap, a respectably married man, a member of a church—but he desires to "taste the forbidden fruit" ("have one last fling") before settling down to the business of being a solid citizen and attaining the good life. He feels that he can do this because he means to retain his religious faith, personified in his wife, who, to reinforce the allegory, is even named Faith. But in order to encounter evil, he must part with his Faith at least temporarily, something he is either willing or compelled to do. It is here that he makes his fatal mistake, for evil turns out to be not some abstraction nor something that can be played with for a while and then put down, but the very pillars of Goodman Brown's world—his ancestors, his earthly rulers, his spiritual overseers, and finally his Faith. In short, so overpowering are the fact and universality of evil in the world that Goodman Brown comes to doubt the existence of any good. By looking upon the very face of evil, he is transformed into a cynic and a misanthrope whose "dying hour was gloom."

Thomas E. Connolly, in "Hawthorne's 'Young Goodman Brown': An Attack on Puritanic Calvinism" (*American Literature* 28 [Nov. 1956]: 370–75), has remarked that Goodman Brown has not *lost* his faith; he has *found* it. That is, Goodman Brown believes that he understands the significance of the Calvinistic teaching of the depravity of humans; this realization makes him doubt and dislike his fellows and in effect paralyzes his moral will so that he questions the motivation of every apparently virtuous act. But this is surely a strange conclusion for Brown to reach, for he has violated the cardinal tenets of Calvinism. If Calvinism stressed anything, it stressed the practical and spiritual folly of placing hope or reliance on human beings and their efforts, which by the very nature of things are bound to fail, whereas God alone never fails. Therefore all trust should be reposed in Him. It is just this teaching that Brown has not learned. On the practical plane, he cannot distinguish between appearance and reality. He takes things and people at face value. If a man *looks* respectable and godly, Brown assumes that he is. And if the man turns out to be a scoundrel, Brown's every standard crumbles. He is in a sense guilty of a kind of idolatry: human institutions in the forms of ministers, church officers, statesmen, and wives have been his god. When they are discredited, he has nothing else to place his trust in and thus becomes a cynic and a misanthrope.

Thus, rather than making a frontal attack on Calvinism, Hawthorne indicted certain reprehensible aspects of Puritanism: the widespread holier-than-thou attitude; the spiritual blindness that led many Puritans to mistake a pious front for genuine religion; the latent sensuality in the apparently austere and disciplined soul (the very capstone of hypocrisy, because sins of the flesh were particularly odious to Puritan orthodoxy).

It will perhaps be argued that Calvinism at its most intense, with its dim view of human nature, is quite likely to produce cynicism and misanthropy. But historically, if paradoxically, Calvinists have been dynamic and full of faith; they have been social and political reformers, educators, enterprisers in business, explorers, foes of tyranny. The religious furnace in which these souls were tempered, however, is too hot for Goodman Brown. He is of a weaker breed, and the sum of his experience with the hard realities of life is disillusion and defeat. He has lost his faith. Whether because his faith was false or because he wished for an objectively verifiable certainty that is the antithesis of faith, Hawthorne does not say. He does not even say whether the whole thing was a dream or reality. Actually, it does not matter. The result remains: faith has been destroyed and supplanted by total despair because Brown is neither a good Calvinist, a good Christian, nor, in the larger sense, a good man.

As we have seen in our discussions of these works, the traditional approach in literary interpretation is neither rigidly dogmatic nor unaesthetic. It is eclectic. And although it has its rationale in the methods discussed in this chapter, it does not eschew insights from any other critical approach; it nonetheless insists on its own fundamental validity. Those other critical approaches, however, do provide insights not stressed in the traditional, such as the appreciation of form, to which we now turn.

· 2 ·

The Formalistic Approach

I. READING A POEM: AN INTRODUCTION TO THE FORMALISTIC APPROACH

Here is the situation: *[handwritten: many paradox]*

The reader is to be presented with a short but complete poem. Its author and its era of composition are unknown. Its language, however, is English; it is not a translation.

Here is the poem:

[handwritten line numbers: 1] A slumber did my spirit seal;
   I had no human fears;
[handwritten: 3] She seemed a thing that could not feel
   The touch of earthly years.

[handwritten: GAP]

[handwritten: 5] No motion has she now, no force;
   She neither hears nor sees;
[handwritten: 7] Rolled round in earth's diurnal course,
   With rocks, and stones, and trees.

[handwritten right margin: youthfulness → on earth movement]

[handwritten right margin: death → heavenly movement]

The poem seems quite simple, easy to grasp and to understand. The speaker—a persona, not necessarily the poet—recalls a frame of mind sometime in the past, when "she" (the female figure) was so active and alive that the speaker (mother? father? lover?) could hardly comprehend any earthly touch to the living female figure. Now, in the present, the speaker tells the reader or listener that the female is dead, but does so by circumlocution, or indirect statement. Only one word, "diurnal," should give even the mildest pause to most readers: it means "daily." Monosyllabic words dominate the poem. The meter is unvarying almost to the point of monotony—alternating soft and strong syllables, usually four of each in the first, third, fifth, and seventh lines; three of each in the other four lines. The rhymes are equally regular and predictable. There is classic restraint and regularity, a tight control.

There is also powerful emotional impact.

Whence comes that impact? Largely from the tightly stated irony and paradox of the poem. The speaker has both gotten what he or she desired and not gotten it: the expectations for the female figure have been realized—and incontrovertibly they have been demolished. Initially the speaker was confident in the eternal life of the female figure. What parent nurtures and enjoys a child while thinking thoughts of death rather than life? What lover thinks constantly if at all that the beloved will die, and prematurely at that? Life seems to ensure continued life. This female figure would somehow transcend earthly normalities, would not even age. The speaker was secure (slumbering) in that assumption. So we know from the first stanza.

But there is a huge gap, and at once a leap beyond that gap, between the first and second stanza. Something happened. Somehow the child or woman died. She already has been buried. The "slumber" of line one has become the eternal sleep of death. The "seal" of the "spirit" has become the coffin seal of the body. Even more poignantly, the life of the dynamic person in lines three and four, where sense perceptions of touching and feeling seem to be transmuted into ethereal or angelic dimensions, is now the unfeeling death of one who has no energy, no vitality, no sense of hearing or seeing. She is no more and no less than a rock or a stone or a tree fixed to the earth. The final irony, that paradox, is that the once motion-filled person is still in motion—but not the vital motion of a human person; she now moves daily a huge distance, a full turn of the earth itself, rotating with a motion not her own, but only that of rocks and stones—gravestones—and rooted trees.

The essential structure, or form, of the poem is the irony

that the speaker got precisely what he or she wanted—but hardly in the way anticipated—a structure that at a fairly obvious level contrasts by means of the two stanzas and resolves the paradox by their interaction. A closer look takes the reader beyond this now-evident contrast of two stanzas. The texture of the poem is enriched by the sleep imagery, the sleep of life becoming the sleep of death. The "slumber" of line 1 connotes rest and quiet, even that of a baby or young child. The sibilant sounds of "s" at first suggest that quiet contentment, but they appear throughout the poem, taking on the irony of the second stanza almost like mournful echoes of the first. "Spirit" and "seal" not only continue the sibilant quality but also in retrospect are ambiguous terms, for "spirit" suggests death as well as life, and "seal" suggests not only security but finality: the coffin and the grave. In the third line the word "thing" at first seems to be a noncommital, simply denotative word: perhaps the poet was not even able to think of a better word, and used a filler. But in retrospect the female figure now is indeed a "thing," like a rock or a stone, a mere thing—in truth, dust. Furthermore, "thing" contrasts with its bluntness of sound with the sibilant sounds of so much of the rest of the poem, and anticipates the alternating sounds of the last line, the "s" sounds alternating with the harder sounds of "r" and the consonant clusters "st" and "tr" in "rocks," "stones," and "trees."

Like the reference to sleep, the references to the senses ("feel," "touch") in the first stanza are expanded in the second: motion, or its lack, involves the muscles in kinetics and kinesthesia; hearing and seeing are explicitly mentioned. But in each of the three cases a negative word precedes the sense word—"no," "neither," "nor." Then in line 7, we meet the awesome reality of kinetic motion without kinesthesia. In "Rolled round" we have the forced motion of the inert body. In a striking change of metrics, we realize that the seeming monotony of the alternating soft and hard syllables is broken here by a spondee in place of the dominant iamb, and the spondee in turn is strengthened by the alliterating "r" and the consonance of the "d" at the end of each of the two words, echoed in the initial "d" of "diurnal." Once having noticed that pounding spondee, we might in retrospect reconsider the two uses of "no" in line five, for they can be read almost as strongly as the stressed syllables of the line, giving still greater impact to the negative effect of the whole statement. Finally, the contrast between lines 7 and 8 is devastating. If we lift the line totally from its context, we can hear almost an ebullient sound in the seventh line, a glorious sweeping rhythm, aided by the vowels or assonance in the middle several words: "Rolled round in earth's diurnal course." But that sweeping, soaring quality comes up against the finality and slowed pace of the heavily impeded line 8, where the punctuation and the three accented monosyllabic words join to give the impact of three strong chords at the end of the symphony.

We have read a poem. Unless we know from other contexts, we still do not know the name of the author, the nationality, or the era of composition. We do not know who the speaker is, not even the sex of the speaker. We do not know if the poem concerns a real-life situation or a totally fictive one. We do not know whether the author took some similar real-life situation or incident that he or she then adapted and transmuted into a poem. We know only the poem itself, a short piece of richly textured literary art that bears up well under close analysis and resolves its tensions by means of irony and paradox, showing them not only in the contrast between two stanzas but also in seemingly minute details. We have read a poem and have analyzed it by using the formalistic approach to literature.

II. THE PROCESS OF FORMALISTIC ANALYSIS: MAKING THE CLOSE READER

What we demonstrated in the preceding page or two is a close reading in practice. The reading stands on its own. Others, perhaps many, have read the poem in much the same way. Indeed, perhaps most readers of the poem in the middle third of the century would have read this poem in something of this way. That is so because the approach to the poem is what we call formalistic, an approach with a methodology, with a history, with practitioners, and with some detractors. Let us now learn more about this formalistic approach.

Intensive reading begins with a sensitivity to the words of the text and all their denotative and connotative values and implications. An awareness of multiple meanings, even the etymologies of words, as traced in dictionaries will offer significant guidelines to what the work says. Usually adequate for most readers is one of the standard collegiate dictionaries. For expanded information on meanings and origins of words, one occasionally may need to check one of the huge unabridged dictionaries such as *Webster's Third New International Dictionary of the English Language* (Merriam). Details and examples of historical changes in word meanings are recorded in the most recent edition of *The Oxford English Dictionary*. Most of these larger dictionaries are usually available in college and public libraries.

Moreover, we must be alert to any allusion to mythology, history, or literature. Eliot's *Waste Land*, Pound's *Cantos*, and Joyce's *Ulysses* are scarcely comprehensible (or at least seem hardly worth the trouble) without some understanding of the many rich allusions in them.

After we have mastered the individual words in the literary text, we look for *structures* and *patterns*, interrelationships, ideally as intensively as Keats contemplated the Grecian urn. We will begin to see relationships of *reference* (pronouns to nouns, a voice to a speaker, an appositive to a name or place, time to a process, etc.), of *grammar* (sentence patterns and their modifiers, parallel words and phrases, agreement of subjects and verbs, etc.), of *tone* (choice of words, manner of speaking, attitudes toward subject and audience, etc.), and of *systems* (related metaphors, symbols, myths, images, allusions, etc.). Such internal relationships gradually reveal a *form*, a principle by which all subordinate patterns can be accommodated and accounted for. When all the words, phrases, metaphors, images, and symbols are examined in terms of each other and of the whole, any literary text will display its own internal logic. When that logic has been established, the reader is very close to identifying the overall form of the work.

The *context* (for example, the nature and personality of the speaker in a poem) must be identified also. In Robert Browning's "My Last Duchess" we must understand not only the personality of the Duke, who is speaking, but also something of the nature of the man to whom he addresses his remarks. Only to the extent that we understand that *what* the Duke is saying is revealed largely by *how* he says it can we really fathom the full implications of the Duchess's story. One of the beautiful ironies of the poem, after all, is the reader's awareness of implications that the Duke does not consciously intend. In Eliot's *Waste Land* one must constantly remember the presence of the Tiresias figure, from whom all other characters in the poem take their being, as the ironically disposed observer-participant-commentator of each episode or context. When the intricate links of the many shards of experience drawn from both the ancient and the modern worlds (and they are linked by the pervasive Tiresias) are perceived, the form of the poem is becoming apparent.

III. A BRIEF HISTORY OF FORMALISTIC CRITICISM

A. The Course of a Half Century

The formalistic approach, as we use the term in this book, emphasizes the manner of reading literature that was given its special dimensions and emphases by English and American critics in the first two-thirds of this century. To many, indeed to most, students of literature during that era, this approach came to be called the New Criticism.

In the last third of this century, the New Criticism came to be called by other names, not always favorable—and some epithets bordered on the vitriolic. At the least it has come to be called by many the *old* New Criticism, for even "newer" approaches have gained popularity and have had little or nothing in common with the old New Criticism. For that matter, the word *formalistic* needs some small qualification as well, for here it will be used more or less synonymously with the methodology of the New Critics, and it is not directly concerned with the Russian formalists, though the methodologies share some principles.

Regardless of shifting attitudes toward English and American formalist criticism (more about that shortly), we are quite content to sail against the winds of change and to assert that being a good reader of literature necessitates our reading closely and reading well. Reading well is what the New Critics helped us to do.

They taught us to look at the individual work of literary art as an organic form. They articulated the concept that in an organic form there is a consistency and an internal vitality that we should look for and appreciate. In doing so, we would appropriate the work to ourselves and make it part of our consciousness in the same way that we might when we study Mahler's *Ninth Symphony* or Michelangelo's *David*, or in the same way that the persona in Keats's ode studied the Grecian urn.

They taught us. But how new were the New Critics when they were called that by John Crowe Ransom in *The New Critics* in 1941? Actually—and this should come as no surprise—there were forebears of great note. The New Critics did not spring suddenly from Zeus's head. We should not be surprised at this because in a form of human endeavor so basic as the creation of literary art we can expect a continuity in the way that art is created or becomes art. Nor should we be surprised that criticism, the informed reading of that art, should have a continuity as well.

More specifically, we should not be surprised because one of the most salient considerations of the New Critics was emphasis on form, on the work of art as an object. Can we imagine any art—whether it be literary, musical, plastic, or dramatic, and regardless of its era, even our own, when formlessness is sometimes important—that does not have some sense of form? The form need not be geometric or physical or otherwise perceptible to the eye, and indeed often it is not, but it is there. To be sure, it might be most easily perceived at a physical level at first: the external and obvious shape of a statue, the geometric pattern of arches and of horizontal and vertical lines in a building, the four-line stanza of Sappho or the pattern of strophes in Pindar, or the careful physical shape of a sonnet, a sestina, or a haiku. The New Critics did not invent these obvious forms.

But they helped us to read better by reminding us of what was there eons earlier. Art entails form; form takes many forms. *(Even one that a critic might not be aware of —)*

So let us consider further some of the background elements of formalistic theory.

B. Backgrounds of Formalistic Theory

Classical art and aesthetics amply testify to a preoccupation with form. Plato exploits dialectic and shapes movement toward Socratic wisdom by his imagery, metaphor, dramatic scenes, characterization, setting, and tone. Aristotle's *Poetics* recommends an "orderly arrangement of parts" that form a beautiful whole or "organism." Horace admonishes the would-be poet: "In short, be your subject what it will, let it be simple and unified." And some awareness of formalism is at least implicit in many other classical, medieval, and Renaissance treatises on art or poetics.

But the Romantic movement in Europe in the late eighteenth and nineteenth centuries intensified speculations about form in literature. Samuel Taylor Coleridge (1772–1834) brought to England (and thus to America) the conception of a dynamic *imagination* as the shaping power and unifier of vision—a conception he had acquired from his studies of the German philosophical idealists: Kant, Hegel, Fichte, and Schelling. Such a conception encouraged discrimination between a poem and other forms of discourse by stressing the poem's power to elicit delight as a "whole" and "distinctive gratification from each component *part.*" In a "legitimate poem," Coleridge declared, the parts "mutually support and explain each other; all in their proportion harmonizing with, and supporting the purpose and known influences of metrical arrangement."

This interrelationship between the whole and the parts was manifested in a consistently recurring image among the Romantics—the image of growth, particularly of vegetation. Perhaps because of the Romantics' infatuation with nature, the analogy usually likened the internal life of a painting or poem to the quintessential unity of parts within a tree, flower, or plant: as the seed determines, so the organism develops and lives. In a letter to John Taylor (February 27, 1818) Keats wrote that one of his "axioms" was "That if Poetry comes not as naturally as the Leaves to a tree it had better not come at all." Shelley uses imagery of growth and of vegetation several times in his "Defence of Poetry." In talking of the relationship of sounds in poetry, he counsels against "the vanity of translation; it were as wise to cast a violet into a crucible that you might discover the formal principle of its colour and odour, as to seek to transfuse from one language into another the creations of a poet. The plant must spring again from its seed, or it will bear no flower. . . ." He calls the thoughts of the poet "the germs of the flower and the fruit of latest time," claiming, "All high poetry is infinite; it is as the first acorn, which contained all oaks potentially." And again of poetry,

> . . . this power arises from within, like the colour of a flower which fades and changes as it is developed. . . . The instinct and intuition of the poetical faculty is still more observable in the plastic and pictorial arts; a great statue or picture grows under the power of the artist as a child in the mother's womb. . . .

In America, Edgar Allan Poe (1809–1849), extending Coleridge's theory, asserted the excellence of short lyric poems and short tales because they can maintain and transmit a single, unitary effect more successfully than can long works like *Paradise Lost*. In "The Philosophy of Composition" Poe demonstrated how the parts of his "The Raven" allegedly developed from the single effect he desires. Poe also reprimanded certain contemporary poets like Henry Wadsworth Longfellow for committing what he called the "heresy of the didactic" by tacking on obtrusive (thus inorganic) moral lessons and accordingly violating the lyric effects of their poems.

Later in the nineteenth century and on into the twentieth, Henry James (1843–1916), in "The Art of Fiction" and the prefaces to his tales and novels, argued for fiction as a "fine art" and for the intricate, necessary interrelationships of parts and the whole:

> There are bad and good novels, as there are bad pictures and good pictures; but that is the only distinction in which I can see any meaning, and I can as little imagine speaking of a novel of character as I can imagine speaking of a picture of character. When one says picture one says of character, when one says novel one says of incident, and the terms seem to be transposed at will. What is character but the determination of incident? What is incident BUT the illustration of character? What is either a picture or a novel that is *not* of character? What else do we seek in it and find in it? It is an incident for a woman to stand up with her hand resting on a table and look at you in a certain way; or if it not be an incident, I think it will be hard to say what it is.

James implies the same interdependence and kinship for all other aspects of a work of fiction—setting, theme, scene and narrative, image and symbol. When the artist is attending to his craft, nothing that goes into the work will be wasted, and form will be present: "Form alone *takes,* and holds and preserves, substance—saves it from the welter of helpless verbiage that we swim in as in a sea of tasteless tepid pudding." When the work achieves "organic form," everything will count.

C. The New Criticism

Although there were antecedents from Plato through James, a systematic and methodological formalistic approach to literary criticism appeared only with the rise in the 1930s of what came to be called the New Criticism. Coming together originally at Vanderbilt University in the years following World War I, the New Critics included a teacher-scholar-poet, John Crowe Ransom, and several bright students—Allen Tate, Robert Penn Warren, and Cleanth Brooks. Associated at first in an informal group that discussed literature, they in time adopted the name of Fugitives and published an elegant literary magazine called *The Fugitive* in Nashville from 1922 to 1925. When the poetry and critical essays of T. S. Eliot came to their attention, they found sturdy reinforcement for ideas that were emerging from their study and writing of lyric poetry. Ideas thus shared and promoted included literature viewed as an organic tradition, the importance of strict attention to form, a conservatism related to classical values, the ideal of a society that encourages order and tradition, a preference for ritual, and the rigorous and analytical reading of literary texts. Eliot was particularly influential in his formulation of the objective correlative ("a set of objects, a situation, a chain of events which shall be the formula of [a] *particular* emotion; such that when the external facts are given, the emotion is immediately invoked"). Eliot was also influential in his endorsement of the English Metaphysical poets of the seventeenth century for their success in blending "states of mind and feeling" in a single "verbal equivalent." Such developments strengthened the emergent New Criticism, which by the 1950s had become the dominant critical system in such influential journals as *Sewanee Review, The Kenyon Review,* and *The Hudson Review* and in college and university English departments.

The New Critics sought precision and structural tightness in the literary work; they favored a style and tone that tended toward irony; they insisted on the presence within the work of everything necessary for its analysis; and they called for an end to a concern by critics and teachers of English with matters outside the work itself—the life of the author, the history of his times, or the social and economic implications of the literary work. In short, they turned the attention of teachers, students, critics, and readers to the essential matter: *what* the work says and *how* it says it as inseparable issues. To their great credit they influenced at least one generation of college students to become more careful and serious readers than they otherwise would have been.

Members and disciples of the group advanced their critical theory and techniques through a series of brilliant college textbooks on literary analysis: *Understanding Poetry* (1939) and *Understanding Fiction* (1943) by Brooks and Warren; *Understanding Drama* (1945) by Brooks and Robert B. Heilman; *The Art of Modern Fiction* by Ray B. West, Jr., and Robert W. Stallman; and *The House of Fiction* (1950) by Caroline Gordon and Allen Tate. After 1942, *The Explicator*, a monthly publication, published hundreds of short textual explications of great varieties of literary works; and prestigious literary journals and quarterlies still publish articles that show the continuing influence of the New Criticism. But even as the formalistic approach of the New Critics was influencing readers, teachers, and students throughout the universities of the United States, well into the second half of the century, others were pointing to what they perceived to be deficiencies or worse in that approach. Frank Lentricchia, in *After the New Criticism* (1980) offers a helpful overview of what was happening. He uses 1957 and the publication of three books that year to give one benchmark for the turn to other approaches and emphases. The three are Northrop Frye's *Anatomy of Criticism*, Cleanth Brooks and W. K. Wimsatt's *Literary Criticism: A Short History*, and Frank Kermode's *Romantic Image*. Coming hard upon Murray Krieger's *New Apologists for Poetry* (1956), they seem to fulfill, Lentricchia says, Krieger's prediction "that the New Criticism had done all it could do for American literary critics . . ." (3). "By about 1957," Lentricchia says, "the moribund condition of the New Criticism and the literary needs it left unfulfilled placed us in a critical void. Even in the late 1940s, however, those triumphant times of the New Criticism, a theoretical opposition was already gathering strength" (4). Lentricchia goes on to cite a number of the works that show that gathering strength, and the reader is referred to his overview.

One article that he does not cite might earn a place here because it provides (witness its title) a kind of synopsis of the reaction setting in even as the vogue of the New Criticism was still gaining strength: "Cleanth Brooks; or the Bankruptcy of Critical Monism" by Ronald S. Crane, a neo-Aristotelian (*Modern Philology* 45 [1948]). Like others in the 1940s, Crane faults the reduction of pieces of literature to one or a few rhetorical devices that brings about a diminution of their potential. Whether it be irony or paradox or tension or texture: these, alone or together, do not a poem make.

However, it is not our present purpose to treat thoroughly the attack on or the divergence from the formalistic approach of the New Criticism. More of that can be seen in works such as those cited by Lentricchia and in the chapters that follow. Our present purpose is to show the enduring contribution of the formalistic approach, even as we call attention to some of its deficiencies.

IV. CONSTANTS OF THE FORMALISTIC APPROACH: SOME KEY CONCEPTS, TERMS, AND DEVICES

We shall draw attention now to several of the constants of the formalistic approach, even though some may have been disparaged by the differing critical emphases of other writers. Keeping in mind the overview we gained from the analysis of "A Slumber Did My Spirit Seal" (we can now reveal that the poem is by William Wordsworth) and keeping in mind that these devices may recur in the analyses later in this chapter, let us look more carefully at these constants.

We must, of course, begin with form. In systems of the past, the word *form* usually meant what we would call *external form*. Thus, when we identify a poem with fourteen lines of iambic pentameter, a conventional pattern of rhymes, and a conventional division into two parts, as a sonnet, we are defining its external form. The same kind of description takes place when we talk about couplets, tercets, *ottava rima*, quatrains, Spenserian stanzas, blank verse, or even free verse. But the formalistic critic is only moderately interested in external forms (in fact, only when external form is related to the work's total form, when stanzaic or metrical pattern is integral to internal relationships, reverberations, patterns, and systems). The process of formalistic analysis is complete only when everything in the work has been accounted for in terms of its overall form.

Organic form is a particular concept important to the New Critics, inherited as we have noted from the English Romantics. In the Romantics, we find the emphasis on organicism not just in literary forms but in a broader, philosophical context, where the world itself is organic; objects within it are organisms that interact with each other in a larger organic universe. This notion may go so far as Wordsworthian pantheism, or what some thought to be pantheism, where a breeze in nature may awaken within the persona of the *Prelude* (in this case the poet himself) a "correspondent breeze" (1.35). Similarly there is the Romantic emphasis on the Aeolian lyre or harp, as in Coleridge's poem "The Eolian Harp," and the reference to the lyre in the imagery and symbolism of Shelley's "Ode to the West Wind," a notion that recurs in the second paragraph of Shelley's "Defence of Poetry." The vegetation imagery, mentioned earlier, is of course part of this organicism. Now the question for us is how this concept of organicism came into formalistic criticism of this century, especially among critics many of whom expressed no fondness for English Romanticism.

In the formalistic approach, the assumption is that a given literary experience takes a shape proper to itself, or at the least that the shape and the experience are functions of each other. This may mean at a minimum that a precise metrical form couples with a complex of sounds in a line of verse to present one small bit of the experience (recall the treatment of the short lyric at the beginning of this chapter). Or it may mean that a generic form, like that of the sonnet, is used repeatedly in a sonnet cycle to show the interrelationship of thoughts to images, or problems to comment or solution. In such a case, even though the overt structure of the sonnet is repetitive, still the experience in any one Italian sonnet is structured across the octave and sestet or in the English form across the three quatrains and the concluding couplet. In a larger work, a full-length play or a novel might adopt much more complex and subtle forms to communicate the experience, such as the interrelationships of plot and subplot in Shakespeare's *Hamlet*, in *Henry IV, Part 1* or in *The Tempest;* or in the complex stream of consciousness of Joyce's *Ulysses* or Faulkner's *Sound and the Fury*. Indeed, the fragmentation of story line and of time line in modern fiction and in some absurdist drama is a major formalistic device used not only to generate within the reader the sense of the immediacy and even the chaos of experience but also to present the philosophical notion of nonmeaning and nihilism. Thus we have the seeming paradox that in some cases the absence of form *is* the form, precisely.

Statements that follow discovery of form must embrace what Ransom called local texture and logical structure (*The World's Body* [New York: Scribner's, 1938]: 347). The logical structure refers to the argument or the concept within the work; local texture comprises the particular details and devices of the work (for example, specific metaphors and images). However, such a dualistic view of a literary work has its dangers, for it might encourage the reduction of logical structure to précis or summary—what Brooks has called the "heresy of paraphrase." In *Understanding Poetry*, Brooks and Warren simply include "idea," along with rhythm and imagery, as a component of form: "the form of a poem is the organization of the material . . . for the creation of the total effect" (3rd ed. [New York: Holt, 1960]: 554). The emphasis, in any case, is upon accounting for all aspects of the work in seeking to name or define its form and effect. Mark Schorer pressed the distinction further between the critic's proper concentration on *form* and an improper total concern with *content only:* "Modern [i.e., formalistic] criticism has shown that to speak of content as such is not to speak of art at all, but of experience; and that it is only when we speak of the *achieved* content, the form, the work of art as a work of art, that we speak as critics. The difference between content, or experience, and achieved content, or art, is technique" ("Technique as Discovery," *The Hudson Review* 1 [Spring 1948]: 67). He goes on to say that "technique is the only means [an author] has of discovering, exploring, developing his subject, of conveying its meaning, and, finally, of evaluating it."

As we turn more specifically to texture, we find that as with form and its potential to embody meaning, imagery and metaphor are an integral part of the work, especially in the poem. Once again, the formalistic critics—obviously—did not invent metaphor: Aristotle, very much a formalist, discussed metaphor in his *Poetics*. But the New Critics delighted in close analysis of imagery and metaphor, and they laid stress on a careful working out of imagery. The consistency of imagery in a lyric, whether it be a single dominant image throughout the poem or a pattern of multiple but related images, became for some an index to the quality of a given poem. Such consistency of imagery helped to create what John Crowe Ransom among others called texture. It was for such reasons that there was much interest in Metaphysical poetry and in the Metaphysical conceit. The interest was aided by publication of Herbert Grierson's collection *Metaphysical Lyrics and Poems of the Seventeenth Century* (1921). It was furthered by the attention of T. S. Eliot, Ransom, and Allen Tate. Critics praised the Metaphysical conceit because of its carefully worked out ("wrought") images that were elaborated over a number of lines, richly textured and endowed with a complexity of meanings, as in John Donne's "The Flea" or in the "stiff twin compasses" of his "Valediction: Forbidding Mourning." Donne's image of the "well-wrought urn" in "The Canonization" is cogent here, not only because of the working out of the image but also because the phrase gave Cleanth Brooks the title to one of his contributions to the rich library of New Criticism, *The Well-Wrought Urn* (1947). By way of contrast, a poem like Shelley's "Lines: 'When the Lamp is Shattered'" was disparaged by formalist critics for its allegedly loose imagery; indeed, much of Shelley, along with other Romantics, was disparaged (but for a defense, see Frederick A. Pottle, "The Case of Shelley," *PMLA* 67 [1952]: 589–608).

When an image (or an incident or other discrete item) takes on meaning beyond its objective self, it moves into the realm of symbol. Here is a dilemma for some formalistic critics, those who espouse the autonomous and autotelic concept of a literary work so strenuously that anything outside it becomes a problem. Symbols may sometimes remain within the work, as it were; but it is the nature of symbols to have extensional possibilities, to open out to the world beyond the art object itself. When meaning and value outside the work of literature are the real purpose of the symbol, some formalistic critics may find fault with the work. On the other hand, such a restriction may well be one of the more limiting concerns of the New Critics (we recall Poe's denunciation of the didactic in favor of beauty), and we take the cautious position that even in a formalistic reading we must go sometimes beyond the pure aestheticism of the work in itself to the extended meaning of the work as suggested by its symbols.

We have already said something of this sort when we alluded to the form of some modern novels or absurdist plays: form can embody theme, and theme transcends the individual work. Symbol is a way of using something integral to the work to reach beyond the work and engage the world of value outside the work. It might be an incident that takes on meaning, such as the apparent happenstance of events in a naturalistic writer like Thomas Hardy; it might be the conventional object or device—a crucifix, a color, a tree—that becomes symbolic of meanings within and without the poem, story, or play. When that happens, the formalistic approach must study such symbols as aspects of form, as exponents of meaning both within and without the work. Not to do so would be to turn the work too much within itself, making it overly centripetal. If a work is too centripetal because of the limiting notion that it should exist in and of and for itself alone, the work becomes an objet d'art, suitable for a shelf but in danger of losing the very life that makes it important to the reader. One must question this restriction, this reductionism, just as one questions Keats's Grecian urn as to whether beauty and truth are indeed the same, and as one questions Emerson's speaker in "The Rhodora," who said that "beauty is its own excuse for being."

Another formalistic term that has brought mixed responses is the intentional fallacy, along with its corollary the affective fallacy. In the intentional fallacy, we are told, the critic or the reader makes the mistake of not divorcing the literary work from any intention that the author might have had for the work. Instead, say Wimsatt and Beardsley in *The Verbal Icon* (1954), the work must give us from within itself any intention that might be garnered, and we must not go to the author for his or her intention: at the very least the author is not a reliable witness. Wimsatt and Beardsley review the arguments of some of the intentionists, and there are legitimate considerations on both sides of the question. For us a proper middle ground would be to take note of external evidence when it seems worthy, but to accept the caution that the work itself must first and always be seen as a work unto itself, having now left the author's care. Wimsatt and Beardsley also warned against the affective fallacy, wherein the work is judged by its effect on the reader or viewer, particularly its emotional effect. Again, however, those avoiding the reductionist tendency of formalistic criticism would note that no work of literary art can be divorced from the reader and therefore from the reader's response. For that matter, no less a critic than Aristotle gave us the concept of catharsis, the purging of the audience at a tragedy that cleanses the emotions. But we admit that the relationship is complex and the formalistic approach is correct in urging caution. (For a helpful and balanced assessment, see "Intentions, problems of" and "Affective fallacy" in the *Princeton Encyclopedia of Poetry and Poetics,* enl. ed., 1974).

Another device that a formalistic approach must heed is the point of view, which, like consistency of imagery, is generally considered a virtue in the work of literary art, for it preserves the internal form, the organic quality of the work. Conversely, a nonexistent point of view (that is, one in which several points of view are not clearly demarcated from each other) flaws the work, for the work then may go in several directions and therefore have no integrity: the center does not hold. Such a fragmentation may be avoided if we grant the narrator the privilege of knowing all, seeing all, from a perspective that in theological terms would have to be called divine. In the great epics and in most traditional novels of an earlier day, the omniscient narrator possessed that godlike quality and narrated from a third-person perspective.

But in more restricted points of view, the very form of the work is conditioned by the point of view to which the author limits the narrator. Telling the story of *The Ambassadors* from the limited point of view of Lambert Strether (this restricted use of the third person is sometimes called central intelligence), James can produce for the reader a depth of moral discrimination and intricate analysis of character and situation because of Strether's New England conscience and determination to live intensively while he is in Paris. As Wayne Booth reminds us in *The Rhetoric of Fiction* (Chicago: U of Chicago P, 1961), however, narrators may be either reliable (if they support the explicit or implicit moral norms of the author) or unreliable (if they do not). Thus Jake Barnes in *The Sun Also Rises* is a completely reliable narrator, for he is the very embodiment of what is often called the "Hemingway code"; on the other hand, the lawyer in Melville's *Bartleby the Scrivener* is unreliable in his early evaluations of himself because he is not involved with humanity. Whatever the point of view we encounter, it has to be recognized as a basic means of control over the area or scope of the action, the quality of the fictional world offered to the reader, and even the reactions of the reader.

In a first-person narration the author may condition the form even more. Thus a young boy named Huckleberry Finn, who narrates his own story, must not be allowed to know more than a young boy such as he would know. His view is limited to what he sees and reports. Nor does he understand all that he reports, not—at least—as a mature person devoid of cultural bias and prejudice might understand. In this first-person point of view, the narration is limited to that person's telling. If the author wishes to communicate anything beyond that to the reader, that wish becomes a challenge in technique, for the information must be reported naively by Huck Finn and interpreted maturely by the reader on the basis of what the author has Huck Finn say (again we must heed the admonitions about the intentional fallacy and the affective fallacy). In this sense Huck Finn is honest on the one hand, but an unreliable narrator on the other. To stretch the point a bit further, we may imagine a psychotic telling a story in a seemingly straightforward way—but the real story may be *about* the psychotic, and what he or she tells us at the obvious level may not have any credibility at all. In some circumstances the author may choose to have a shifting point of view to achieve different effects at different times (possibly this is what Chaucer the author did to Chaucer the pilgrim). Or there may be multiple points of view, as in Faulkner's *The Sound and the Fury*. Still another is the point of view that would claim total objectivity—the scenic or dramatic: we read only the dialogue of characters, with no hints of a narrator to intrude any perspective other than what we get from the dialogue itself. All these points of view condition the form of literature, and a formalistic approach must study them for the reader to appreciate the fullness of the work.

Failure to note such an aspect of form will result in a mis-reading or in an inadequate reading of the work. This challenge to the reader may be further illustrated by turning briefly to lyric poetry, where tone of voice is analogous to point of view. Although we do not usually think of point of view as an aspect of lyric poetry, the fact is that in a lyric there is a speaker—that is, a first-person situation. This immediately sets a context and a set of circumstances, for the speaker is doing something, somewhere. Possibly there is also a hearer, a second person (we readers only overhear the speaker), so that the hearer also conditions the experience. In Robert Browning's "Andrea del Sarto" it means much to know that Andrea is addressing a woman and that they are among his paintings at a certain time of day. Consequently it is even more important to know what Andrea feels about his inadequacy as a painter and as a man: his tone of voice, as much as details revealed to us, will largely reveal those feelings. Conversely, another painter, Fra Lippo Lippi in the poem of that name, responds ebulliently to his world and his confidence about his ability to capture and interpret it in his painting: his colloquial, jovial tone communicates this attitude. In Browning's "Porphyria's Lover" the reader will go totally astray if he does not understand that the lover is a madman—and that the beloved though present has been murdered by him. In a more traditional love lyric or in one that describes a beautiful scene in nature, the speaker may reasonably be trusted to speak the truth. But how does one interpret the speaker's voice in Donne's "Song" ("Go and Catch a Falling Star")? What is the mixture of genial satire, sardonicism, mere playfulness? The way the reader hears the speaker will condition the poem, give it its form, indeed may make the poem into poems by varying the voice. So the formalist critic ends up with a problem: one poem, or several? Perhaps, finally, there is only one, and that one is the resolution of all the possibilities in one reality, a kind of super-form that resolves and incorporates all the several forms.

This resolution is like the principle of the arch. In an arch the way down is the way up: the arch stands because the force of gravity pulls the several stones down while at the same time pushing them against the keystone. Gravity therefore counteracts itself to keep the entire arch standing; for that matter, the arch can carry great weight—just as a piece of literature might.

This aspect of formalist criticism might be called tension, the resolution of opposites, often in irony and paradox. Coleridge enunciated at least part of this notion early on; the New Critics laid great stress on the terminology, sometimes almost to the exclusion of other elements. The basic terms—tension, irony, paradox—are often nearly indistinguishable, so closely do they work together. C. Hugh Holman summarizes tension as "A term introduced into contemporary criticism by Allen Tate, by which he means the integral unity of a poem, a unity which results from the successful resolution in the work of the conflicts of abstraction and concreteness, of general and particular, of denotation and connotation. . . . Good poetry, Tate asserts, is the 'full, organized body of all the extension and intension that we can find in it'" (*A Handbook to Literature*, 5th ed., 1986). Holman further notes that "This concept has been widely used by the New Critics, particularly of poetry as a pattern of paradox or as a form of irony."

One could hardly find a better demonstration of the interrelationships of tension, irony, and paradox than what Robert Penn Warren has provided in "Pure and Impure Poetry," which first appeared in *The Kenyon Review* (1943) and later in *Selected Essays of Robert Penn Warren* (Random, 1958). In making a case for impure poetry—poetry of inclusiveness—Warren not only analyzes the arguments of purists but also provides excellent analyses of poems and passages that include the impure and thereby prove themselves as poetry. Regularly, the ironies and paradoxes—the tensions—are at the heart of the success of the items he studies. Near the conclusion of the essay he says:

> Can we make any generalizations about the nature of the poetic structure? First, it involves resistances, at various levels. There is the tension between the rhythm of the poem and the rhythms of speech . . . ; between the formality of the rhythm and the informality of the language; between the particular and the general, the concrete and the abstract; between the elements of even the simplest metaphor; between the beautiful and the

ugly; between ideas . . . ; between the elements involved in irony; between prosaisms and poeticisms. . . . This list is not intended to be exhaustive; it is intended to be merely suggestive. But it may be taken to imply that the poet is like the jiujitsu expert; he wins by utilizing the resistance of his opponent—the materials of the poem. In other words, a poem, to be good, must earn itself. It is a motion toward a point of rest, but if it is not a resisted motion, it is motion of no consequence.

We may now turn to the formalist approach in practice, applying some of its methods to the four literary works that we analyzed in chapter 1.

V. THE FORMALISTIC APPROACH IN PRACTICE

A. Word, Image, and Theme: Space-Time Metaphors in "To His Coy Mistress"

August Strindberg, the Swedish novelist and playwright, said in the preface to *Miss Julie* that he "let people's minds work irregularly, as they do in real life." As a consequence, "The dialogue wanders, gathering in the opening scenes material which is later picked up, worked over, repeated, expounded and developed like the theme in a musical composition."

Tracing such thematic patterns in a literary composition assumes that significant literature does attempt to communicate, or at least to embody, meaningful experience in an aesthetically appealing form. This is not to say that literature merely sugarcoats a beneficial pill. Rather, in the creation of any given work, a literary artist has an idea, or an actual experience, or an imagined experience that he or she wishes to communicate or to embody. Consciously or otherwise, the artist then chooses a means of doing so, selecting or allowing the unconscious mind to present specific devices, and arranging them so that they can embody or communicate that experience. Once the author has created such a work for us, we readers must re-create the experience, in part by carefully tracing the motifs used to communicate it. If Strindberg has given us material in *Miss Julie* that he later picked up, worked over, and developed "like the theme in a musical composition," then our role is to seek out the indications of that theme. Bit by bit as we notice instances of a pattern, we work our way into the experience of the story, poem, or play. As we follow the hints of thematic statement, recognize similar but new images, or identify related symbols, we gradually come to live the experience inherent in the work. The evocative power of steadily repeated images and symbols makes the experience a part of our own consciousness and sensibility. Thus the image satisfies our senses, the pattern our instinctive desire for order, and the thematic statement our intellect and our moral sensibility.

Andrew Marvell's poem "To His Coy Mistress" presents us with a clear instance of how a particular set of images can open out to themes in the way just described. The opening line of the poem—"Had we but world enough and time"—introduces us to the space-time continuum. Rich in possibilities of verbal patterns, the motif is much more, for the structure of the poem depends on the subjunctive concept, the condition contrary to fact, which gives the whole poem its meaning: "Had we," the speaker says, knowing that they do not. From that point on, the hyperbole, the playfulness, the grim fear of annihilation are all based on the feeling of the speaker that he is bound by the dimensions of space and time.

Clearly, this poem is a proposition made by the eternal male to the eternal female. Just as clearly, and in a wholly different realm, the motif of space and time shows this poem to be a philosophical consideration of time, of eternity, of man's pleasure (hedonism) and of salvation in an afterlife (traditional Christianity). In this way Marvell includes in one short poem the range between man's lust and man's philosophy.

On the other hand, we find that the words used to imply this range tend to be suggestive, to shift their meanings so as to demand that they be read at different levels at the same time. Let us begin with instances of the space motif. The space motif appears not only in obvious but also in veiled allusions. In the first section of the poem we find "world," "sit down," "which way / To walk," the suggested distance between "Indian Ganges" and the Humber, the distance implicit in the allusions to the Flood and to the widespread Jews of the Diaspora, "vaster than empires," the sense of spatial movement as the speaker's eyes move over the woman's body, and the hint of spatial relationship in "lower rate." The word "long" (line 4) refers to time, but has spatial meaning, too. Several other words ("before," "till," "go," "last") also have overlapping qualities, but perhaps we strain too far to consider them.

Space and time are clearly related in the magnificent image of the opening lines of the second stanza: "But at my back I always hear / Time's winged chariot hurrying near." The next couplet provides "yonder," "before," "deserts," and, again, a phrase that suggests both space and time: "vast eternity." In the third stanza the word "sits" echoes the earlier use of the word, and several words suggest movement or action in space: "transpires," "sport," "birds of prey," "devour," "languish in slow-chapped power," "roll," "tear . . . / Thorough." The space motif climaxes in an image that again incorporates the time motif: the sun, by which the man measures time and which will not stand still in space, will be forced to run.

The time motif also appears in its own right, and not only by means of imagery. The word itself appears once in each stanza: near the beginning of stanzas 1 and 2 (lines 1 and 22), and in the third stanza as a central part of the lover's proposition (39). Clustering around this basic unifying motif are these phrases and allusions from the first stanza: the "long love's day," the specific time spans spent in adoring the woman's body and the vaster if less specific "before the Flood" and "Till the conversion of the Jews," and the slow growth of "vegetable love" and the two uses of "age" (lines 17 and 18). At the beginning of stanza 2, the powerful image of time's winged chariot as it moves across a desert includes the words "always" and "eternity." Other time words are "no more" and "long-preserved." There is also the sense of elapsed time in the allusions to the future decomposition of the lovers' bodies. The third stanza, although it delays the use of the word "time," has for its first syllable the forceful, imperative "now." The word appears twice more in the stanza (lines 37 and 38). It is strengthened by "instant," "at once," and "languish in [Time's] slow-chapped power." The phrase "thorough the iron gates of life," though it has more important meanings, also may suggest the passing from temporal life into the not so certain eternity mentioned earlier. The concluding couplet of the poem, as already shown, combines space and time. Further, it may extend time backward to suggest Old Testament days and classic mythology: Joshua stopped the sun so that the Israelites could win a battle, and, even more pertinently, Zeus lengthened the night he spent with Amphitryon's wife.

For the poem is also a love poem, both in its traditional context of the courtly love complaint and in the simple fact of its subject matter: fearing that the afterlife may be a vast space without time, the speaker looks for a means of enjoying whatever he can. This *carpe diem* theme is not uncommon, nor is the theme of seduction. What gives the poem unusual power, however, is the overbearing sense of a cold, calculated drive to use the pleasures of sex to counterbalance the threats of empty eternity. Thus a second major motif—after the space-time relationship—used to present the theme, is the sexual motif.

We can follow this theme beginning with the title, which immediately sets up the situation. In the second line the word "coyness" leads us into the poem itself; even the word "crime" suggests the unconventional (though crime and conventional morality are reversed in the context of the lover's address). The motif gradually emerges, romantically at first, but more frankly, even brutally, as the speaker continues. In the first stanza the distant Ganges and the redness of rubies are romantic enough; the word "complain," in the sense of the courtly lover's song, echoes the whole courtly tradition. The word "love" appears twice before the courtly catalogue of the lady's beautiful body. The catalogue in turn builds to a climax with the increasing time spans and the veiled suggestiveness of "rest" and "part."

The second stanza, though it continues to be somewhat veiled, is less romantic, and becomes gruesome even while insisting upon sexual love. The lady's beauty will disappear in the marble vault. We may associate the word "marble" with the texture and loveliness of the living woman's skin, but here the lover stresses the time when that loveliness will be transferred to stone. The same type of transference of the lover's song, which finds no echoes in that vault, occurs in a veiled image of unrealized sexual union in life: worms will corrupt the woman in a way that the lover could not. "Quaint honor" is an ironic play on words to suggest the pudendum (*quaint* as in Middle English *queynte*; see Chaucer's "Miller's Tale"). The fires of lust will become ashes (with an implicit comparison to the coldness of marble), and the stanza closes with puns on "private" and "embrace."

The third stanza resumes the romantic imagery of the first ("youthful hue," "morning dew"), but it continues the bolder imagery of the second section. "Pore" is a somewhat unromantic allusion to the woman's body, and "instant fires" recalls the lust and ashes of the preceding stanza. "Sport" takes still a different tack, though it reminds us of the playfulness of the first stanza. After this line, the grimness of the second stanza is even more in evidence. The amorous birds are not turtledoves, but birds of prey, devouring time—and each other. Although the romantic or sentimental is present in the speaker's suggestion that they "roll all our strength and all / Our sweetness up into one ball," the emphasis on the rough and violent continues in the paradoxical "tear our pleasures with rough strife." Once the coy lady's virginity is torn away, the lover will have passed not through the pearly gates of eternity, but through the iron gates of life. Thus the lover's affirmation of life, compounded of despair and defiance, is produced by his suggestion that the birth canal of life and procreation is preferable to the empty vault and deserts of vast eternity. On the one hand, the instances of the sexual motif point to a degeneration from romantic convention in the first section to scarcely veiled explicitness in the last. But on the other hand, the speaker has proceeded from a question about the nature of eternity and the meaning of the space-time relationship in this world to an affirmation of what he suspects is one of the few realities left him. The very concreteness, the physicality of the sexual motif, provides an answer to the philosophic speculation about space, time, and eternity. Obviously different, the motifs just as obviously fuse to embody the theme of the poem.

There are other, lesser motifs that we could trace had we ourselves space enough and time, such as wings and birds, roundness, and minerals and other things of earth (rubies, marble, iron, ashes, and dust). Each of these serves as a means to greater insights into the poem.

In sum, a formalist reading of "To His Coy Mistress" can originate in a study of images and metaphors—here, space-time images. It can then lead to complexes of other images—precious stones and marble vaults, chariots and rivers, worms and dust. Finally, it is the nature of a formalist approach to lead us to see how images and metaphors form, shape, confect a consideration of philosophic themes—in this case a speculation on whether love and even existence itself can extend beyond the time we know, and, if they cannot, whether instant gratification is a sufficient response to the question raised.

B. The Dark, the Light, and the Pink: Ambiguity as Form in "Young Goodman Brown"

In short fiction, as in a poem, we can look for the telling word or phrase, the recurring or patterned imagery, the symbolic object or character, the hint of or clue to meaning greater than that of the action or plot alone. Because we can no more justify stopping with a mere summary of what happens outwardly in the story than we can with a mere prose paraphrase of a lyric poem's content, we must look for the key to a story's form in one or more devices or images or motifs that offer a pattern that leads us to larger implications. In short, we seek a point at which the structure of the story coincides with and illuminates its meaning.

As we approach a formalistic reading of Hawthorne's story, we should make another point or two of comparison and contrast. The lyric poem generally embraces a dramatic situation. That is, a speaker reacts to an experience, a feeling, an idea, or even a physical sensation. Only one voice is ordinarily present in the lyric poem, but in other literary genres there is usually a group of characters. In fiction the story is told by the author, by one of the characters in the story, or by someone who has heard of an episode. Unlike the novel, the other major fictional type, the short story is characteristically concerned with relatively few characters and with only one major situation, which achieves its climax and solution and thus quickly comes to an end. The short story is restricted in scope, like a news story, for example, but unlike the news story, the short story possesses balance and design—the polish and finish, the completeness that we associate with the work of art. A principle of unity operates throughout to give that single effect that Poe emphasized as necessary. In brief, like any other imaginative literary work, the short story possesses form.

Paradoxical as it may seem, we wish to suggest that ambiguity is a formal device in "Young Goodman Brown." One sure way to see this ambiguity is to trace the relationships between light and dark in the story, for the interplay of daylight and darkness, of town and (dark) forest, is important. For evidence of that importance the reader is urged to consult Richard Harter Fogle's classic study, *Hawthorne's Fiction: The Light and the Dark* (Norman: U of Oklahoma P, 1952). We shall not neglect the interplay of dark and light—indeed we assume it—but we wish to focus on another device of ambiguity.

In our formalistic reading of Marvell's "To His Coy Mistress" we stressed the recurrent pattern of words, images, and metaphors of space and time as a means of seeing the form that embodies meaning in that poem. In "Young Goodman Brown" we can start with a clearly emphasized image that almost immediately takes on symbolic qualities. That is the set of pink ribbons that belongs to Faith, young Brown's wife. Whatever she is (and much of the effect of the story centers on that "whatever"), the pink ribbons are her emblem as much as the scarlet letter is Hester Prynne's. They are mentioned three times in the first page or so of the story. Near the center of the story, a pink ribbon falls, or seems to fall, from a cloud that Goodman Brown sees, or thinks he sees, overhead. At the end of the story, when Faith eagerly greets her returning husband, she still wears her pink ribbons. Like the admixture of light and dark in the tale—as in much of

Hawthorne—the ribbons are neither red nor white. They are somewhere between: they are ambiguity objectified. Clearly Hawthorne meant them to be suggestive, to be an index to one or more themes in the tale. But suggestive of what? Are they emblematic of love, of innocence, of good? Conversely, do they suggest evil, or hypocrisy, or the ambiguous and puzzling blend of good and evil? Are they symbolic of sex, of femininity, or of Christian faith? Should we even attempt to limit the meaning to one possibility? Would we be wise—or slovenly—to let the ribbons mean more than one thing in the story?

Of this we can be sure: to follow this motif as it guides us to related symbols and patterns of relationships is to probe the complex interweaving of ideas within the story. Specifically, in the interpretation that follows we suggest that the mysterious pink ribbons are—at least among other things—an index to elements of theology. To see that relationship let us first consider the theological matrix of the story.

Because the Puritan setting of "Young Goodman Brown" is basic to the story, we can expect that some of its thematic patterns derive from traditional Christian concepts. For example, readers generally assume that Goodman Brown loses his faith—in Christ, in human beings, or in both. But the story is rich in ambiguities, and it is therefore not surprising that at least one reader has arrived at the opposite conclusion. Thomas E. Connolly (*American Literature* 28 [1956]: 370–75) has argued that the story is an attack on Calvinism, and that Faith (that is, faith) is not lost in the story. On the contrary, he says, Goodman Brown is confirmed in his faith, made aware of "its full and terrible significance." Either way—loss of faith or still firmer belief—we see the story in a theological context. Although we do not have to accept either of these views, we do not have to deny them either. Instead, let us accept the theological matrix within which both views exist. As a matter of fact, let us pursue this theological view by following the pattern of relationships of faith, hope, and love, and their opposed vices, in other words the form that this pattern creates in the mind of the reader.

We can assume that Hawthorne was familiar with some of the numerous passages from the Bible that bear upon the present interpretation. Twice in the first epistle to the Thessalonians, Saint Paul mentions the need for faith, hope, and love (1:3 and 5:8). In 1 Corinthians 13, after extolling love as the most abiding of the virtues, Paul concludes his eloquent description with this statement: "So there abide faith, hope, and love, these three; but the greatest of these is love." The author of the first epistle of Peter wrote, "But above all things have a constant mutual love among yourselves; for love covers a multitude of sins" (4:8). To these may be added the telling passages on love of God and neighbor (Matt. 22:36–40 and Rom. 13:9–10) and related passages on love (such as Col. 3:14 and 1 Tim. 1:5). Faith, hope, and love, we should note, have traditionally been called the theological virtues because they have God (*theos*) for their immediate object.

Quite possibly Hawthorne had some of these passages in mind, for it appears that he wove into the cloth of "Young Goodman Brown" a pattern of steady attention to these virtues. Surely he provided a clue for us when he chose Faith as the name for Goodman Brown's wife. Hawthorne thereby gave faith first place in the story, not necessarily because faith is the story's dominant theme (indeed, love may well be the dominant theme), but because faith is important in Puritan theology and because it is traditionally listed as the first of the three virtues. Allusions to faith could be made explicit in so many passages in the story and implicit in so many others that they would provide an evident pattern to suggest clearly the other two virtues. (Similarly, the epithet *goodman* could take on symbolic qualities and function almost as Brown's given name, not simply as something comparable to modern *mister*.)

An analysis of these passages, for example, shows not only explicit mentions of faith but also implicit allusions to the virtues of faith, hope, and love, and to their opposed vices, doubt, despair, and hatred. The first scene includes these: "And Faith, as the wife was aptly named"; "My love and my Faith"; "dost thou doubt me already . . . ?"; "he looked back and saw the head of Faith still peeping after him with a melancholy air"; "Poor little Faith!"; and "I'll cling to her skirts and follow her to heaven." Both Goodman Brown and the man he meets in the forest make similar allusions in the second scene, where we read: "Faith kept me back a while"; "We have been a race of honest men and good Christians"; "We are a people of prayer, and good works to boot" (a hint of the theological debate on faith and good works); "Well, then, to end the matter at once, there is my wife, Faith"; "that Faith should come to any harm"; and "why I should quit my dear Faith and go after [Goody Cloyse]." In the episode after the older man leaves Goodman Brown, we have these passages: "so purely and sweetly now, in the arms of Faith!"; "He looked up to the sky, doubting whether there really was a heaven above him"; "With heaven above and Faith below, I will yet stand firm against the devil!"; "a cloud," "confused and doubtful of voices," "he doubted"; "'Faith!' shouted Goodman Brown, in a voice of agony and desperation"; and "'My Faith is gone! . . . Come, devil. . . .' And, maddened with despair. . . ." The last scenes, the forest conclave and young Goodman Brown's return home, offer these: "'But where is Faith?' thought Goodman Brown; and, as hope came"; "the wretched man beheld his Faith . . . before that unhallowed altar"; "'Faith! Faith!' cried the husband, 'look up to heaven . . .'"; "the head of Faith . . . gazing anxiously"; "a distrustful, if not a desperate man"; "he shrank from the bosom of Faith . . . and turned away"; and "no hopeful verse . . . , for his dying hour was gloom."

With these passages in mind, let us recall that there may be both symbolical and allegorical uses of the word "faith." Such ambivalence can complicate a reading of the story. If the tale is allegorical, for example, it may be that Goodman Brown gained his faith (that is, the belief that he is one of the elect) only three months before the action of the story, when he and Faith were married. The fall of the pink ribbon may be a sin or a fall, just as Adam's fall was the original sin, a lapse from grace. The allegory may further suggest that Goodman Brown shortly loses his new faith, for "he shrank from the bosom of Faith." But allegory is difficult to maintain, often requiring a rigid one-to-one equivalence between the surface meaning and a "higher" meaning. Thus if Faith is faith, and Goodman Brown loses the latter, how do we explain that Faith remains with him and even outlives him? Strict allegory would require that she disappear, perhaps even vanish in that dark cloud from which the pink ribbon apparently falls. On the other hand, a pattern of symbolism centering on Faith is easier to handle, and may even be more rewarding by offering us more pervasive, more subtly interweaving ideas that, through their very ambiguity, suggest the difficulties of the theological questions in the story. Such a symbolic view also frees the story from a strict adherence to the Calvinistic concept of election and conviction in the faith, so that the story becomes more universally concerned with Goodman Brown as Everyman Brown.

Whether we emphasize symbol or allegory, however, Goodman Brown must remain a character in his own right, one who progressively loses faith in his ultimate salvation, in his forebears as members of the elect or at least as "good" people, and in his wife and fellow townspeople as holy Christians. At a literal level, he does not lose Faith, for she greets him when he returns from the forest, she still wears her pink ribbons, she follows his corpse to the grave. Furthermore, she keeps her pledge to him, for it is he who shrinks from her. In other words, Brown has not completely lost Faith; rather he has lost faith, a theological key to heaven.

But when faith is lost, not all is lost, though it may very nearly be. Total loss comes later and gradually as Brown commits other sins. We can follow this emerging pattern when we recall that the loss of faith is closely allied to the loss of hope. We find that, in the story, despair (the vice opposed to hope) can be easily associated with doubt (the vice opposed to faith). For example, the two vices are nearly allied when Goodman Brown recognizes the pink ribbon: "'My Faith is gone!' cried he, after one stupefied moment. 'There is no good on earth; and sin is but a name. Come, devil; for to thee is this world given.' And, maddened with *despair*, so that he laughed loud and long, did Goodman Brown grasp his staff and set forth again . . ." (our emphasis).

Doubt, although surely opposed to belief, here leads to despair as much as to infidelity. Similarly, many passages that point to faith also point to hope. When Goodman Brown says, "'I'll follow her to heaven,'" he expresses hope as well as belief. When he says, "'With heaven above and Faith below,'" he hopes to "'stand firm against the devil.'" When he cries, "'Faith, look up to heaven,'" he utters what may be his last hope for salvation. Once again we see how motifs function in a formal structure. It is easy to touch the web at any one point and make it vibrate elsewhere.

Thus we must emphasize that Brown's hope is eroded by increasing doubt, the opposite of faith. We recall that the passages already quoted include the words "desperate," "despair," and "no hopeful verse." When Goodman Brown reenters the town, he has gone far toward a complete failure to trust in God. His thoughts and his actions when he sees the child talking to Goody Cloyse border on the desperate, both in the sense of despair and in the sense of frenzy. Later, we know that he has fully despaired, "for his dying hour was gloom."

"But the greatest of these is love," and "love covers a multitude of sins," the Scriptures insist. Goodman Brown sins against this virtue too, and as we follow these reiterations of the structural components we may well conclude that Hawthorne considered this sin the greatest sin in Brown's life. Sins against love of neighbor are important in other Hawthorne stories. It is a sin against love that Ethan Brand and Roger Chillingworth commit. It is a sin against love of which Rappaccini's daughter accuses Guasconti: "Farewell, Giovanni! Thy words of hatred are like lead within my heart; but they, too, will fall away as I ascend. Oh, was there not, from the first, more poison in thy nature than in mine?" In *The House of the Seven Gables,* it is love that finally overcomes the hate-engendered curse of seven generations.

In "Young Goodman Brown" perhaps the motif of love-hate is first suggested in the opening scene, when Goodman Brown refuses his wife's request that he remain: "'My love and my Faith,' replied young Goodman Brown, 'of all nights in the year, this one night must I tarry away from thee. . . . What, my sweet, pretty wife, dost thou doubt me already, and we but three months married?'" Significantly, the words "love" and "Faith" are used almost as synonyms. When the pink ribbons are mentioned in the next paragraph almost as an epithet ("Faith, with the pink ribbons"), they are emblematic of one virtue as much as the other. Later, Goodman Brown's love of others is diminished when he learns that he is of a family that has hated enough to lash the "Quaker woman so smartly through the streets of Salem" and "to set fire to an Indian village." Instead of being concerned for his own neighbor, he turns against Goody Cloyse, resigning her to the powers of darkness: "What if a wretched old woman do choose to go to the devil . . . ?" He turns against Faith and against God Himself when, after the pink ribbon has fallen from the cloud, he says, "'Come, devil; for to thee is this world given.'" To be sure, he still loves Faith enough at the forest conclave to call upon her yet to look to heaven; but next morning when she almost kisses her husband in front of the whole village, "Goodman Brown looked sternly and sadly into her face, and passed on without a greeting." By this time he is becoming guilty of the specific sin called rash judgment, for he rashly makes successive judgments on his neighbors. He shrinks from the blessing of "the good old minister," he disparages the prayers of old Deacon Gookin, he snatches a child away from the catechizing of "Goody Cloyse, that excellent old Christian." Thenceforth he stubbornly isolates himself from his fellow men and from his own wife. On the Sabbath day he questions their hymns and their sermons, at midnight he shrinks from his wife, at morning or eventide he scowls at family prayers. Having given his allegiance to the devil, he cannot fulfill the injunction of the second great commandment any more than he can fulfill that of the first. Unable to love himself, he is unable to love his neighbor.

"Faith, hope, and love: these three" he has lost, replacing them with their opposed vices, and the pink ribbons serve as emblems for them all and lead to a double pattern of virtues and vices. We can explore further, for the two counterrunning patterns exist in a context of other religiously oriented motifs. Thus, apart from the formal relation of virtues and their opposing vices, we have other biblical motifs. Let us now focus on a few echoes of the Old Testament.

The forest into which Goodman Brown ventures, dark and forbidding as it is, is equated generally with temptation and sin. Clearly he is uneasy about venturing upon this temptation, about leaving Faith behind him. But as any sinner might think, he seems to say, "Just this once, and then. . . ." Specifically, however, the forest equates with the Garden of Eden, where grew the Tree of Knowledge of Good and Evil. Thus it is not surprising that the "good man" meets there the Puritan dark man, that is, an embodiment of the dark forces that Eve had encountered in the serpent. To be sure, this time the avatar is a man "in grave and decent attire" and he has an "indescribable air of one who knew the world and who would not have felt abashed at the governor's table or in King William's court. . . ." But he is soon identified with specific evil actions in the past, and when Goody Cloyse meets him, she screams, "The devil!" Though he appears as a man this time, not as the serpent of Genesis, he carries a staff that writhes like a snake; and we recall not only the snake of the Garden of Eden, but Aaron's staff, which, when thrown down before Pharaoh, turned into a snake (Exod. 7:9–12). Should we miss the motif in its early instance, Hawthorne soon draws the explicit comparison of the staff to the "rods which its owner had formerly lent to the Egyptian Magi." When Goodman Brown approaches the central temptation in the forest congregation of devotees, he—like Adam, like Everyman—is initiated into the knowledge of his race. We should recall a maxim in *The New England Primer*: "In Adam's fall, we sinned all."

Still following these echoes of the Old Testament story, we can now extend our pattern to include the Calvinist setting of early New England. Hawthorne and Goodman Brown, we must note here, are not necessarily thinking in identical ways. Instead, Hawthorne permits Brown's speeches and those of the dark man to carry forward the theological implications of man's depraved nature. It is Goodman Brown who sees evil in all he sees, including himself and Faith. As he and Faith, like Adam and Eve before them, stand at the place of temptation (where not two but four trees mark the place of sin), it is the dark man who insists upon their identity with their race, their knowledge of evil:

> Welcome, my children, to the communion of your race. Ye have found thus young your nature and your destiny.

Yet here are they all in my worshipping assembly.

By the sympathy of your human hearts for sin . . .

It shall be yours to penetrate, in every bosom, the deep mystery of sin . . .

Evil is the nature of mankind. Evil must be your only happiness. Welcome again, my children, to the communion of your race.

Thus when Goodman Brown calls upon Faith to "look up to heaven," he cannot forget that the pink ribbon has already fallen from heaven. For the Calvinist Brown, original sin is a reality. His cry to Faith is quite literally "after the Fall" (postlapsarian), and the Fall cannot be wished away.

But whether Hawthorne or his reader is Calvinist or not, whether the fall of the ribbon is dream or not, the damage is done to Goodman Brown. On this Hawthorne and the reader can agree, for faith, hope, and love come near to vanishing along with the forest conclave when it dissolves into ambiguous shadow. In this way, Hawthorne need not commit himself to Calvinism or, conversely, to an attack upon it. As with the dream, "be it so if you will." The effect on Goodman Brown remains: Whether because of his own fault or because of the depraved nature that is a consequence of Adam's fall, Goodman Brown has lost faith, and hope, and love.

In orthodox Christianity the principles of theology are presented both explicitly and implicitly in the Scriptures. In "Young Goodman Brown" the motifs of faith, hope, and love, summed up in the pink ribbons, blend each into each in a context that derives from the Scriptures. If the blend sometimes confuses us, like the alternating light and dark of the forest conclave, and more particularly like the mystery of the pink ribbons, it is perhaps no less than Hawthorne intended when he presented Goodman Brown's initiation into the knowledge of good and evil, a knowledge that rapidly becomes confusion. For Goodman Brown it is a knowledge by which he seems to turn the very names and epithets of Goodman, Goody, and Gookin into variant spellings of "evil," just as Brown transmutes faith, hope, and love into their opposed vices. For the reader the pink ribbons, like the balance of town and country, like the interplay of light and dark, remain in the mind an index to ambiguity, which is, paradoxically, as we have said, a formalistic device in the story.

C. Romance and Reality, Land and River: The Journey as Repetitive Form in *Huckleberry Finn*

In the preceding section, on formalistic qualities in "Young Goodman Brown," we noted that the short story is generally concerned with relatively few characters and with only one major situation. The short story achieves its climax and solution, and quickly concludes. The novel, however, contains more characters; and its plot, a number of episodes or situations. Its ampler space provides opportunity for creation of a world, with the consequent opportunity for the reader to be immersed in that world. But because the novel is ample, in comparison with a lyric poem or a short story, it offers a further challenge to its creator to give it its form. In fact, historically the formalistic approach in criticism has focused more on lyric poetry and short stories than on the novel. Nevertheless, the novel, too, is an art form, and a close reading will present one or more ways of seeing its form and how the author controls that form.

It will become clear as we approach the form of *Huckleberry Finn* that at one level its form can be simplistically diagrammed as a capital letter "I" lying on its side. At each end there is a block of chapters set on the land and in a world where Tom Sawyer can exist and even dominate. In the middle are chapters largely related to the river as Huck and Jim travel down that river; here realism, not a Tom Sawyer romanticism, dominates. Further, in the central portion there is a pattern of alternations between land and river. Taking the novel as a whole, then, there is a pattern of departures and returns.

But Twain was not limited to a pattern that can be charted, as it were, on graph paper. In a master stroke of the creative art, he chose Huck Finn himself as the point-of-view character. In doing so, Twain abandoned the simpler omniscient (or authorial) point of view that he had very successfully used in *The Adventures of Tom Sawyer* for a relatively sophisticated technique. He allowed the central character to relate his adventures in his own way—the point of view called first-person narrator. T. S. Eliot refers to the difference in points of view as indicative of the major qualitative distinction between *Tom Sawyer* and *Huckleberry Finn*: Tom's story is told by an adult looking at a boy and his gang; Huck's narrative requires that "we see the world through his eyes" (Eliot, "Introduction," *The [sic] Adventures of Huckleberry Finn* [London: Cresset, 1950]; rpt. in *Adventures of Huckleberry Finn: An Annotated Text, Backgrounds and Sources, Essays in Criticism*, ed. Sculley Bradley, Richmond Croom Beatty, and E. Hudson Long [New York: Norton, 1962]: 321). Granted that Twain sometimes allows us to see beyond Huck's relatively simple narrative manner some dimensions of meaning not apparent to Huck, the point of view has been so contrived (and controlled) that we do not see anything that is not at least implicit in Huck's straightforward narration.

Several questions can be raised. What is the character of Huck like? How does his manner of telling his story control our responses to that story? Finally, how does this point of view assist us in perceiving the novel's form?

First, Huck is an objective narrator. He is objective about himself, even when that objectivity tends to reflect negatively upon himself. He is objective about the society he repeatedly confronts, even when, as he often fears, that society possesses virtues and sanctions to which he must ever remain a stranger. He is an outcast, he knows that he is an outcast, and he does not blame the society that has made and will keep him an outcast. He always assumes in his characteristic modesty that he must somehow be to blame for the estrangement. His deceptions, his evasions, his lapses from conventional respectability are always motivated by the requirements of a given situation; he is probably the first thoroughgoing, honest pragmatist in American fiction. When he lies or steals, he assumes that society is right and that he is simply depraved. He does not make excuses for himself, and his conscience is the stern voice of a pietistic, hypocritical backwoods society asserting itself within that sensitive and wistful psyche. We know that he is neither depraved nor dishonest, because we judge that society by the damning clues that emerge from the naive account of a boy about thirteen years old who has been forced to lie in order to get out of trouble but who never lies to himself or to his reader. In part, his lack of subtlety is a measure of his reliability as narrator: he has mastered neither the genteel speech of "respectable" folks nor their deceit, evasions of truth, and penchant for pious platitudes. He is always refreshingly himself, even when he is telling a tall tale or engaging in one of his ambitious masquerades to get out of a jam.

Thus the point of view Twain carefully establishes from the first words of the narrative offers a position from which the reader must consider the events of the narrative. That position never wavers from the trustworthy point of view of the hero-narrator's clear-eyed gaze. He becomes at once the medium and the norm for the story that unfolds. By him we can measure (although he never overtly does it himself) the hypocrisy of Miss Watson, perceive the cumulative contrast between Huck and the incorrigible Tom Sawyer, and finally judge the whole of society along the river. Eliot makes this important discrimination: "Huck has not imagination, in the sense in which Tom has it; he has, instead, vision. He sees the real world; and he does not judge it—he allows it to judge itself" (321).

Huck's characteristic mode of speech is ironic and self-effacing. Although at times he can be proud of the success of his tall tales and masquerades, in the things that matter he is given to understatement. Of his return to "civilized" life with the Widow Douglas, he tersely confides, "Well, then, the old thing commenced again." Of the senseless horror with which the Grangerford-Shepherdson feud ends, Huck says with admirable restraint: "I ain't a-going to tell *all* that happened—it would make me sick again if I was to do that. I wished I hadn't ever come ashore that night to see such things. I ain't ever going to get shut of them—lots of times I dream about them." And in one of the most artfully conceived, understated, but eloquent endings in all fiction, Huck bids his reader and civilization goodbye simultaneously: "But I reckon I got to light out for the Territory ahead of the rest, because Aunt Sally she's going to adopt me and sivilize me, and I can't stand it. I been there before."

The movement of the novel likewise has an effect on the total shape of the work. The apparently aimless plot with its straightforward sequence—what happened, what happened next, and then what happened after that, to paraphrase Gertrude Stein—is admirably suited to the personality of Huck as narrator. In the conventional romantic novel, of course, we expect to find a more or less complex central situation, in which two lovers come together by various stratagems of the novelist, have their difficulties (they disagree about more or less crucial matters or they must contend against parents, a social milieu, deprivations of war or cattle rustlers, or dozens of other possible impediments to their union), resolve their problems, and are destined to live happily ever afterward. Even in such a classic novel as Jane Austen's *Pride and Prejudice*, the separate chapters and the pieces of the plot concern the manifestations, against the background of early nineteenth-century English provincial life, of the many facets of Mr. Darcy's insuperable pride and Elizabeth Bennet's equally tenacious prejudice, but everything works toward the happy union of two very attractive young people.

In *Huckleberry Finn*, however, there is no real center to the plot as such. Instead we have what Kenneth Burke has called repetitive form: "the consistent maintaining of a principle under new guises . . . a restatement of the same thing in different ways. . . . A succession of images, each of them regiving the same lyric mood; a character repeating his identity, his 'number,' under changing situations; the sustaining of an attitude as in satire . . ." (*Counter-Statement* [Los Altos, CA: Hermes, 1953]: 125). The separate situations or episodes are loosely strung together by the presence of Huck and Jim as they make their way down the Mississippi River from St. Petersburg, while the river flows through all, becoming really a vast highway across backwoods America. In the separate episodes there are new characters who, after Huck moves on, usually do not reappear. There are new settings and always new situations. At the beginning, there are five chapters about the adventures of Huck and Tom and the gang in St. Petersburg; at the end, there are twelve chapters centering on the Phelps farm that chronicle the high jinks of the boys in trying to free Jim; in between, there are twenty-six chapters in which Huck and Jim pursue true freedom and in which Tom Sawyer does not appear. This large midsection of the book includes such revealing experiences as Jim and Huck's encounter with the "house of death" (ch. 9); the dual masquerades before the perceptive Mrs. Judith Loftus (ch. 11); Huck's life with the Grangerfords (chs. 17 and 18); the performance of the Duke and Dauphin at Pokeville (ch. 20); the Arkansas premiere of Shakespeare and the shooting of Boggs by Colonel Sherburn (chs. 21 and 22); and, finally, the relatively lengthy involvement with the Wilks family (chs. 24–29).

Despite changes in settings and dramatis personae, the separate episodes share a cumulative role (their repetitive form): Huck learns bit by bit about the depravity that hides beneath respectability and piety. He learns gradually and unwillingly that society or civilization is vicious and predatory and that the individual has small chance to assert himself against a monolithic mass. Harmless as the sentimental tastes of the Grangerfords or their preference for the conventionally pretty may seem, Twain's superb sense for the objective correlative allows us to *realize* (without being *told*) that conventional piety and sentimentality hide depravity no more effectually than the high coloring of the chalk fruit compensates for the chips that expose the underlying chalk. Likewise, elaborate manners, love of tradition, and "cultivated" tastes for graveyard school poetry and lugubrious drawings are merely genteel facades for barbarism and savagery. Mrs. Judith Loftus, probably the best-developed minor character in the entire novel, for all her sentimental response to the hackneyed story of a mistreated apprentice, sees the plight of the runaway slave merely in terms of the cash reward she and her husband may win. Even the Wilks girls, as charming as they seem to Huck, are easily taken in by the grossest sentimentality and pious clichés. A review of the several episodes discloses that, for all their apparent differences, they are really reenactments of the same insistent revelation: the mass of humanity is hopelessly depraved, and the genuinely honest individual is constantly being victimized, betrayed, and threatened.

The framework of the plot is, then, a journey—a journey

from north to south, a journey from relative innocence to horrifying knowledge. Huck tends to see people for what they are, but he does not suspect the depths of evil and the pervasiveness of sheer meanness, of man's inhumanity to man, until he has completed his journey. The relative harmlessness of Miss Watson's lack of compassion and her devotion to the letter rather than the spirit of religious law or of Tom's incurable romanticism does not become really sinister until Huck reenters the seemingly good world at the Phelps farm, a world that is really the same as the "good" world of St. Petersburg—a connection that is stressed by the kinship of Aunt Sally and Aunt Polly. Into that world the values of Tom Sawyer are once more injected, but Huck discovers that he has endured too much on his journey down the river to become Tom's foil again.

Much has been written about the end of the novel. In *The Green Hills of Africa* Hemingway complained that the section about the Phelps farm is "cheating," that the novel should have ended with Huck and Jim floating down the river to an inevitable and tragic end. But precisely because Huck has come full circle, back into the world of St. Petersburg, there also comes his shocking realization that "you can't go home again." Therefore, Huck must set off on another flight from an oppressive civilization into the temporary freedom of the Territory.

The characters, except for Huck and Jim, are stereotypes. The women, except perhaps for Mrs. Judith Loftus, seem flat to most readers—either humor characters like the elderly women or pure virgins like Sophia Grangerford or Mary Jane Wilks. The men represent the frontier types Twain knew—the braggart, the superman like Colonel Sherburn, the vagrant actors like the Duke and the Dauphin, the lawyers, the itinerant preachers, and backwoodsmen generally. But distinction in characterization is another formal suggestion that the stereotype is another integer of the anti-individualism that pervades the society represented in this novel.

One can imagine something like a descending scale of viciousness operating in the characters in the novel. The relatively harmless good people like the Widow Douglas, the Wilks girls, and Aunt Sally and Uncle Silas rank near the top.

Somewhere along the middle are the Grangerfords, whose basic kindness and devotion to what they suppose is an aristocratic tradition of manners and forms must be set against their overweening sentimentality, their canting theological commitments, and their single-minded obsession with a senseless and brutal blood feud whose origin has long before been lost to memory. Alongside the Grangerfords is Mrs. Judith Loftus, with her paradoxical combination of sympathy and generosity for the homeless, ostensibly mistreated white waif, and unmitigated avarice and indifference to the humanity of the miserable runaway slave. And the bottom of the scale is crowded with animalistic, predatory characters like the Duke and Dauphin, pap Finn, and the mob that pursues Colonel Sherburn.

Even such a simple classification of the characters enables us to see a significance in the arrangement of episodes. Huck's flight into freedom, although its origin lies in his need to escape both the civilization of Miss Watson and possible death at the hands of pap, starts out as a kind of lark—even a Tom Sawyer type of adventure. But the adventures become increasingly sinister for Huck, and as he and Jim drift down the river, involvements with society take on a darker and darker cast. By the time he and Jim arrive at Pikesville and have been victimized by the Duke and Dauphin, they have seen virtually every kind of depravity posssible on the frontier, once the sanctuary of the seekers after freedom.

Thus the narrative moves in a circular pattern. Huck has left St. Petersburg to escape from conventional gentility and morality. But he sees the same conventions shaping life at the Phelps farm. Tom Sawyer returns to play more of his tricks, this time with Jim as victim. Although Huck has gone home again (back to the world of St. Petersburg), he knows now that there is really no home for him as human beings define home. He was merely a social outcast before; now he is a moral outcast because he cannot accept conventional morality, which is after all part of the texture of civilization. At the end of the novel, then, he is preparing to set forth once again, on another journey that time and history will probably defeat. Huck had been subjected to the same kind of cruel, senseless captivity at the hands of pap that Jim must endure on the Phelps farm. Now Huck realizes, finally and tragically, that all society is constantly arranging to capture the individual in one way or another—that, indeed, all members of society are really captives of an oppressive sameness that destroys any mode of individualism. Huck and Jim alone have a capacity to pursue freedom.

Only the great, flowing river defines the lineaments of otherwise elusive freedom; that mighty force of nature opposes and offers the only possible escape from the blighting tyranny of towns and farm communities. The Mississippi is the novel's major symbol. It is the one place where a person does not need to lie to himself or to others. Its ceaseless flow mocks the static, stultifying society on its banks. There are lyrical passages in which Huck communicates, even with all his colloquial limitations, his feelings about the river, its symbolic functions, as in the image-packed description that follows the horrors of the Grangerford-Shepherdson carnage (ch. 19). In that memorable passage Huck extols the freedom and contemplation that the river encourages. In contrast to the oppressive places on land, the raft and the river promise release: "We said there warn't no home like a raft, after all. Other places do seem so cramped up and smothery, but a raft don't. You feel mighty free and easy and comfortable on a raft."

Like the river, Huck's narrative flows spontaneously and ever onward. Around each bend lies a possible new adventure; in the eddies, a lyrical interlude. But the river always carries Huck and Jim out of each adventure toward another uncertain try for freedom. That freedom is never really achieved is a major irony, but the book's structure parallels the river's flow. The separate adventures become infinite variations upon (repetitive forms of) the quest for freedom. That the final thwarting of freedom is perpetrated by the forces of St. Petersburg, of course, is no fault of the river or its promise of freedom; it simply seems that membership in humanity generates what we have elsewhere called the circular pattern of flight and captivity.

D. Dialectic as Form: The Trap Metaphor in *Hamlet*

> My stronger guilt defeats my strong intent;
> And like a man to double business bound,
> I stand in pause where I shall first begin,
> And both neglect. (III.iii)

The words are not those of Hamlet. They are spoken by Claudius, as he tries to pray for forgiveness, even as he knows that he cannot give up those things for which he murdered his brother—his crown, his fulfilled ambition, and his wife. But the words may easily have been Hamlet's, for he too is by "double business bound." Indeed, much of the play centers on doubleness. A consequence of that doubleness for many of the characters is that they are apparently caught in a trap—a key metaphor in the play—or, in another image, "Hoist with [one's] own petar" (III.iv).

Let us examine that metaphor of the trap, for it leads clearly to our seeing how dialectic provides form in *Hamlet.* Several times in the play, but in varying images, we find allusions to different kinds of entanglement. Polonius injudiciously uses the metaphor to warn Ophelia away from Hamlet's "holy vows of heaven," vows that he says are "springes to catch woodcocks" (I.iii). More significant is Hamlet's deliberate misnaming of "The Murder of Gonzago"; he calls it "The Mousetrap" (III.ii) because it is, as he says elsewhere, "the thing / Wherein I'll catch the conscience of the King" (II.ii). Claudius feels that he is trapped: "O limed soul, that, struggling to be free, / Art more engag'd" (III.iii). Hamlet, in the hands of plotters, finds himself "thus be-netted round with villainies" and one for whom Claudius has "Thrown out his angle [fish hook] for my proper life" (V.ii). The dying Laertes echoes his father's metaphor when he tells Osric that he is "as a woodcock to mine own springe" (V.ii). Here we have a pattern of trap images—springes, lime, nets, mousetraps, and angles or hooks. Now traps are usually for animals, but we are dealing with human beings, persons who are trapped in their own dilemmas, in their own questions, in the very questioning of the universe.

Let us expand our formal approach to *Hamlet* by characterizing once again the world of the work. The setting, we know, is the castle of Elsinore in medieval Denmark. But what is this Denmark? It is, to be sure, largely a Denmark of Shakespeare's imagination—an imaginatively apprehended Denmark that has only tenuous relationships to a historical Denmark. It is a world, however, peopled by Shakespeare's characters that takes on definition from what happens and what is said in the play and that in turn determines what can be done and said. We need go no further than the first scene of act I to realize that it is a disturbed world, that a sense of mystery and deep anxiety preoccupies the soldiers of the watch. The ghost has appeared already and is expected to appear again. The guards instinctively assume that the apparition of the former king has more than passing import; and, in their troubled questions to Horatio about the mysterious preparations for war, the guards show how closely they regard the connection between the unnatural appearance of the dead king and the welfare of the state. The guards have no answers for the mystery, their uncertainty, or their premonitions; their quandary is mirrored in abundant questions and minimal answers—a rhetorical phenomenon that recurs throughout the play, even in the soliloquies of Hamlet. The sense of cosmic implication in the special situation of Denmark emerges strongly in the exchange between Hamlet and his friends Rosencrantz and Guildenstern:

> *Hamlet.* Denmark's a prison.
> *Rosencrantz.* Then is the world one.
> *Hamlet.* A goodly one; in which there are many confines, wards, and dungeons, Denmark being the one o' th' worst.
> (II.ii)

These remarks recall the assertion of Marcellus as Hamlet and the ghost go offstage: "Something is rotten in the state of Denmark" (I.iv). Indeed, Hamlet's obsession is both the product and the cure of the rottenness of Denmark, but the rottenness, he acknowledges, pervades all of nature: ". . . this goodly frame the earth seems to me a sterile promontory; this most excellent canopy, the air, look you, this brave o'erhanging firmament, this majestical roof fretted with golden fire— why, it appeareth nothing to me but a foul and pestilent

congregation of vapors" (II.ii). Much earlier, before his encounter with the ghost, Hamlet has expressed his extreme
pessimism at man's having to endure earthly existence
within nature's unwholesome realm:

> How weary, stale, flat and unprofitable
> Seem to me all the uses of this world!
> Fie on't, ah, fie, 'tis an unweeded garden
> That grows to seed. Things rank and gross in nature
> Possess it merely. (I.ii)

As he speaks these lines, Hamlet apparently has no idea of
the truth of his father's death but is dismayed over his
mother's hasty marriage to the new king. He has discovered a
seeming paradox in being: the fair, in nature and humanity,
inevitably submits to the dominion of the foul. His obsession
with the nature and cause of the paradox focuses his attention on Denmark as the model of nature and human frailty.
Thus a pattern of increasing parallels of Denmark to the
cosmos and of man to nature develops. Question and answer,
dialogue and soliloquy, become a verbal unity of repeated
words and phrases, looking forward to larger thematic assertion and backward to earlier adumbration.

The play constitutes a vast poem in which speculation
about nature, human nature, the health of the state, and human destiny intensifies into a passionate dialectic. Mystery,
riddle, enigma, and metaphysical question complicate the
dialogue. Particularly in his soliloquies Hamlet confronts
questions that have obsessed protagonists from Sophocles's
Oedipus to Tom Stoppard's Rosencrantz and Guildenstern.
What begins with the relatively simple questions of the soldiers of the watch in act 1 is magnified and complicated as
the play moves on. Increasingly tenuous and rarified probes
of the maddening gulf between reality and appearance prolif-
erate. Moreover, the contrast between what the simple man
cheerfully accepts at face value and what the thoughtful man
is driven to question calls into doubt every surface of utterance, act, or thing. In the world of *Hamlet* the cosmic
implications of myriad distinctions between "seem" and
"be" confront us at every hand.

An index to form looms in the crucial qualitative differences between Hamlet's mode of speech and that of the other inhabitants of his strange world. Because Hamlet's utterances and manners are characteristically unconventional, the other major characters (except Horatio, of course) assume that he is mad or at least temporarily deranged. Conversely, because they do speak the simple, relatively safe language of ordinary existence, he assumes that they are hiding or twisting the truth. No one who easily settles for seeming is quite trustworthy to the man obsessed with the pursuit of being. Even the ghost's nature and origin (he may be a diabolical agent, after all) must be tentative for Hamlet until he can settle the validity of the ghost's revelations by the "play within the play." Even Ophelia must be treated as the possible tool of Claudius and Polonius. The presence of Rosencrantz and Guildenstern, not to mention their mission on the journey to England, arouses Hamlet's deepest suspicions. Only Horatio is exempt from distrust, and even to him Hamlet cannot divulge the full dimension of his subversion. Yet though Hamlet seems to speak only in riddle and to act solely with evasion, his utterances and acts always actually bespeak the full measure of his feelings and his increasingly single-minded absorption with his inevitable mission. The important qualification of his honesty lies in his full knowledge that others do not (or cannot) comprehend his real meanings and that others are hardly vitally concerned with deep truths about the state, mankind, or themselves.

For our purposes, of course, the important fact is that these contrasting levels of meaning and understanding achieve formal expression. When the king demands some explanation for his extraordinary melancholy, Hamlet replies, "I am too much in the sun" (I.ii). The reply thus establishes, although Claudius does not perceive it, Hamlet's judgment of and opposition to the easy acceptance of "things as they are." And when the queen tries to reconcile him to the inevitability of death in the natural scheme and asks, "Why seems it so particular with thee?" he responds with a revealing contrast between the seeming evidences of mourning and real woe— an unequivocal condemnation of the queen's apparently easy acceptance of his father's death as opposed to the vindication of his refusal to view that death as merely an occasion for ceremonial mourning duties. To the joint entreaty of Claudius and Gertrude that he remain in Denmark, he replies only to his mother: "I shall in all my best obey you, madam" (I.ii). But in thus disdaining to answer the king, he has promised really nothing to his mother, although she takes his reply for complete submission to the royal couple. Again we see that every statement of Hamlet is dialectic: that is, it tends toward double meaning—a kind of countermeaning for the world of Denmark and subtler meaning for Hamlet and the reader.

As we have observed, Hamlet's overriding concern, even before he knows of the ghost's appearance, is the frustration of living in a world attuned to imperfection. He sees, wherever he looks, the pervasive blight in nature, especially human nature. Man, outwardly the acme of creation, is susceptible to "some vicious mole of nature," and no matter how virtuous he otherwise may be, the "dram of evil" or the "stamp of one defect" adulterates nobility (I.iv). Hamlet finds that "one may smile and smile, and be a villain" (I.v). To the uncomprehending Guildenstern, Hamlet emphasizes his basic concern with the strange puzzle of corrupted and corrupting man:

> What a piece of work is a man, how noble in reason, how infinite in faculties, in form and moving how express and admirable, in action how like an angel, in apprehension how like a god: the beauty of the world, the paragon of animals! And yet to me what is this quintessence of dust? Man delights not me— no, nor woman either, though by your smiling you seem to say so. (II.ii)

This preoccupation with the paradox of man, recurring as it does throughout the play, obviously takes precedence over the revenge ordered by the ghost. From the beginning of Hamlet's inquiry into the world that surrounds him— whether he is considering his father's death, his mother's remarriage, the real or supposed defection of his friends, or the fallen state of man—we see a developing pattern of meaning in which man's private world of marriage bed and lust for power becomes part of the larger dimension of identity and worth. The implications of the dangers inherent in this "man's" view of the world in *Hamlet* are explored in chapter 5 ("Feminist Approaches").

Reams have been written about Hamlet's reasons for the delay in carrying out his revenge; for our purpose, however, the delay is not particularly important, except insofar as it emphasizes Hamlet's greater obsession with the pervasive blight within the cosmos. From almost every bit of verbal evidence, he considers as paramount the larger role of investigator and punitive agent of all humankind (for example, his bristling verbal attack on the queen, his accidental—but to his mind quite justifiable—murder of Polonius, his indignation about the state of the theater and his instructions to the players, his bitter castigation of Ophelia, his apparent delight not only in foiling Rosencrantz and Guildenstern but also in arranging their destruction, and his fight with Laertes over the grave of Ophelia). Hamlet, in living up to what he conceives to be a higher role than that of mere avenger, recurrently broods about his self-imposed mission, although he characteristically avoids naming it. In his warfare against bestiality, however, he asserts his allegiance to heaven-sent reason and its dictates:

> What is a man,
> If his chief good and market of his time
> Be but to sleep and feed? A beast, no more.
> Sure he that made us with such large discourse,
> Looking before and after, gave us not
> That capability and godlike reason
> To fust in us unused. Now, whether it be
> Bestial oblivion, or some craven scruple
> of thinking too precisely on th'event—
> A thought which, quartered, hath but one part wisdom
> And ever three parts coward—I do not know
> Why yet I live to say, "This thing's to do,"
> Sith I have cause, and will, and strength, and means
> To do't. (IV.vi)

With some envy he regards the active competence of Fortinbras as opposed to his own "craven scruple / Of thinking too precisely on th'event" (that is, his obligation to act to avenge his father's death). Although he promises himself to give priority to that obligation, he goes to England on what he suspects to be a ruse, presumably because for the moment he prefers outwitting Rosencrantz, Guildenstern, and their kind. In short, almost from his first appearance in the play, Hamlet obviously is convinced that to him is given a vast though somewhat general (even ambiguous) task:

> The time is out of joint. O cursed spite
> That ever I was born to set it right! (I.v)

The time, like the place of Denmark, has been corrupted by men vulnerable to natural flaws. And once again Hamlet's statement (this time in the philosophic, lyrical mode of *soliloquy*) offers formal reinforcement for the dialectic of the play—the opposition of two attitudes toward human experience that must achieve resolution or synthesis before the play's end.

To the ideal of setting things right, then, Hamlet gives his allegiance. The order he supports transcends the expediency of Polonius, the apostle of practicality, and of Claudius, the devotee of power and sexuality. Again and again we see Hamlet's visionary appraisal of an order so remote from the ken of most people that he appears at times inhuman in his refusal to be touched by the scales of ordinary joy or sorrow. He will set straight the political and social order by ferreting out bestiality, corruption (of state, marriage bed, or theater), trickery, and deceit. He is obsessed throughout the play by the "dusty death" to which all must come, and his speeches abound in images of sickness and death. But if he has finally gotten the king, along with his confederates, "Hoist with his own petar" (III.iv), Hamlet also brings himself, through his own trickery, deceit, perhaps even his own ambitions, to the fate of Yorick.

Thus does the play turn upon itself. It is no simple morality play. It begins in an atmosphere of mourning for the late king and apprehensions about the appearance of the ghost, and it ends in a scene littered with corpses. The noble prince, like his father before him, is, despite his best intentions, sullied by the "foul crimes done in my days of nature" (I.v). All men apparently are, as Laertes says of himself, "as a woodcock to mine own springe" (V.ii) (that is, like a fool caught in his own snare). And though all beauty and aspiration (a counterpoint theme) are reduced ultimately to a "quintessence of dust," it is in Hamlet's striving, however imperfectly and destructively, to bend the order of nature to a higher law that we must see the play's tragic assertion in the midst of an otherwise pervasive and unrelieved pessimism.

The design of the play can be perceived in part by the elaborate play upon the words "see" and "know" and their cognates. Whereas the deity can be understood as "Looking before and after" (IV.iv), the player king points out to his queen that there is a hiatus between what people intend and what they do: "Our thoughts are ours, their ends none of our own" (III.ii). Forced by Hamlet to consider the difference between her two husbands, Gertrude cries out in anguish against having to see into her own motivations:

> O Hamlet, speak no more.
> Thou turn'st mine eyes into my very soul,
> And there I see such black and grained spots
> As will not leave their tinct. (III.iv)

But she does not see the ghost of her former husband, nor can she see the metaphysical implications of Hamlet's reason in madness. The blind eye sockets of Yorick's skull once saw their quota of experience, but most people in Denmark are quite content with the surface appearances of life and refuse even to consider the ends to which mortality brings everyone. The intricate weavings of images of sight thus become a kind of tragic algebra for the plight of a man who "seemed to find his way without his eyes" (II.i) and who found himself at last "placed to the view" of the "yet unknowing world" (V.ii).

The traveling players had acted out the crime of Denmark on another stage, but their play seemed to most of the audience only a diversion in a pageant of images designed to keep them from really knowing themselves or their fellows to be corrupted by nature and doomed at last to become "my Lady Worm's, chapless and knocked about the mazzard with a sexton's spade" (V.i). The contexts of these words assert a systematic enlargement of the play's tragic pronouncement of human ignorance in the midst of appearances. Formally, the play progresses from the relatively simple speculations of the soldiers of the watch to the sophisticated complexity of metaphysical inquiry. There may not be final answers to the questions Hamlet ponders, but the questions assume a formal order as their dimensions are structured by speech and action—in miniature, by the play within the play; in extension, by the tragedy itself.

Ophelia, in her madness, utters perhaps the key line of the play: "Lord, we know what we are, but we know not what we may be" (IV.v). Hamlet has earlier said that if the king reacts as expected to the play within the play, "I know my course" (II.ii). But he is not sure of his course, nor does he even know himself—at least not until the final act. In the prison of the world he can only pursue his destiny, which, as he realizes before the duel, inevitably leads to the grave. The contest between human aspiration and natural order in which Hamlet finds himself is all too unequal: idealism turns out to be a poor match for the prison walls of either Denmark or the grave.

VI. LIMITATIONS OF THE FORMALISTIC APPROACH

By the 1950s, dissent was in the air. Still outraged by the award of the Bollingen Prize for Poetry to Ezra Pound in 1949, some voices thought they detected a pronounced elitism, if not more sinister rightist tendencies, in the New Critics, their disciples, and the poets to whom they had granted the favor of their attention. The details of this political argument need not compel our attention here. What does concern us is the realization that by 1955, some doubters were pointing to the formalistic critics' absorption with details, their greater success with intensive than with extensive criticism, their obvious preference for poets like Eliot and Yeats, and their lack of success with the novel and the drama (C. Hugh Holman, "The Defense of Art: Criticism Since 1930," *The Development of American Literary Criticism*, ed. Floyd Stovall [Chapel Hill: U of North Carolina P, 1955]: 238–39).

Less general caveats have emphasized the restriction of

formalistic criticism to a certain kind of literature simply because that kind proved itself especially amenable—lyric poetry generally but especially English poetry of the seventeenth century and the "modernist" poetry that stems from Pound and Eliot, and some virtually self-selecting fiction that significantly displays poetic textures (for example, *Moby-Dick* and *Ulysses*). New Critics tended to ignore or undervalue some poetry and other genres that do not easily respond to formalistic approaches (for example, the poetry of Wordsworth and Shelley, philosophical and didactic verse generally, and the essay). Apparently the problems increase whenever the language of the literary work tends to approach that of the philosopher, or even that of the critic himself. The formalistic approach sometimes seems to lapse into a treasure hunt for objective correlatives, conceits, the image, or ironic turns of phrase. It has not seemed to work particularly well for most American poetry written since 1950; as students often point out, it tends to overlook feeling and appears heartless and cold in its absorption with form.

Robert Langbaum has pronounced the New Criticism "dead—dead of its very success." For, says he, "We are all New Critics nowadays, whether we like it or not, in that we cannot avoid discerning and appreciating wit in poetry, or reading with close attention to words, images, ironies, and so on" (*The Modern Spirit: Essays on the Continuity of Nineteenth- and Twentieth-Century Literature* [New York: Oxford UP, 1970]: 11). There is more to criticism than "understanding the text, [which] is where criticism begins, not where it ends" (14). Langbaum believes that the New Criticism took us for a time outside the "main stream of criticism" (represented by Aristotle, Coleridge, and Arnold), but now we should return, with the tools of explication and analysis given us by the New Critics, to that main stream. That is, instead of insisting upon literature's autonomy, we must resume relating it to life and ideas.

· 3 ·

The Psychological Approach: Freud

I. AIMS AND PRINCIPLES

Having discussed two of the basic approaches to literary understanding, the traditional and the formalistic, we now examine a third interpretive perspective, the psychological. Of all the critical approaches to literature, this has been one of the most controversial, the most abused, and—for many readers—the least appreciated. Yet, for all the difficulties involved in its proper application to interpretive analysis, the psychological approach can be fascinating and rewarding. Our purpose in this chapter is threefold: (1) to account briefly for the misunderstanding of psychological criticism; (2) to outline the psychological theory most commonly used as an interpretive tool by modern critics; and (3) to show by examples how readers may apply this mode of interpretation to enhance their understanding and appreciation of literature.

The idea of *enhancement* must be understood as a preface to our discussion. It is axiomatic that no single approach can exhaust the manifold interpretive possibilities of a worthwhile literary work; each approach has its own peculiar limitations. As we have already discovered, the limitations of the traditional approach lie in its tendency to overlook the structural intricacies of the work. The formalistic approach, on the other hand, often neglects historical and sociological contexts that may provide important insights into the mean-

116

ing of the work. In turn, the crucial limitation of the psychological approach is its aesthetic inadequacy: psychological interpretation can afford many profound clues toward solving a work's thematic and symbolic mysteries, but it can seldom account for the beautiful symmetry of a well-wrought poem or of a fictional masterpiece. Though the psychological approach is an excellent tool for reading beneath the lines, the interpretive craftsman must often use other tools, such as the traditional and the formalistic approaches, for a proper rendering of the lines themselves.

A. Abuses and Misunderstandings of the Psychological Approach

In the general sense of the word, there is nothing new about the psychological approach. As early as the fourth century B.C., Aristotle used it in setting forth his classic definition of tragedy as combining the emotions of pity and terror to produce catharsis. The "compleat gentleman" of the English Renaissance, Sir Philip Sidney, with his statements about the moral effects of poetry, was psychologizing literature, as were such romantic poets as Coleridge, Wordsworth, and Shelley with their theories of the imagination. In this sense, then, virtually every literary critic has been concerned at some time with the psychology of writing or responding to literature.

During the twentieth century, however, psychologcial criticism has come to be associated with a particular school of thought: the psychoanalytic theories of Sigmund Freud (1856–1939) and his followers. (The currently most significant of these followers, Jacques Lacan, will be discussed in chapter 5.) From this association have derived most of the abuses and misunderstandings of the modern psychological approach to literature. Abuses of the approach have resulted from an excess of enthusiasm, which has been manifested in several ways. First, the practitioners of the Freudian approach often push their critical theses too hard, forcing literature into a Procrustean bed of psychoanalytic theory at the expense of other relevant considerations (for example, the work's total thematic and aesthetic context). Second, the literary criticism of the psychoanalytic extremists has at times degenerated into a special occultism with its own mystique and jargon exclusively for the in-group. Third, many critics of the psychological school have been either literary scholars who have understood the principles of psychology imperfectly or professional psychologists who have had little feeling for literature as art: the former have abused Freudian insights through oversimplification and distortion; the latter have bruised our literary sensibilities.

These abuses have given rise to a widespread mistrust of the psychological approach as a tool for critical analysis. Conservative scholars and teachers of literature, often shocked by such terms as *anal eroticism, phallic symbol,* and *Oedipal complex,* and confused by the clinical diagnoses of literary problems (for example, the interpretation of Hamlet's character as a "severe case of hysteria on a cyclothymic basis"—that is, a manic-depressive psychosis), have rejected all psychological criticism, other than the commonsense type, as pretentious nonsense. By explaining a few of the principles of Freudian psychology that have been applied to literary interpretation and by providing some cautionary remarks, we hope to introduce the reader to a balanced critical perspective that will enable him or her to appreciate the instructive possibilities of the psychological approach while avoiding the pitfalls of either extremist attitude.

B. Freud's Theories

The foundation of Freud's contribution to modern psychology is his emphasis on the unconscious aspects of the human psyche. A brilliant creative genius, Freud provided convincing evidence, through his many carefully recorded case studies, that most of our actions are motivated by psychological forces over which we have very limited control. He demonstrated that, like the iceberg, the human mind is structured so that its great weight and density lie beneath the surface (below the level of consciousness). In "The Anatomy of the Mental Personality, Lecture XXI," in *New Introductory Lec-*

tures on Psychoanalysis (New York: Norton, 1964), Freud discriminates between the levels of conscious and unconscious mental activity:

> The oldest and best meaning of the word "unconscious" is the descriptive one; we call "unconscious" any mental process the existence of which we are obligated to assume—because, for instance, we infer it in some way from its effects—but of which we are not directly aware. . . . If we want to be more accurate, we should modify the statement by saying that we call a process "unconscious" when we have to assume that it was active *at a certain time*, although *at that time* we knew nothing about it. (99–100)

Freud further emphasizes the importance of the unconscious by pointing out that even the "most conscious processes are conscious for only a short period; quite soon they become *latent*, though they can easily become conscious again" (100). In view of this, Freud defines two kinds of unconscious:

> one which is transformed into conscious material easily and under conditions which frequently arise, and another in the case of which such a transformation is difficult, can only come about with a considerable expenditure of energy, or may never occur at all. . . . We call the unconscious which is only latent, and so can easily become conscious, the "preconscious," and keep the name "unconscious" for the other. (101)

That most of the individual's mental processes are unconscious is thus Freud's first major premise. The second (which has been rejected by a great many professional psychologists, including some of Freud's own disciples—for example, Carl Gustav Jung and Alfred Adler) is that all human behavior is motivated ultimately by what we would call sexuality. Freud designates the prime psychic force as libido, or sexual energy. His third major premise is that because of the powerful social taboos attached to certain sexual impulses, many of our desires and memories are repressed (that is, actively excluded from conscious awareness).

Starting from these three premises, we may examine several corollaries of Freudian theory. Principal among these is

Freud's assignment of the mental processes to three psychic zones: the id, the ego, and the superego. An explanation of these zones may be illustrated with Freud's own diagram:

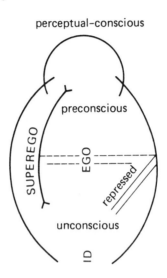

The diagram reveals immediately the vast portion of the mental apparatus that is not conscious. Furthermore, it helps to clarify the relationship between ego, id, and superego, as well as their collective relationship to the conscious and the unconscious. We should note that the id is entirely unconscious and that only a small portion of the ego and the superego is conscious. With this diagram as a guide, we may define the nature and functions of the three psychic zones.

1. The *id* is the reservoir of libido, the primary source of all psychic energy. It functions to fulfill the primordial life principle, which Freud considers to be the *pleasure principle*. Without consciousness or semblance of rational order, the id is characterized by a tremendous and amorphous vitality. Speaking metaphorically, Freud explains this "obscure inaccessible part of our personality" as "a chaos, a cauldron of seething excitement [with] no organization and no unified will, only an impulse to obtain satisfaction for the instinctual needs, in accordance with the pleasure principle" (103–4). He further stresses that the "laws of logic—above all, the

law of contradiction—do not hold for processes of the id. Contradictory impulses exist side by side without neutralizing each other or drawing apart. . . . Naturally, the id knows no values, no good and evil, no morality" (104–5). The id is, in short, the source of all our aggressions and desires. It is lawless, asocial, and amoral. Its function is to gratify our instincts for pleasure without regard for social conventions, legal ethics, or moral restraint. Unchecked, it would lead us to any lengths—to destruction and even self-destruction—to satisfy its impulses for pleasure. Safety for the self and for others does not lie within the province of the id; its concern is purely for instinctual gratification, heedless of consequence. For centuries before Freud, this force was recognized in human nature but often attributed to supernatural and external rather than to natural and internal forces: the id as defined by Freud is identical in many respects to the Devil as defined by theologians. Thus there is a certain psychological validity in the old saying that a rambunctious child (whose id has not yet been brought under control by ego and superego) is "full of the devil." We may also see in young children (and neurotic adults) certain uncontrolled impulses toward pleasure that often lead to excessive self-indulgence and even to self-injury.

2. In view of the id's dangerous potentialities, it is necessary that other psychic agencies protect the individual and society. The first of these regulating agencies, that which protects the individual, is the *ego*. This is the rational governing agent of the psyche. Though the ego lacks the strong vitality of the id, it regulates the instinctual drives of the id so that they may be released in nondestructive behavioral patterns. And though a large portion of the ego is unconscious, the ego nevertheless comprises what we ordinarily think of as the conscious mind. As Freud points out, "In popular language, we may say that the ego stands for reason and circumspection, while the id stands for the untamed passions." Whereas the id is governed solely by the pleasure principle, the ego is governed by the *reality principle*. Consequently, the ego serves as intermediary between the world within and the world without.

3. The other regulating agent, that which primarily func-

tions to protect society, is the *superego*. Largely unconscious, the superego is the moral censoring agency, the repository of conscience and pride. It is, as Freud says in "The Anatomy of the Mental Personality," the "representative of all moral restrictions, the advocate of the impulse toward perfection, in short it is as much as we have been able to apprehend psychologically of what people call the 'higher' things in human life" (95). Acting either directly or through the ego, the superego serves to repress or inhibit the drives of the id, to block off and thrust back into the unconscious those impulses toward pleasure that society regards as unacceptable, such as overt aggression, sexual passions, and the Oedipal instinct. Freud attributes the development of the superego to the parental influence that manifests itself in terms of punishment for what society considers to be bad behavior and reward for what society considers good behavior. An overactive superego creates an unconscious sense of guilt (hence the familiar term *guilt complex* and the popular misconception that Freud advocated the relaxing of all moral inhibitions and social restraints). Whereas the id is dominated by the pleasure principle and the ego by the reality principle, the superego is dominated by the *morality principle*. We might say that the id would make us devils, that the superego would have us behave as angels (or, worse, as creatures of absolute social conformity), and that it remains for the ego to keep us healthy human beings by maintaining a balance between these two opposing forces. It was this balance that Freud advocated—not a complete removal of inhibiting factors.

One of the most instructive applications of this Freudian tripartition to literary criticism is the well-known essay "In Nomine Diaboli" by Henry A. Murray (*New England Quarterly* 24 [1951]: 435–52), a knowledgeable psychoanalyst and a sensitive literary critic as well. In analyzing Herman Melville's masterpiece *Moby-Dick* with the tools provided by Freud, Murray explains the White Whale as a symbolic embodiment of the strict conscience of New England Puritanism (that is, as a projection of Melville's own superego). Captain Ahab, the monomaniac who leads the crew of the *Pequod* to destruction through his insane compulsion to pursue and strike back at the creature who has injured him, is interpreted as the symbol of a rapacious and uncontrollable id. Starbuck, the sane Christian and first mate who struggles to mediate between the forces embodied in Moby-Dick and Ahab, symbolizes a balanced and sensible rationalism (that is, the ego). Though many scholars are reluctant to accept Freud's tripartition of the human psyche, they have not reacted against this aspect of psychoanalytic criticism so strongly as against the application of his sexual theories to the symbolic interpretation of literature. Let us briefly examine the highlights of such theories. Perhaps the most controversial (and, to many persons, the most offensive) facet of psychoanalytic criticism is its tendency to interpret imagery in terms of sexuality. Following Freud's example in his interpretation of dreams, the psychoanalytic critic tends to see all concave images (ponds, flowers, cups or vases, caves, and hollows) as female or yonic symbols, and all images whose length exceeds their diameter (towers, mountain peaks, snakes, knives, lances, and swords) as male or phallic symbols. Perhaps even more objectionable to some is the interpretation of such activities as dancing, riding, and flying as symbols of sexual pleasure: for example, in *The Life and Works of Edgar Allan Poe: A Psycho-Analytic Interpretation* (London: Imago, 1949), Marie Bonaparte interprets the figure of Psyche in "Ulalume" as an ambivalent mother figure, both the longed-for mother and the mother as superego who shields her son from his incestuous instincts, concluding with the following startling observation: "Psyche's drooping, trailing wings in this poem symbolise in concrete form Poe's physical impotence. We know that flying, to all races, unconsciously symbolises the sex act, and that antiquity often represented the penis erect and winged." For the skeptical reader Bonaparte provides this explanation:

> Infinite are the symbols man has the capacity to create, as indeed, the dreams and religions of the savage and civilized well show. Every natural object may be utilised to this end yet, despite their multiple shapes, the objects and relations to which they attach are relatively few: these include the beings we loved first, such as mother, father, brothers or sisters and their bodies, but mainly our own bodies and genitals, and theirs. Almost all symbolism is sexual, in its widest sense,

taking the word as the deeply-buried primal urge behind all expressions of love, from the cradle to the grave. (294)

Although such observations as these may have a sound psychoanalytic basis, their relevance to sound critical analysis has been questioned by many scholars. We may sympathize with their incredulousness when we encounter the Freudian essay that interprets even a seemingly innocent fairy tale like "Little Red Riding Hood" as an allegory of the age-old conflict between male and female in which the plucky young virgin, whose red cap is a menstrual symbol, outwits the ruthless, sex-hungry "wolf" (Erich Fromm, *The Forgotten Language* [New York: Grove, 1957]: 235–41).

Perhaps even more controversial than Freudian dream symbolism are Freud's theories concerning child psychology. Contrary to traditional beliefs, Freud found infancy and childhood a period of intense sexual experience, sexual in a sense much broader than is commonly attached to the term. During the first five years of life, the child passes through a series of phases in erotic development, each phase being characterized by emphasis on a particular *erogenous zone* (that is, a portion of the body in which sexual pleasure becomes localized). Freud indicated three such zones: the *oral*, the *anal*, and the *genital*. (Note that the uninitiated layman, unfamiliar with the breadth of Freud's term, generally restricts the meaning of "sexuality" to "genital sexuality.") These zones are associated not only with pleasure in stimulation but also with the gratification of our vital needs: eating, elimination, and reproduction. If for some reason the individual is frustrated in gratifying these needs during childhood, the adult personality may be warped accordingly (that is, development may be arrested or *fixated*). For example, adults who are compulsively fastidious may suffer, according to the psychoanalyst, from an anal fixation traceable to overly strict toilet training during early childhood. Likewise, compulsive cigarette smoking may be interpreted as a symptom of oral fixation traceable to premature weaning. Even among "normal" adults, sublimated responses occur when the individual is vicariously stimulated by images associated with one of the major erogenous zones. In his *Fiction and the*

Unconscious (Boston: Beacon, 1957), Simon O. Lesser suggests that the anal-erotic quality in *Robinson Crusoe* (manifested in the hero's scrupulous record keeping and orderliness) accounts at least partially for the unconscious appeal of Defoe's masterpiece (306).

According to Freud, the child reaches the stage of genital primacy around age five, at which time the Oedipus complex manifests itself. In simple terms, the Oedipus complex derives from the boy's unconscious rivalry with his father for the love of his mother. Freud borrowed the term from the classic Sophoclean tragedy in which the hero unwittingly murders his father and marries his mother. In *The Ego and the Id* (New York: Norton, 1962), Freud describes the complex as follows:

> . . . the boy deals with his father by identifying himself with him. For a time these two relationships [the child's devotion to his mother and identification with his father] proceed side by side, until the boy's sexual wishes in regard to his mother become more intense and his father is perceived as an obstacle to them; from this the Oedipus complex originates. His identification with his father then takes on a hostile colouring and changes into a wish to get rid of his father in order to take his place with his mother. Henceforward his relation to his father is ambivalent; it seems as if the ambivalence inherent in the identification from the beginning had become manifest. An ambivalent attitude to his father and an object-relation of a solely affectionate kind to his mother make up the content of the simple positive Oedipus complex in a boy. (21–22)

Further ramifications of the Oedipus complex are a fear of castration and an identification of the father with strict authority in all forms; subsequent hostility to authority is therefore associated with the Oedipal ambivalence to which Freud refers. (The Oedipus complex figures strongly in Jacques Lacan's psychoanalytic theory [see chapter 5].) A story like Nathaniel Hawthorne's "My Kinsman, Major Molineux," for instance, has been interpreted by Lesser as essentially a symbolic rebellion against the father figure. And with this insight we may find meaning in the young hero's disturbing outburst of laughter as he watches the cruel tarring and feathering of his once-respected relative: The youth is expressing his unconscious joy in being released from parental authority. Now he is free, as the friendly stranger suggests, to make his own way in the adult world without the help (and restraint) of his kinsman.

II. THE PSYCHOLOGICAL APPROACH IN PRACTICE

A. *Hamlet:* The Oedipus Complex

Although Freud himself made some applications of his theories to art and literature, it remained for an English disciple, the psychoanalyst Ernest Jones, to provide the first full-scale psychoanalytic treatment of a major literary work. Jones's *Hamlet and Oedipus*, originally published as an essay in *The American Journal of Psychology* in 1910, was later revised and enlarged in a paperback edition (Garden City, NY: Doubleday [Anchor], 1949).

Jones bases his argument on the thesis that Hamlet's much-debated delay in killing his uncle, Claudius, is to be explained in terms of internal rather than external circumstances and that the "play is mainly concerned with a hero's unavailing fight against what can only be called a disordered mind." In his carefully documented essay Jones builds a highly persuasive case history of Hamlet as a psychoneurotic who suffers from manic-depressive hysteria combined with an abulia (an inability to exercise will power and come to decisions)—all of which may be traced to the hero's severely repressed Oedipal feelings. Jones points out that no really satisfying argument has ever been substantiated for the idea that Hamlet avenges his father's murder as quickly as practicable. Shakespeare makes Claudius's guilt as well as Hamlet's duty perfectly clear from the outset—if we are to trust the words of the ghost and the gloomy insights of the hero himself. The fact is, however, that Hamlet does not fulfill this duty until absolutely forced to do so by physical circumstances—and even then only after Gertrude, his mother, is dead. Jones also elucidates the strong misogyny that Hamlet displays throughout the play, especially as it is directed against Ophelia, and his almost physical revulsion from sex. All of this adds up to a classic example of the neurotically repressed Oedipus complex.

The ambivalence that typifies the child's attitude toward his father is dramatized in the characters of the ghost (the good, lovable father with whom the boy identifies) and Claudius (the hated father as tyrant and rival), both of whom are dramatic projections of the hero's own conscious-unconscious ambivalence toward the father figure. The ghost represents the conscious ideal of fatherhood, the image that is socially acceptable:

> See, what a grace was seated on this brow:
> Hyperion's curls, the front of Jove himself,
> An eye like Mars, to threaten and command,
> A station like the herald Mercury
> New-lighted on a heaven-kissing hill,
> A combination and a form indeed,
> Where every god did seem to set his seal,
> To give the world assurance of a man:
> This was your husband. (III.iv)

His view of Claudius, on the other hand, represents Hamlet's repressed hostility toward his father as a rival for his mother's affection. This new king-father is the symbolic perpetrator of the very deeds toward which the son is impelled by his own unconscious motives: murder of his father and incest with his mother. Hamlet cannot bring himself to kill Claudius because to do so he must, in a psychological sense, kill himself. His delay and frustration in trying to fulfill the ghost's demand for vengeance may therefore be explained by the fact that, as Jones puts it, the "thought of incest and parricide combined is too intolerable to be borne. One part of him tries to carry out the task, the other flinches inexorably from the thought of it" (78–79).

Norman N. Holland neatly summed up the reasons both for Hamlet's delay and also for our three-hundred-year delay in comprehending Hamlet's true motives:

> Now what do critics mean when they say that Hamlet cannot act because of his Oedipus complex? The argument is very simple, very elegant. One, people over the centuries have been unable to say why Hamlet delays in killing the man who mur-

dered his father and married his mother. Two, psychoanalytic experience shows that every child wants to do just exactly that. Three, Hamlet delays because he cannot punish Claudius for doing what he himself wished to do as a child and, unconsciously, still wishes to do: he would be punishing himself. Four, the fact that this wish is unconscious explains why people could not explain Hamlet's delay. (*The Shakespearean Imagination* [Bloomington: Indiana UP, 1968]: 158)

A corollary to the Oedipal problem in *Hamlet* is the pronounced misogyny in Hamlet's character. Because of his mother's abnormally sensual affection for her son, an affection that would have deeply marked Hamlet as a child with an Oedipal neurosis, he has in the course of his psychic development repressed his incestuous impulses so severely that this repression colors his attitude toward all women: "The total reaction culminates in the bitter misogyny of his outburst against Ophelia, who is devastated at having to bear a reaction so wholly out of proportion to her own offense and has no idea that in reviling her Hamlet is really expressing his bitter resentment against his mother" (Jones 96). The famous "Get thee to a nunnery" speech has even more sinister overtones than are generally recognized, explains Jones, when we understand the pathological degree of Hamlet's conditions and read "nunnery" as Elizabethan slang for brothel.

> The underlying theme relates ultimately to the splitting of the mother image which the infantile unconscious effects into two opposite pictures: one of a virginal Madonna, an inaccessible saint towards whom all sensual approaches are unthinkable, and the other of a sensual creature accessible to everyone. . . . When sexual repression is highly pronounced, as with Hamlet, then both types of women are felt to be hostile: the pure one out of resentment at her repulses, the sensual one out of the temptation she offers to plunge into guiltiness. Misogyny, as in the play, is the inevitable result. (97–98)

Although it has been attacked by the anti-Freudians and occasionally disparaged as "obsolete" by the neo-Freudians, Jones's critical tour de force has nevertheless attained the status of a modern classic. "Both as an important seminal work which led to a considerable re-examination of *Hamlet*, and as an example of a thorough and intelligent application of psychoanalysis to drama," writes Claudia C. Morrison, "Jones's essay stands as the single most important Freudian study of literature to appear in America . . ." (*Freud and the Critic* [Chapel Hill: U of North Carolina P, 1968]: 175).

B. Rebellion against the Father in *Huckleberry Finn*

Mark Twain's great novel has this in common with Shakespeare's masterpiece: both are concerned with the theme of rebellion—with a hostile treatment of the father figure. In both works the father figure is finally slain, and knowledge of his death brings a curious sense of relief—and release—for the reader. As we have seen, from the psychoanalytic viewpoint all rebellion is in essence a rejection of parental, especially paternal, authority. Sociologically speaking, Huck rebels against the unjust, inhumane restrictions of a society that condones slavery, hypocrisy, and cruelty. However, Mark Twain showed a remarkable pre-Freudian insight when he dramatized this theme of rebellion in the portrayal of Huck's detestable father as the lowest common denominator of social authority. The main plot of the novel is launched with Huck's escape from pap Finn ("pap," in keeping with the reductive treatment of this father figure, is not capitalized), a flight that coincides with Jim's escape from Miss Watson.

Symbolically, Huck and Jim, in order to gain freedom and to regain prelapsarian bliss (the happiness enjoyed by Adam before the Fall), must escape whatever is represented by Miss Watson and pap Finn (who reminds Huck of Adam all covered with mud—that is, Adam after the Fall). Despite their superficial and rather melodramatic differences, Miss Watson and pap Finn have much in common. They represent extremes of authority: authority at its most respectable and at its most contemptible. What is more, they both represent social and legal morality, again in the extremes of the social spectrum. Notwithstanding his obvious worthlessness, pap

Finn is still Huck's sole guardian by law and holds near-absolute power over him, an authority condoned by society, just as Miss Watson has a similar power over Jim. In the light of such authority both Miss Watson and pap Finn may be said to represent the superego (for example, when Huck goes against his conscience by refusing to turn Jim in to the authorities, it is the letter to Miss Watson that he tears up). In this sense, then, it is to escape the oppressive tyranny and cruel restraints of the superego that Huck and Jim take flight on the river.

Huckleberry Finn cannot by any means be read as a psychological allegory, and it would be foolish to set up a strict one-to-one relationship of characters and events to ideas, particularly because Mark Twain wrote the book with no notion of Freudian concepts. But like most great writers, Twain knew human nature; and from the psychoanalytic perspective, a "linked analogy" can be seen between the structure of his novel and the Freudian structure of the human psyche. Water in any form is generally interpreted by the psychoanalysts as a female symbol, more specifically as a maternal symbol. From the superegoistic milieu of society Huck and Jim flee to the river, where they find freedom. Except when invaded by men, the river is characterized by a strange, fluid, dreamlike peacefulness; Huck's most lyrical comments are those describing the beauty of the river:

Two or three days and nights went by; I reckon I might say they swum by, they slid along so quiet and smooth and lovely. . . . Not a sound anywheres—perfectly still—just like the whole world was asleep. . . . [Then] the nice breeze springs up, and comes fanning you from over there, so cool and fresh and sweet to smell on account of the woods and flowers; but sometimes not that way, because they've left dead fish laying around, gars and such, and they do get pretty rank. . . . [And] we would watch the lonesomeness of the river, and kind of lazy along, and by and by lazy off to sleep. . . . It's lovely to live on a raft. We had the sky up there, all speckled with stars, and we used to lay on our backs and look up at them, and discuss about whether they was made or only just happened. . . . Jim said the moon could 'a' *laid* them; well, that looked kind of reasonable, so I didn't say nothing against it, because

I've seen a frog lay most as many, so of course it could be done. (ch. 19)

The foregoing passage is redolent with female-maternal imagery; it also suggests the dark, mysterious serenity associated with the prenatal state, as well as with death, in psychoanalytic interpretation. The tension between land and water may be seen as analogous to that between the conscious and the unconscious in Freudian theory. Lacking a real mother, Huck finds his symbolic mother in the river; in Freudian terms, he returns to the womb. From this matrix he undergoes a series of symbolic deaths and rebirths, punctuated structurally by the episodes on land. As James M. Cox (*Sewanee Review* 62 [1954]: 389–405) has pointed out, Huck's fake murder in escaping from pap Finn is crucial to our understanding the central informing pattern of death and rebirth: "Having killed himself, Huck is 'dead' throughout the entire journey down the river. He is indeed the man without identity who is reborn at almost every river bend, not because he desires a new role, but because he must re-create himself to elude the forces which close in on him from every side. The rebirth theme which began with pap's reform becomes the driving idea behind the entire action." Enhancing this pattern is the hermaphroditic figure of Jim, Huck's adopted friend and parent, whose blackness coincides with the darkness associated with death, the unconscious, and the maternal. (We are reminded of Whitman's celebration of death as the Dark Mother in such famous poems as "Out of the Cradle Endlessly Rocking" and "When Lilacs Last in the Dooryard Bloom'd.") Jim's qualities are more maternal than paternal. He possesses the gentleness, unquestioning loyalty, and loving kindness that we traditionally ascribe to the mother, in sharp contrast to the brutal authoritarianism of pap.

Viewed from a slightly different psychological angle, *Huckleberry Finn* is a story of the child as victim, embodying the betrayal-of-innocence theme that has become one of the chief motifs in American fiction. Philip Young, in *Ernest Hemingway* (New York: Holt, 1952), has detected similarities between Huck's plight and that of the Hemingway hero.

Young sees Huck as the wounded child, permanently scarred by traumas of death and violence; he has counted thirteen corpses in the novel and observes that virtually every major episode in the book ends with violence or death. Young makes explicit the causal relationship between the traumatic experiences suffered by Huck (and later by Hemingway's protagonists) and the growing preoccupation with death that dominates much modern literature:

> [Huck] is a wounded and damaged boy. He will never get over the terror he has seen and been through, is guilt-ridden and can't sleep at night for his thoughts. When he is able to sleep he is tortured with bad dreams. . . . This is a boy who has undergone an unhappy process of growing up, and has grown clean out of his creator's grasp. . . . Precisely as Clemens could never solve his own complications, save in the unmitigated but sophomoric pessimism of his last books, so he could not solve them for Huck, who had got too hot to handle and was dropped. What the man never realized was that in his journey by water he had been hinting at a solution all along: an excessive exposure to violence and death produced first a compulsive fascination with dying, and finally an ideal symbol for it. (200–201)

This ideal symbol is the dark river itself, which is suggestive of the Freudian death instinct, the unconscious instinct in all living things to return to the nonliving state and thereby achieve permanent surcease from the pain of living.

Our recognition of these symbolic implications does not, by any means, exhaust the interpretive potential of Twain's novel, nor does it preclude insights gained from other critical approaches. Such recognition should *enhance* our appreciation of the greatness of *Huckleberry Finn* by revealing that Mark Twain produced a masterwork that, intentionally or not, has appealed in a profound psychological way to many generations of readers. The Freudian reading—particularly in its focus on the death of the Father and the search for the Feminine—has enjoyed renewed attention from feminist psychoanalytic critics (see chapter 5, "Feminist Approaches").

C. "Young Goodman Brown": Id versus Superego

The theme of innocence betrayed is also central to Nathaniel Hawthorne's "Young Goodman Brown," the tale of the young bridegroom who leaves his wife Faith to spend a night with Satan in the forest. The events of that terrifying night are a classic traumatic experience for the youth. At the center of the dark wilderness he discovers a witches' Sabbath involving all the honored teachers, preachers, and friends of his village. The climax is reached when his own immaculate bride is brought forth to stand by his side and pledge eternal allegiance to the Fiend of Hell. Following this climactic moment in which the hero resists the diabolical urge to join the fraternity of evil, he wakes to find himself in the deserted forest wondering if what has happened was dream or reality. Regardless of the answer, he is a changed man. He returns in the morning to the village and to his Faith, but he is never at peace with himself again. Henceforth he can never hear the singing of a holy hymn without also hearing echoes of the anthem of sin from that terrible night in the forest. He shrinks even from the side of Faith. His dying hour is gloom, and no hopeful epitaph is engraved upon his tombstone.

Aside from the clearly intended allegorical meanings discussed elsewhere in this book, it is the story's underlying psychological implications that concern us here. We start with the assumption that, through symbolism and technique, "Young Goodman Brown" means more than it says. In this respect our task is one of extrapolation, an inferring of the unknown from the known. Our first premise is that Brown's journey is more than a physical one; it is a psychological one as well. To see what this journey means in psychological terms, we need to examine the setting, the time and place. Impelled by unmistakably libidinal force, the hero moves from the village of Salem into the forest. The village is a place of light and order, both social and spiritual order. Brown leaves Faith behind in the town at sunset and returns to Faith in the morning. The journey into the wilderness is taken in the night: "My journey . . . forth and back again," explains the young man to his wife, "must needs be done 'twixt now and sunrise." It is in the forest, a place of darkness and unknown terrors, that Brown meets the Devil. On one level, then, the village may be equated with consciousness, the forest with the dark recesses of the unconscious. But, more precisely, the village, as a place of social and moral order (and inhibition) is analogous to Freud's superego, conscience, the morally inhibiting agent of the psyche; the forest, as a place of wild, untamed passions and terrors, has the attributes of the Freudian id. As mediator between these opposing forces, Brown himself resembles the poor ego, which tries to effect a healthy balance and is shattered because it is unable to do so.

Why can't he reconcile these forces? Is his predicament that of all human beings, as is indicated by his common, nondistinctive surname? If so, are we all destined to die in gloom? Certainly, Hawthorne implies, we cannot remain always in the village, outside the forest. And sooner or later, we must all confront Satan. Let us examine this diabolical figure for a moment. When we first see him (after being prepared by Brown's expressed fear, "What if the devil himself should be at my very elbow!"), he is "seated at the foot of an old tree"—an allusion to the "old tree" of forbidden fruit and the knowledge of sin. He is described as "bearing a considerable resemblance" to the hero himself. He is, in short, Brown's own alter ego, the dramatic projection of a part of Brown's psyche, just as Faith is the projection of another part of his psyche. The staff Satan is carrying, similar to the maple stick he later gives to Brown, is like a "great black snake . . . a living serpent"—a standard Freudian symbol for the uncontrollable phallus. As he moves on through the forest, Brown encounters other figures, the most repected of his moral tutors: old Goody Cloyse, Deacon Gookin, and, at last, even Faith herself, her pink ribbon reflecting the ambiguity that Brown is unable to resolve, for pink is the mixture of white (for purity) and red (for passion). Thoroughly unnerved—then maddened—by disillusionment, Brown capitulates to the wild evil in this heart of darkness and becomes "himself the chief horror of the scene, [shrinking] not from its other horrors." That the whole lurid scene may be interpreted as the projection of Brown's formerly repressed impulses is indicated in Hawthorne's description of the transformed protagonist:

> In truth, all through the haunted forest there could be nothing more frightful than the figure of Goodman Brown. On he flew among the black pines, brandishing his staff with frenzied gestures, now giving vent to an inspiration of horrid blasphemy, and now shouting forth such laughter as set all the echoes of the forest laughing like demons around him. *The fiend in his own shape is less hideous than when he rages in the breast of man.* (our italics)

Though Hawthorne implies that Brown's problem is that of Everyman, he does not suggest that all humans share Brown's gloomy destiny. Like Freud, Hawthorne saw the dangers of an overactive suppression of libido and the consequent development of a tyrannous superego, though he thought of the problem in his own terms as an imbalance of head versus heart. Goodman Brown is the tragic victim of a society that has shut its eyes to the inevitable "naturalness" of sex as a part of humankind's physical and mental constitution, a society whose moral system would suppress too severely natural human impulses.

Among Puritans the word "nature" was virtually synonymous with "sin." In Hawthorne's *The Scarlet Letter*, little Pearl, illegitimate daughter of Hester Prynne and the Reverend Mr. Arthur Dimmesdale, is identified throughout as the "child of nature." In his speech to the General Court in 1645, Governor John Winthrop defined "natural liberty"—as distinguished from "civil liberty"—as a "liberty to do evil as well as good . . . the exercise and maintaining of [which] makes men grow more evil, and in time to be worse than brute beasts. . . ." Hawthorne, himself a descendant of Puritan witch hunters and a member of New England society, the moral standards of which had been strongly conditioned by its Puritan heritage, was obsessed with the nature of sin and with the psychological results of violating the taboos imposed by this system. Young Goodman Brown dramatizes the neurosis resulting from such a violation.

After his night in the forest he becomes a walking guilt complex, burdened with anxiety and doubt. Why? Because he has not been properly educated to confront the realities of the external world or of the inner world, because from the cradle on he has been indoctrinated with admonitions against tasting the forbidden fruit, and because sin and Satan have been inadvertently glamorized by prohibition, he has developed a morbid compulsion to taste of them. He is not necessarily evil; he is, like most young people, curious. But because of the severity of Puritan taboos about natural impulses, his curiosity has become an obsession. His dramatic reactions in the forest are typical of what happens in actual cases of extreme repression. Furthermore, the very nature of his wilderness fantasy substantiates Freud's theory that our repressed desires express themselves in our dreams, that dreams are symbolic forms of wish fulfillment. Hawthorne, writing more than a generation before Freud, was a keen enough psychologist to be aware of many of the same phenomena Freud was to systematize through clinical evidence.

D. *The Turn of the Screw:* The Consequences of Sexual Repression

Perhaps the most famous story dealing with the theme of sexual repression is Henry James's *Turn of the Screw*. One of the most celebrated ghost stories of our literature, this many-faceted gem has been the focus of critical controversy since 1924, when Edna Denton published her Freudian analysis of the tale, "Henry James to the Ruminant Reader: The Turn of the Screw" (*The Arts* 6 [1924]: 245–55). This interpretation was buttressed ten years later by the highly respected critic Edmund Wilson in "The Ambiguity of Henry James" (*Hound and Horn* 7 [1934]: 385–406). It was exhaustively reinforced a generation later by Thomas M. Cranfill and Robert L. Clark, Jr., in *An Anatomy of "The Turn of the Screw"* (Austin: U of Texas P, 1965).

In *The Turn of the Screw* a young English governess takes a position as tutor and protectress of two beautiful children living in a magnificent old country mansion. The children's parents are dead, and their legal guardian is a debonair bachelor uncle who lives in London and does not want to be bothered with looking after his wards. He hires the young governess (the narrator of the story) with the proviso that she is to be in complete charge at Bly, his country estate, and that she will under no circumstances disturb him with appeals or complaints about her problems there. Though she is only twenty years old, she is to become governess of the estate as well as of the children, and the uncle is to be left alone, disburdened of worries about the welfare of his wards. The children, Miles and Flora, are two perfectly well-mannered youngsters with whom the governess immediately falls in love.

All seems to be well at Bly, except for the ugly mystery surrounding the relationship between the governess's predecessor, Miss Jessel, and the uncle's former valet, Peter Quint, both of whom are now dead. Also, there is the puzzling dismissal of little Miles from his school on the grounds that he "was an injury" to his fellow students. As best the governess can discover from rumor and from the scanty information given her by the housekeeper, Mrs. Grose, there had been "an affair" between Miss Jessel and Quint, carried on in the presence of the children, which had left some subtle mark of corruption on Miles and Flora. On several occasions after her arrival at Bly the governess sees the ghosts of Miss Jessel and Quint and deduces that they are somehow diabolically intent upon ensnaring the children's young souls. The actions of the children themselves, though superficially normal, suggest to the governess that her apprehensions are not without foundation. At the end of the narrative, little Flora turns against the governess and is taken off to the city by Mrs. Grose as a means, presumably, of preserving her from further corruption by Miss Jessel. The governess stays at Bly with Miles and fights for his soul against the apparition of Peter Quint. In this final climactic struggle the governess seems to triumph in driving off the evil spirit, but the little boy dies from the terrible ordeal of being dispossessed.

No brief summary can do justice to the complexities and the exquisite horror of James's tale; our primary concern here is with its interpretive possibilities. In his preface to volume 12 of the New York Edition of his collected works, James himself disavows all psychical implications in *The Turn of the Screw*, designating it as a pure and simple *amusette* intended "to catch those not easily caught (the 'fun' of the capture of the merely witless being ever but small), the jaded, the disillusioned, the fastidious"—a "Christmas-tide toy" designed "to rouse the dear old sacred terror."

Two questions may be asked about James's statements. Is he serious in disavowing a clinical intent? And does it really matter whether or not he intended the story to be no more than a simple *amusette*? The first question is unanswerable. We cannot be sure about his stated purpose; perhaps it, too, is designed to catch the literal-minded reader. To the second question we must answer a qualified no. In the strictest interpretive sense, James's conscious intentions are not directly relevant to our critical analysis of his story. Because the mind of the artist is structured essentially like other human minds and is therefore influenced by a welter of unconscious forces, the author may write more profoundly than he or she realizes. The important thing is not so much what writers intend as what we as careful, informed readers find in their works. The fact is, a very strong case can be made for the clinical implications in James's story.

In his essay on *The Turn of the Screw* (revised for *The Triple Thinkers* [New York: Oxford UP, 1948]), Edmund Wilson pointed out that no one except the governess ever admits to seeing the ghosts of Peter Quint and Miss Jessel. We assume that the children see them, as we infer this from their curious behavior, but in truth we have only the governess's word. Mrs. Grose, the simple, illiterate housekeeper whose name signifies her down-to-earthness, never sees either ghost, despite several opportunities to do so. She, too, relies only upon the word of the highly sensitive governess.

What, then, is the significance of the ghosts, and why does only the young governess see them? To the psychoanalytic observer, the answer is fairly obvious. The governess is suffering from hallucinations, the result of a severe case of sexual repression; the ghosts are dramatic projections of her own unconscious sexual desires. As James's narrator informs us at the beginning of the story, she has been reared as the "youngest of several daughters of a poor country parson." We may therefore infer that, in such a sheltered, feminine world, her normal libidinous instincts have been powerfully inhibited, like those of Goodman Brown, by her parents and by a Victorian middle-class society even more repressive than the Puritan. She is admittedly infatuated with the children's uncle ("a gentleman, a bachelor in the prime of life, such a figure as had never risen, save in a dream or an old novel, before a fluttered, anxious girl out of a Hampshire vicarage"), and it is dressed in the uncle's clothing that the red-headed Peter Quint first appears. Not only this, but Peter Quint, whose very name is a metonymy of his libidinous function (his first and last names being vulgar terms for the male and female sexual organs), makes his first appearance on the tower, a phallic symbol, just as Miss Jessel first appears beside the lake, a female symbol. Wilson lends further support to his case by pointing out the pieces of wood with which little Flora is playing under the fascinated gaze of the governess at the time of Miss Jessel's initial appearance; the child is attempting to insert the mast of a toy ship into its appropriate hole. To sum up the Freudian case in Wilson's words:

> When we look back in the light of these hints, we are inclined to conclude from analogy that the story is primarily intended as a characterization of the governess: her somber and guilty visions and the way she behaves about them seem to present, from the moment we examine them from the obverse side of her narrative, an accurate and distressing picture of the poor country parson's daughter, with her English middle-class class-consciousness, her inability to admit to herself her natural sexual impulses and the relentless English "authority" which enables her to put over on inferiors even purposes which are totally deluded and not at all in the other people's best interests. . . . We find now that [this story] is a variation of one of [James's] familiar themes: the thwarted Anglo-Saxon spinster. . . . (*The Triple Thinkers* 95)

E. Death Wish in Poe's Fiction

Aside from Ernest Jones's *Hamlet and Oedipus* and Edmund Wilson's essay on *The Turn of the Screw*, one of the most widely known psychoanalytic studies of literature is Marie Bonaparte's *Life and Works of Edgar Allan Poe* (London:

Imago, 1949). A pupil of Sigmund Freud, Bonaparte is, like Jones, one of those rare critics who has combined a thorough professional knowledge of psychoanalysis with a comparable grasp of her literary subject. For the uninitiated her book is as fantastic as it is fascinating. Her main thesis is that Poe's life and works are informed throughout by the Oedipal complex: hatred of father and psychopathic love of mother. The rejection of authority forms the core of Poe's critical writings; the mother fixation (the death wish or longing to return to the womb, manifested, for example, in his obsession with premature burial) is the matrix for Poe's poetry and fiction. Even his fatal weakness for drink is explained as a form of escape that enabled him to remain faithful to his dead mother, through a rigidly enforced chastity that was further ensured by alcoholic overindulgence. As Bonaparte writes,

> Ever since he was three, in fact, Poe had been doomed by fate to live in constant mourning. A fixation on a dead mother was to bar him forever from earthly love, and make him shun health and vitality in his loved ones. Forever faithful to the grave, his imagination had but two ways open before it: the heavens or the tomb according to whether he followed the "soul" or body of his lost one. . . .
>
> Thus, through his eternal fidelity to the dead mother, Poe, to all intents, became necrophilist. . . . Had [his necrophilia] been unrepressed, Poe would no doubt have been a criminal. (83)

Using such psychoanalytic theories as her foundation, Bonaparte proceeds to analyze work after work with a logical consistency that is as unsettling as it is monotonous. "The Cask of Amontillado" and "The Tell-Tale Heart" are seen as tales of revenge against the father. The wine vault in the former story is a symbol of the "interior of the woman's body . . . where the coveted, supreme intoxication dwells, [and] thus becomes the instrument of retribution. . . ." The victim in "The Tell-Tale Heart" is likewise interpreted as a symbol of Poe's hated father, John Allan, and his horrible blind eye is a token of retributive castration. "The Fall of the House of Usher" is a psychoanalytic model of the Oedipal guilt complex. Madeline Usher, the vault in which she is prematurely interred, the house itself are all, according to

Freudian symbology, mother images. The weird tale of Ethelred, read to Roderick by the narrator and climaxed by the slaying of the dragon, is a reenactment of the slaying of the father to gain the mother-treasure.

F. Love and Death in Blake's "Sick Rose"

Though few writers lend themselves so readily as Poe to the psychoanalytic approach, a great deal of serious literature, if we accept Marie Bonaparte's premises, can be interpreted along the same basic lines established by Freud. The Romantic poets especially are susceptible of Freudian interpretations because, as F. L. Lucas has asserted, Romanticism is related to the unconscious—as opposed to Classicism, which, with its emphasis on restraint and order, is oriented toward the conscious, particularly the ego and superego.

A richly symbolic poem like William Blake's "Sick Rose" is exemplary:

> O Rose, thou art sick!
>   The invisible worm,
> That flies in the night,
>   In the howling storm,
>
> Has found out thy bed
>   Of crimson joy;
> And his dark secret love
>   Does thy life destroy.

From the Freudian perspective the sexual implications of Blake's imagery are readily discernible. The rose is a classic symbol of feminine beauty. But this beauty is being despoiled by some agent of masculine sexuality: the worm, symbol of death, of decay, and also of the phallus (worm = serpent = sexual instinct). Again, as in Poe's "Ulalume," we encounter flying as a symbol of sexual intercourse. Images of night, darkness, and howling storm suggest attributes of the unconscious or the id, as in the forest of "Young Goodman Brown." The second stanza sets forth in rather explicit images the idea of sensual destruction. In short, Blake's poem is a vaguely disturbing parable of the death instinct, which psychoanalysts affirm is closely conjoined with sexual passion.

The sharp juxtaposition of "crimson joy" and "destroy" (coupled with "bed" and "his dark secret love") suggests that Eros, unmitigated by higher spiritual love, is the agent of evil as well of mortality.

G. Sexual Imagery in "To His Coy Mistress"

We see a similar juxtaposition in Andrew Marvell's "To His Coy Mistress," one of the most celebrated erotic poems in English literature. The speaker begins his proposition of love by stating an impossible condition: "Had we but world enough, and time, / This coyness, Lady, were no crime." Flattering his prospective mistress as "Lady," he proceeds to outline the "ideal" relationship of the two lovers:

> We would sit down and think which way
> To walk and pass our long love's day.
> . . . .
> For, Lady, you deserve this state,
> Nor would I love at lower rate.

The speaker's argument in this first stanza achieves a fine sublimation. He has managed to refine his seductive motive of all its grossness, yet, ever so subtly, he has not swerved from his main purpose. His objective, despite the contradictory deceptiveness of "vegetable love" (a passion whose burning is so slow as to be imperceptible), is nevertheless the same: it is only a matter of time before the woman must capitulate to his blandishments.

But this "only" makes all the difference in the world, as he demonstrates in his second stanza, shifting dramatically from the allusive persuasion of the first stanza to the overt pressure of the second:

> But at my back I always hear
> Time's winged chariot hurrying near;
> And yonder all before us lie
> Deserts of vast eternity.

The flying chariot of Time (again we find the subtle implication of sexual union in the image of flying) is juxtaposed against an eternity of oblivion, just as the slow but sure fecundity of a vegetable love growing to the vastness of empires is contrasted with the barren deserts of death. After setting forth this prospect, the speaker dares to reveal precisely what all this means in terms of love:

> Thy beauty shall no more be found,
> Nor, in thy marble vault, shall sound
> My echoing song; then worms shall try
> That long preserved virginity,
> And your quaint honor turn to dust,
> And into ashes all my lust.

This statement, in even sharper contrast with the gentle cajolery of the first stanza, is brutal in its explicitness. The "marble vault" is a thinly disguised vaginal metaphor suggesting both rigor mortis and the fleshless pelvis of the skeleton. "My echoing song" and the sensual meanings of the lines following are extremely coarse (cf. "quaint" and James's "Quint" as yonic puns). From the eternal burning of a vegetable passion, in the face of reality, we see that all love must at last end in ashes—just as all chastity must end, the same as sexual profligacy, in dust. The speaker concludes this stanza with a devastating anticlimax:

> The grave's a fine and private place,
> But none, I think, do there embrace.

In the final stanza the speaker relaxes his harsh irony and appeals passionately to his reluctant sweetheart to seize the moment. Again, in contrast with both the vegetable metaphor of the first stanza and the frightening directness of the second stanza, he achieves a sublimation of sensual statement through the bold sincerity of his passion and through the brilliance of his imagery:

> Now therefore, while the youthful hue
> Sits on thy skin like morning dew,
> And while thy willing soul transpires
> At every pore with instant fires,
> Now let us sport us while we may,
> And now, like amorous birds of prey,
> Rather at once our time devour
> Than languish in his slow-chapped power.

Let us roll all our strength and all
Our sweetness up into one ball,
And tear our pleasures with rough strife
Thorough the iron gates of life:
Thus, though we cannot make our sun
Stand still, yet we will make him run.

Here, too, the sexual imagery is overt. The fire image, which smolders in stanza 1 and turns to ashes in stanza 2, explodes into passion in the concluding stanza. ("Fire, in the unconscious," says Marie Bonaparte, "is the classic symbol of urethral eroticism.") Furthermore, in contrast to the tone of Blake's "Sick Rose," here love-as-destruction is set forth rapturously. The poet conveys, instead of sinister corruption, a sense of desperate ecstasy. The eating-biting metaphor (oral eroticism in its primal form) is fused with the flying symbol in "amorous birds of prey" and set with metaphysical brilliance against the alternative of a slow, cannibalistic dissolution within the horrible maw of Time. In his last four lines the lover drives his message home with an orgastic force through the use of harshly rhythmic spondees ("Thus, though" and "Stand still") and strongly suggestive puns ("make our sun" and "Make him run").

To read Marvell's great poem as nothing more than a glorification of sexual activity is, of course, a gross oversimplification. "To His Coy Mistress" is much more, as we have indicated in the preceding chapters and will elaborate in the following chapters. We agree with the formalistic critic that literature is autonomous, but we must also concur with critic Wayne Shumaker that it is "continuous with nonaesthetic life" (*Literature and the Irrational* [Englewood Cliffs, NJ: Prentice-Hall, 1960]: 263). As Simon O. Lesser has said, "Among [literary works] whose artistic authenticity cannot be questioned we give the highest place precisely to those works which ignore no aspect of man's nature, which confront the most disagreeable aspects of life deliberately and unflinchingly. . . ." Great literature has always dealt not merely with those aspects of the human mind that are pleasant and conscious but with the total human psyche, many facets of which are both unpleasant and unconscious. The enduring appeal of Marvell's poem, like that of the other works we have examined, derives from this kind of artistic and honest confrontation.

III. OTHER POSSIBILITIES AND LIMITATIONS OF THE PSYCHOLOGICAL APPROACH

This brings us to a final re-capitulation and a few words of defense as well as of caution about the Freudian approach. First, in defense: incredibly farfetched as some psychoanalytic interpretations seem to many readers, such interpretations, handled by qualified critics, are not unsubstantiated in fact; they are based upon psychological insights often derived from and supported by actual case histories, and they are set forth in such works as those of Ernest Jones and Marie Bonaparte with remarkable cogency. They are—if we accept the basic premises of psychoanalysis—very difficult to refute. Futhermore, regardless of their factual validity, such theories have had a tremendous impact upon modern writing (in the works of such creative artists as James Joyce, Eugene O'Neill, Tennessee Williams, and Philip Roth, to mention only a few) and upon modern literary criticism (for example, in the essays of such major and diverse critics as Edmund Wilson, Lionel Trilling, F. L. Lucas, Frederick Hoffman, Sandra Gilbert, Hélène Cixous, and Julia Kristeva). It is therefore important that the serious student of literature be acquainted with psychoanalytic theory.

The danger is that the serious student may become theory-ridden, forgetting that Freud's is not the only approach to literary analysis. To see a great work of fiction or a great poem primarily as a psychological case study is often to miss its wider significance and perhaps even the essential aesthetic experience it should provide. A number of great works, despite the claims of the more zealous Freudians and post-Freudians, do not lend themselves readily, if at all, to the psychoanalytic approach, and even those that do cannot be studied exclusively from the psychological perspective. Literary interpretation and psychoanalysis are two distinct fields, and though they may be closely associated, they can in no sense be regarded as parts of one discipline. The literary critic who views the masterpiece solely through the lens of Freud is liable to see art through a glass darkly. However, those readers who reject psychoanalysis as neurotic nonsense deprive themselves of a valuable tool in understanding not only literature but human nature and their individual selves as well.

· 4 ·

Mythological and Archetypal Approaches

I. DEFINITIONS AND MISCONCEPTIONS

In *The Masks of God: Primitive Mythology* (New York: Viking, 1959), Joseph Campbell recounts a curious phenomenon of animal behavior. Newly hatched chickens, bits of eggshells still clinging to their tails, will dart for cover when a hawk flies overhead; yet they remain unaffected by other birds. Furthermore, a wooden model of a hawk, drawn forward along a wire above their coop, will send them scurrying (if the model is pulled backward, however, there is no response). "Whence," Campbell asks, "this abrupt seizure by an image to which there is no counterpart in the chicken's world? Living gulls and ducks, herons and pigeons, leave it cold; but *the work of art strikes some very deep chord!*" (31; our italics).

Campbell's hinted analogy, though only roughly approximate, will serve nonetheless as an instructive introduction to the mythological approach to literature. For it is with the relationship of literary art to "some very deep chord" in human nature that mythological criticism deals. The myth critic is concerned to seek out those mysterious elements that inform certain literary works and that elicit, with almost uncanny force, dramatic and universal human reactions. The myth critic wishes to discover how certain works of literature, usually those that have become, or promise to become,

147

[Handwritten margin note (left side): The things that are in "common" w/all human beings.]

[Handwritten margin note: Associates w/ Reader responses Psychological]

"classics," image a kind of reality to which readers give perennial response—while other works, seemingly as well constructed, and even some forms of reality, leave them cold. Speaking figuratively, the myth critic studies in depth the "wooden hawks" of great literature: the so-called archetypes or archetypal patterns that the writer has drawn forward along the tensed structural wires of his or her masterpiece and that vibrate in such a way that a sympathetic resonance is set off deep within the reader.

An obviously close connection exists between mythological criticism and the psychological approach discussed in chapter 3: both are concerned with the motives that underlie human behavior. Between the two approaches are differences of degree and of affinities. Psychology tends to be experimental and diagnostic; it is closely related to biological science. Mythology tends to be speculative and philosophic; its affinities are with religion, anthropology, and cultural history. Such generalizations, of course, risk oversimplification; for instance, a great psychologist like Sigmund Freud ranged far beyond experimental and clinical study into the realms of myth, and his distinguished sometime protégé, Carl Gustav Jung, became one of the foremost mythologists of our time. Even so, the two approaches are distinct, and mythology is wider in its scope than psychology. For example, what psychoanalysis attempts to disclose about the individual personality, the study of myths reveals about the mind and character of a people. And just as dreams reflect the unconscious desires and anxieties of the individual, so myths are the symbolic projections of a people's hopes, values, fears, and aspirations.

According to the common misconception and misuse of the term, myths are merely primitive fictions, illusions, or opinions based upon false reasoning. Actually, mythology encompasses more than grade school stories about the Greek and Roman deities or clever fables invented for the amusement of children (or the harassment of students in college literature courses). It may be true that myths do not meet our current standards of factual reality, but then neither does any great literature. Instead, they both reflect a more profound reality. As Mark Schorer says in *William Blake: The Politics of Vision* (New York: Holt, 1946), "Myth is fundamental, the dramatic representation of our deepest instinctual life, of a primary awareness of man in the universe, capable of many configurations, upon which all particular opinions and attitudes depend" (29). According to Alan W. Watts in *Myth and Ritual in Christianity* (New York: Vanguard, 1954), "Myth is to be defined as a complex of stories—some no doubt fact, and some fantasy—which, for various reasons, human beings regard as demonstrations of the inner meaning of the universe and of human life" (7).

Myths are by nature collective and communal; they bind a tribe or a nation together in common psychological and spiritual activities. In *The Language of Poetry*, edited by Allen Tate (New York: Russell, 1960), Philip Wheelwright explains, "Myth is the expression of a profound sense of togetherness of feeling and of action and of wholeness of living" (11). Moreover, like Melville's famous white whale (itself an archetypal image), myth is ubiquitous in time as well as place. It is a dynamic factor everywhere in human society; it transcends time, uniting the past (traditional modes of belief) with the present (current values) and reaching toward the future (spiritual and cultural aspirations).

It holds humans together as one ; expands to the past ; future as well.

II. SOME EXAMPLES OF ARCHETYPES

Having established the significance of myth, we need to examine its relationship to archetypes and archetypal patterns. Although every people has its own distinctive mythology that may be reflected in legend, folklore, and ideology—although, in other words, myths take their specific shapes from the cultural environments in which they grow—myth is, in the general sense, universal. Furthermore, similar motifs or themes may be found among many different mythologies, and certain images that recur in the myths of peoples widely separated in time and place tend to have a common meaning or, more accurately, tend to elicit comparable psychological responses and to serve similar cultural functions. Such motifs and images are called *archetypes*. Stated simply, archetypes are universal symbols. As Philip Wheelwright explains in *Metaphor and Reality* (Bloomington: Indiana UP, 1962), such symbols are

> those which carry the same or very similar meanings for a large portion, if not all, of mankind. It is a discoverable fact that certain symbols, such as the sky father and earth mother, light, blood, up-down, the axis of a wheel, and others, recur again and again in cultures so remote from one another in space and time that there is no likelihood of any historical influence and causal connection among them. (111)

Examples of these archetypes and the symbolic meanings with which they tend to be widely associated follow (it should be noted that these meanings may vary significantly from one context to another):

A. Images

1. Water: the mystery of creation; birth-death-resurrection; purification and redemption; fertility and growth.
   According to Jung, water is also the commonest symbol for the unconscious.
   a. The sea: the mother of all life; spiritual mystery and infinity; death and rebirth; timelessness and eternity; the unconscious.
   b. Rivers: death and rebirth (baptism); the flowing of time into eternity; transitional phases of the life cycle; incarnations of deities.
2. Sun (fire and sky are closely related): creative energy; law in nature; consciousness (thinking, enlightenment, wisdom, spiritual vision); father principle (moon and earth tend to be associated with female or mother principle); passage of time and life.
   a. Rising sun: birth; creation; enlightenment.
   b. Setting sun: death.
3. Colors
   a. Red: blood, sacrifice, violent passion; disorder.
   b. Green: growth; sensation; hope; fertility; in negative context may be associated with death and decay.

   c. Blue: usually highly positive, associated with truth, religious feeling, security, spiritual purity (the color of the Great Mother or Holy Mother).

   d. Black (darkness): chaos, mystery, the unknown; death; primal wisdom; the unconscious; evil; melancholy.

   e. White: highly multivalent, signifying, in its positive aspects, light, purity, innocence, and timelessness; in its negative aspects, death, terror, the supernatural, and the blinding truth of an inscrutable cosmic mystery (see, for instance, Herman Melville's chapter "The Whiteness of the Whale" in *Moby-Dick*).

4. Circle (sphere): wholeness, unity.

   a. Mandala (a geometric figure based upon the squaring of a circle around a unifying center; see the accompanying illustration of the classic Shri-Yantra mandala): the desire for spiritual unity and psychic integration. Note that in its classic Asian forms the mandala juxtaposes the triangle, the square, and the circle with their numerical equivalents of three, four, and seven.

   b. Egg (oval): the mystery of life and the forces of generation.

   c. Yang-yin: a Chinese symbol (below) representing the union of the opposite forces of the yang (masculine principle, light, activity, the conscious mind) and the yin (female principle, darkness, passivity, the unconscious).

    d. Ouroboros: the ancient symbol of the snake biting its own tail, signifying the eternal cycle of life, primordial unconsciousness, the unity of opposing forces (cf. yang-yin).

5. Serpent (snake, worm): symbol of energy and pure force (cf. libido); evil, corruption, sensuality; destruction; mystery; wisdom; the unconscious.

6. Numbers:

    a. Three: light; spiritual awareness and unity (cf. the Holy Trinity); the male principle.

    b. Four: associated with the circle, life cycle, four seasons; female principle, earth, nature; four elements (earth, air, fire, water)

    c. Seven: the most potent of all symbolic numbers— signifying the union of *three* and *four*, the completion of a cycle, perfect order.

7. The archetypal woman (Great Mother—the mysteries of life death, transformation):

    a. The Good Mother (positive aspects of the Earth Mother): associated with the life principle, birth, warmth, nourishment, protection, fertility, growth, abundance (for example, Demeter, Ceres).

    b. The Terrible Mother (including the negative aspects of the Earth Mother): the witch, sorceress, siren, whore, femme fatale—associated with sensuality, sexual orgies, fear, danger, darkness, dismemberment, emasculation, death; the unconscious in its terrifying aspects.

    c. The Soul Mate: the Sophia figure, Holy Mother, the

princess or "beautiful lady"—incarnation of inspiration and spiritual fulfillment (cf. the Jungian anima).

8. The Wise Old Man (savior, redeemer, guru): personification of the spiritual principle, representing "knowledge, reflection, insight, wisdom, cleverness, and intuition on the one hand, and on the other, moral qualities such as goodwill and readiness to help, which make his 'spiritual' character sufficiently plain. . . . Apart from his cleverness, wisdom, and insight, the old man . . . is also notable for his moral qualities; what is more, he even tests the moral qualities of others and makes gifts dependent on this test. . . . The old man always appears when the hero is in a hopeless and desperate situation from which only profound reflection or a lucky idea . . . can extricate him. But since, for internal and external reasons, the hero cannot accomplish this himself, the knowledge needed to compensate the deficiency comes in the form of a personified thought, i.e., in the shape of this sagacious and helpful old man" (C. G. Jung, *The Archetypes and the Collective Unconscious*, trans. R. F. C. Hull, 2nd ed. [Princeton, NJ: Princeton UP, 1968]: 217ff.)

9. Garden: paradise; innocence; unspoiled beauty (especially feminine); fertility.

10. Tree: "In its most general sense, the symbolism of the tree denotes life of the cosmos: its consistence, growth, proliferation, generative and regenerative processes. It stands for inexhaustible life, and is therefore equivalent to a symbol of immortality" (J. E. Cirlot, *A Dictionary of Symbols*, trans. Jack Sage [New York: Philosophical, 1962]: 328; cf. the depiction of the cross of redemption as the tree of life in Christian iconography).

11. Desert: spiritual aridity; death; nihilism, hopelessness.

These examples are by no means exhaustive, but represent some of the more common archetypal images that the reader is likely to encounter in literature. The images we have listed do not necessarily function as archetypes every time they appear in a literary work. The discreet critic interprets them as such only if the total context of the work logically supports an archetypal reading.

B. Archetypal Motifs or Patterns

1. Creation: perhaps the most fundamental of all archetypal motifs—virtually every mythology is built on some account of how the cosmos, nature, and humankind were brought into existence by some supernatural Being or beings.

2. Immortality: another fundamental archetype, generally taking one of two basic narrative forms:

   a. Escape from time: "return to paradise," the state of perfect, timeless bliss enjoyed by man and woman before their tragic Fall into corruption and mortality.

   b. Mystical submersion into cyclical time: the theme of endless death and regeneration—human beings achieve a kind of immortality by submitting to the vast, mysterious rhythm of Nature's eternal cycle, particularly the cycle of the seasons.

3. Hero archetypes (archetypes of transformation and redemption):

   a. The quest: the hero (savior, deliverer) undertakes some long journey during which he or she must perform impossible tasks, battle with monsters, solve unanswerable riddles, and overcome insurmountable obstacles in order to save the kingdom.

   b. Initiation: the hero undergoes a series of excruciating ordeals in passing from ignorance and immaturity to social and spiritual adulthood, that is, in achieving maturity and becoming a full-fledged member of his or her social group. The initiation most commonly consists of three distinct phases: (1) separation, (2) transformation, and (3) return. Like the quest, this is a variation of the death-and-rebirth archetype.

   ex. Red Badge of Courage.

   c. The sacrificial scapegoat: the hero, with whom the welfare of the tribe or nation is identified, must die to atone for the people's sins and restore the land to fruitfulness.

C. Archetypes as Genres

Finally, in addition to appearing as images and motifs, archetypes may be found in even more complex combinations as genres or types of literature that conform with the major phases of the seasonal cycle. Northrop Frye, in his *Anatomy of Criticism* (Princeton, NJ: Princeton UP, 1957), indicates the correspondent genres for the four seasons as follows:

1. The mythos of spring: comedy
2. The mythos of summer: romance
3. The mythos of fall: tragedy
4. The mythos of winter: irony

With brilliant audacity Frye identifies myth with literature, asserting that myth is a "structural organizing principle of literary form" (341) and that an archetype is essentially an "element of one's literary experience" (365). And in *The Stubborn Structure* (Ithaca, NY: Cornell UP, 1970) he claims that "mythology as a whole provides a kind of diagram or blueprint of what literature as a whole is all about, an imaginative survey of the human situation from the beginning to the end, from the height to the depth, of what is imaginatively conceivable" (102).

III. MYTH CRITICISM IN PRACTICE

Frye's contribution leads us directly into the mythological approach to literary analysis. As our discussion of mythology has shown, the task of the myth critic is a special one. Unlike the traditional critic, who relies heavily on history and the biography of the writer, the myth critic is interested more in prehistory and the biographies of the gods. Unlike the formalistic critic, who concentrates on the shape and symmetry of the work itself, the myth critic probes for the inner spirit which gives that form its vitality and its enduring appeal. And, unlike the Freudian critic, who is prone to look on the artifact as the product of some sexual neurosis, the myth critic sees the work holistically, as the manifestation of vitalizing, integrative forces arising from the depths of humankind's collective psyche.

Despite the special importance of the myth critic's contribution, this approach is, for several reasons, poorly understood. In the first place, only during the present century have the proper interpretive tools become available through the development of such disciplines as anthropology, psychology, and cultural history. Second, many scholars and teachers of literature have remained skeptical of myth criticism because of its tendencies toward the cultic and the occult. Finally, there has been a discouraging confusion over concepts and definitions among the myth initiates themselves, which has caused many would-be myth critics to turn their energies to more clearly defined approaches such as the traditional or formalistic. In carefully picking our way through this maze, we can discover at least three separate though not necessarily exclusive disciplines, each of which has figured prominently in the development of myth criticism. In the following pages we examine these in roughly chronological order, noting how each may be applied to critical analysis.

A. Anthropology and Its Uses

The rapid advancement of modern anthropology since the end of the nineteenth century has been the most important single influence on the growth of myth criticism. Shortly after the turn of the century this influence was revealed in a series of important studies published by the Cambridge Hellenists, a group of British scholars who applied recent anthropological discoveries to the understanding of Greek classics in terms of mythic and ritualistic origins. Noteworthy contributions by members of this group include *Anthropology and the Classics* (New York: Oxford UP, 1908), a symposium edited by R. R. Marett; Jane Harrison's *Themis* (London: Cambridge UP, 1912); Gilbert Murray's *Euripides and His Age* (New York: Holt, 1913); and F. M. Cornford's *Origin of Attic Comedy* (London: Arnold, 1914). But by far the most significant member of the British school was Sir James G. Frazer, whose monumental *The Golden Bough* has exerted an enormous influence on twentieth-century literature, not merely on the critics but also on such creative writers as James Joyce, Thomas Mann, and T. S. Eliot. Frazer's work, a comparative study of the primitive origins of religion in magic, ritual, and myth, was first published in two volumes in 1890, later expanded to twelve volumes, and then published in a one-volume abridged edition in 1922. Frazer's main contribution was to demonstrate the "essential similarity of man's chief wants everywhere and at all times," particularly as these wants were reflected throughout ancient mythologies. He explains, for example, in the abridged edition (New York: Macmillan, 1922), that

> Under the names of Osiris, Tammuz, Adonis, and Attis, the peoples of Egypt and Western Asia represented the yearly decay and revival of life, especially vegetable life, which they personified as a god who annually died and rose again from the dead. In name and detail the rites varied from place to place: in substance they were the same. (325)

The central motif with which Frazer deals is the archetype of crucifixion and resurrection, specifically the myths describing the "killing of the divine king." Among many primitive peoples it was believed that the ruler was a divine or semidivine being whose life was identified with the life cycle in nature and in human existence. Because of this identification, the safety of the people and even of the world was felt to depend upon the life of the god-king. A vigorous, healthy ruler would ensure natural and human productivity; on the other hand, a sick or maimed king would bring blight and disease to the land and its people. Frazer points out that if

> the course of nature is dependent on the man-god's life, what catastrophes may not be expected from the gradual enfeeblement of his powers and their final extinction in death? There is only one way of averting these dangers. The man-god must be killed as soon as he shows symptoms that his powers are beginning to fail, and his soul must be transferred to a vigorous successor before it has been seriously impaired by threatened decay. (265)

Among some peoples the kings were put to death at regular intervals to ensure the welfare of the tribe; later, however, substitute figures were killed in place of the kings themselves, or the sacrifices became purely symbolic rather than literal.

Corollary to the rite of sacrifice was the scapegoat archetype. This motif centered in the belief that, by tranferring the
corruptions of the tribe to a sacred animal or person, then by
killing (and in some instances eating) this scapegoat, the
tribe could achieve the cleansing and atonement thought
necessary for natural and spiritual rebirth. Pointing out that
food and children are the primary needs for human survival,
Frazer emphasizes that the rites of blood sacrifice and purification were considered by ancient peoples as a magical
guarantee of rejuvenation, an assurance of life, both vegetable and human. If such customs strike us as incredibly primitive, we need only to recognize their vestiges in our own
civilized world—for example, the irrational satisfaction that
some people gain by the persecution of such minority groups
as blacks and Jews as scapegoats, or the more wholesome
feelings of renewal derived from our New Year's festivities
and resolutions, the homely tradition of spring-cleaning, our
celebration of Easter and even the Eucharist. Modern writers
themselves have employed the scapegoat motif with striking
relevance (for example, Shirley Jackson in "The Lottery,"
Robert Heinlein in *Stranger in a Strange Land*, and Tom
Tryon in *Harvest Home*).

The insights of Frazer and the Cambridge Hellenists have
been extremely helpful in myth criticism, especially in the
mythological approach to drama. Many scholars theorize
that tragedy originated from the primitive rites we have described. The tragedies of Sophocles and Aeschylus, for
example, were written to be played during the festival of
Dionysos, annual vegetation ceremonies during which the
ancient Greeks celebrated the deaths of the winter-kings and
the rebirths of the gods of spring and renewed life.

Sophocles's *Oedipus* is an excellent example of the fusion
of myth and literature. Sophocles produced a great play, but
the plot of *Oedipus* was not his invention. It was a well-
known mythic narrative long before he immortalized it as
tragic drama. Both the myth and the play contain a number of
familiar archetypes, as a brief summary of the plot indicates.
The king and queen of ancient Thebes, Laius and Jocasta, are
told in a prophecy that their newborn son, after he has grown

up, will murder his father and marry his mother. To prevent this catastrophe, the king orders one of his men to pierce the infant's heels and abandon him to die in the wilderness. But the child is saved by a shepherd and taken to Corinth, where he is reared as the son of King Polybus and Queen Merope, who lead the boy to believe that they are his real parents. After reaching maturity and hearing of a prophecy that he is destined to commit patricide and incest, Oedipus flees from Corinth to Thebes. On his journey he meets an old man and his servants, quarrels with them and kills them. Before entering Thebes he encounters the Sphinx (who holds the city under a spell), solves her riddle, and frees the city; his reward is the hand of the widowed Queen Jocasta. He then rules a prosperous Thebes for many years, fathering four children by Jocasta. At last, however, a blight falls upon his kingdom because Laius's slayer has gone unpunished. Oedipus starts an intensive investigation to find the culprit—only to discover ultimately that he himself is the guilty one, that the old man whom he had killed on his journey to Thebes was Laius, his real father. Overwhelmed by this revelation, Oedipus blinds himself with a brooch taken from his dead mother-wife, who has hanged herself, and goes into exile. Following his sacrificial punishment, Thebes is restored to health and abundance.

Even in this bare summary we may discern at least two archetypal motifs: (1) In the quest motif, Oedipus, as the hero, undertakes a journey during which he encounters the Sphinx, a supernatural monster with the body of a lion and the head of a woman; by answering her riddle, he delivers the kingdom and marries the queen. (2) In the king-as-sacrificial-scapegoat motif, the welfare of the state, both human and natural (Thebes is stricken by both plague and drought), is bound up with the personal fate of the ruler; only after Oedipus has offered himself up as a scapegoat is the land redeemed.

Considering that Sophocles wrote his tragedy expressly for a ritual occasion, we are hardly surprised that *Oedipus* reflects certain facets of the fertility myths described by Frazer. More remarkable, and more instructive for the student interested in myth criticism, is the revelation of similar facets in the great tragedy written by Shakespeare two thousand years later.

1. The Sacrificial Hero: Hamlet

One of the first modern scholars to point out these similarities was Gilbert Murray. In his "Hamlet and Orestes," delivered as a lecture in 1914 and subsequently published in *The Classical Tradition in Poetry* (Cambridge, MA: Harvard UP, 1927), Murray indicated a number of parallels between the mythic elements of Shakespeare's play and those in *Oedipus* and in the *Agamemnon* of Aeschylus. The heroes of all three works derive from the *Golden Bough* kings; they are all haunted, sacrificial figures. Furthermore, as with the Greek tragedies, the story of Hamlet was not the playwright's invention but was drawn from legend. As literary historians tell us, the old Scandinavian story of Amlehtus or Amlet, Prince of Jutland, was recorded as early as the twelfth century by Saxo Grammaticus in his *History of the Danes*. Murray cites an even earlier passing reference to the prototypal Hamlet in a Scandinavian poem composed about A.D. 980. Giorgio de Santillana and Hertha von Dechend in *Hamlet's Mill* (Boston: Gambit, 1969) have traced this archetypal character back through the legendary Icelandic Amlodhi to Oriental mythology. It is therefore evident that the core of Shakespeare's play is mythic. In Murray's words,

> The things that thrill and amaze us in *Hamlet* . . . are not any historical particulars about mediaeval Elsinore . . . but things belonging to the old stories and the old magic rites, which stirred and thrilled our forefathers five and six thousand years ago; set them dancing all night on the hills, tearing beasts and men in pieces, and giving up their own bodies to a ghastly death, in hope thereby to keep the green world from dying and to be the saviours of their own people. (236)

By the time Sophocles and Aeschylus were producing their tragedies for Athenian audiences, such sacrifices were no longer performed literally but were acted out symbolically on stage; yet their mythic significance was the same. Indeed, their significance was very similar in the case of

Shakespeare's audiences. The Elizabethans were a myth-minded and symbol-receptive people. There was no need for Shakespeare to interpret for his audience: they *felt* the mythic content of his plays. And though myth may smolder only feebly in the present-day audience, we still respond, despite our intellectual sophistication, to the archetypes in *Hamlet.*

Such critics as Murray and Francis Fergusson have provided clues to many of Hamlet's archetypal mysteries. In *The Idea of Theater* (Princeton, NJ: Princeton UP, 1949), Fergusson discloses point by point how the scenes in Shakespeare's play follow the same ritual pattern as those in Greek tragedy, specifically in *Oedipus;* he indicates that

> in both plays a royal sufferer is associated with pollution, in its very sources, of an entire social order. Both plays open with an invocation for the well-being of the endangered body politic. In both, the destiny of the individual and of society are closely intertwined; and in both the suffering of the royal victim seems to be necessary before purgation and renewal can be achieved.
> (118)

To appreciate how closely the moral norms in Shakespeare's play are related to those of ancient vegetation myths, we need only to note how often images of disease and corruption are used to symbolize the evil that has blighted Hamlet's Denmark. The following statement from Philip Wheelwright's *The Burning Fountain* (Bloomington: Indiana UP, 1954), explaining the organic source of good and evil, is directly relevant to the moral vision in *Hamlet,* particularly to the implications of Claudius's crime and its disastrous consequences. From the natural or organic standpoint,

> Good is life, vitality, propagation, health; evil is death, impotence, disease. Of these several terms *health* and *disease* are the most important and comprehensive. Death is but an interim evil; it occurs periodically, but there is the assurance of new life ever springing up to take its place. The normal cycle of life and death is a healthy cycle, and the purpose of the major seasonal festivals (for example, the Festival of Dionysos) was at least as much to celebrate joyfully the turning wheel of great creative Nature as to achieve magical effects. Disease and

blight, however, interrupt the cycle; they are the real destroyers; and health is the good most highly to be prized. (197)

Wheelwright continues by pointing out that because murder (not to be confused with ritual sacrifice) does violence to both the natural cycle of life and the social organism, the murderer is symbolically diseased. Furthermore, when the victim is a member of the murderer's own family, an even more compact organism than the tribe or the political state, the disease is especially virulent.

We should mention one other myth that relates closely to the meaning of *Hamlet,* the myth of divine appointment. This was the belief, strongly fostered by such Tudor monarchs as Henry VII, Henry VIII, and Elizabeth I, that not only had the Tudors been divinely appointed to bring order and happiness out of civil strife but also any attempt to break this divine ordinance (for example, by insurrection or assassination) would result in social, political, and natural chaos. We see this Tudor myth reflected in several of Shakespeare's plays (for example, in *Richard III, Macbeth,* and *King Lear*) where interference with the order of divine succession or appointment results in both political and natural chaos, and where a deformed, corrupt, or weak monarch epitomizes a diseased political state. This national myth is, quite obviously, central in *Hamlet.*

The relevance of myth to *Hamlet* should now be apparent. The play's thematic heart is the ancient, archetypal mystery of the life cycle itself. Its pulse is the same tragic rhythm that moved Sophocles's audience at the festival of Dionysos and moves us today through forces that transcend our conscious processes. Through the insights provided us by anthropological scholars, however, we may perceive the essential archetypal pattern of Shakespeare's tragedy. Hamlet's Denmark is a diseased and rotten state because Claudius's "foul and most unnatural murder" of his king-brother has subverted the divinely ordained laws of nature and of kingly succession. The disruption is intensified by the blood kinship between victim and murderer. Claudius, whom the ghost identifies as "The Serpent," bears the primal blood curse of Cain. And because the state is identified with its ruler, Denmark shares and suffers also from his blood guilt. Its natural cycle interrupted, the nation is threatened by chaos: civil strife within and war without. As Hamlet exclaims, "The time is out of joint; O cursed spite, / That ever I was born to set it right!"

Hamlet's role in the drama is that of the prince-hero who, to deliver his nation from the blight that has fallen upon it, must not only avenge his father's murder but also offer himself up as a royal scapegoat. As a member of the royal family, Hamlet is infected with the regicidal virus even though he is personally innocent. We might say, using another metaphor from pathology, that Claudius's murderous cancer has metastasized so that the royal court and even the nation itself is threatened with fatal deterioration. Hamlet's task is to seek out the source of this malady and to eliminate it. Only after a thorough purgation can Denmark be restored to a state of wholesome balance. Hamlet's reluctance to accept the role of cathartic agent is a principal reason for his procrastination in killing Claudius, an act that may well involve his self-destruction. He is a reluctant but dutiful scapegoat, and he realizes ultimately that there can be no substitute victim in this sacrificial rite—hence his decision to accept Laertes's challenge to a dueling match that he suspects has been fixed by Claudius. The bloody climax of the tragedy is therefore not merely spectacular melodrama but an essential element in the archetypal pattern of sacrifice-atonement-catharsis. Not only must all those die who have been infected by the evil contagion (Claudius, Gertrude, Polonius, Rosencrantz and Guildenstern—even Ophelia and Laertes), but the prince-hero himself must suffer "crucifixion" before Denmark can be purged and reborn under the healthy new regime of Fortinbras.

Enhancing the motif of the sacrificial scapegoat is Hamlet's long and difficult spiritual journey—his initiation, as it were—from innocent, carefree youth (he has been a university student) through a series of painful ordeals to sadder, but wiser, maturity. His is a long night's journey of the soul, and Shakespeare employs archetypal imagery to convey this thematic motif: *Hamlet* is an autumnal, nighttime play dominated by images of darkness and blood, and the hero appropriately wears black, the archetypal color of melancholy. The superficial object of his dark quest is to solve the riddle of his father's death. On a deeper level, his quest leads him down the labyrinthine ways of the human mystery, the mystery of human life and destiny. (Observe how consistently his soliloquies turn toward the puzzles of life and of self.) As with the riddle of the Sphinx, the enigmatic answer is "man," the clue to which is given in Polonius's glib admonition, "To thine own self be true." In this sense, then, Hamlet's quest is the quest undertaken by all of us who would gain that rare and elusive philosopher's stone, self-knowledge.

2. Archetypes of Time and Immortality: "To His Coy Mistress"

Even though the mythological approach lends itself more readily to the interpretation of drama and the novel than to shorter literary forms such as the lyric poem, it is not uncommon to find elements of myth in these shorter works. In fact, mythopoeic poets like William Blake, William Butler Yeats, and T. S. Eliot carefully structured many of their works on myth. Even those poets who are not self-appointed myth-makers often employ images and motifs that, intentionally or not, function as archetypes. Andrew Marvell's "To His Coy Mistress" seems to fit into this latter category.

Because of its strongly suggestive (and suggested) sensuality and its apparently cynical theme, "To His Coy Mistress" is sometimes dismissed as an immature if not immoral love poem. But to see the poem as little more than a clever proposition is to miss its greatness. No literary work survives because it is merely clever, or merely well written. It must partake somehow of the universal and, in doing so, may contain elements of the archetypal. Let us examine "To His Coy Mistress" with an eye to its archetypal content.

Superficially a love poem, "To His Coy Mistress" is, in a deeper sense, a poem about time. As such, it is concerned with immortality, a fundamental motif in myth. In the first two stanzas we encounter an inversion or rejection of traditional conceptions of human immortality. Stanza 1 is an ironic presentation of the "escape from time" to some paradisal state in which lovers may dally for an eternity. But such a state of perfect, eternal bliss is a foolish delusion, as the speaker suggests in his subjunctive "Had we . . ." and in his description of love as some kind of monstrous vegetable growing slowly to an infinite size in the archetypal garden. Stanza 2 presents, in dramatic contrast, the desert archetype in terms of another kind of time, naturalistic time. This is the time governed by the inexorable laws of nature (note the sun archetype imaged in "Time's winged chariot"), the laws of decay, death, and physical extinction. Stanza 2 is as extreme in its philosophical realism as the first stanza is in its impracticable idealization.

The concluding stanza, radically altered in tone, presents a third kind of time, an escape into cyclical time and thereby a chance for immortality. Again we encounter the sun archetype, but this is the sun of "soul" and of "instant fires"— images not of death but of life and creative energy, which are fused with the sphere ("Let us roll all our strength and all / Our sweetness up into one ball"), the archetype of primal wholeness and fulfillment. In *Myth and Reality* (New York: Harper, 1963), Mircea Eliade indicates that one of the most widespread motifs in immortality myths is the *regressus ad uterum* (a "return to the origin" of creation or to the symbolic womb of life) and that this return is considered to be symbolically feasible by some philosophers (for example, the Chinese Taoists) through alchemical fire:

> During the fusion of metals the Taoist alchemist tries to bring about in his own body the union of the two cosmological principles, Heaven and Earth, in order to reproduce the primordial chaotic situation that existed before the Creation. This primordial situation . . . corresponds both to the egg (that is, the archetypal sphere) or the embryo and to the paradisal and innocent state of the uncreated World. (83–84)

We are not suggesting that Marvell was familiar with Taoist philosophy or that he was consciously aware of immortality archetypes. However, in representing the age-old dilemma of time and immortality, Marvell employed a cluster of images charged with mythic significance. His poet-lover seems to offer the alchemy of love as a way of defeating the laws of naturalistic time; love is a means of participating in, even intensifying, the mysterious rhythms of nature's eternal cycle. If life is to be judged, as some philosophers have suggested, not by duration but by intensity, then Marvell's lovers, at least during the act of love, will achieve a kind of immortality by "devouring" time or by transcending the laws of clock time ("Time's winged chariot"). And if this alchemical transmutation requires a fire hot enough to melt them into one primordial ball, then it is perhaps also hot enough to melt the sun itself and "make him run." Thus we see that the overt sexuality of Marvell's poem is, in a mythic sense, suggestive of a profound metaphysical insight, an insight that continues to fascinate those philosophers and scientists who would penetrate the mysteries of time and eternity.

B. Jungian Psychology and Its Archetypal Insights

The second major influence on mythological criticism is the work of C. G. Jung, the great psychologist-philosopher and onetime student of Freud who broke with the master because of what he regarded as a too-narrow approach to psychoanalysis. Jung believed libido (psychic energy) to be more than sexual; also, he considered Freudian theories too negative because of Freud's emphasis on the neurotic rather than the healthy aspects of the psyche.

Jung's primary contribution to myth criticism is his theory of racial memory and archetypes. In developing this concept, Jung expanded Freud's theories of the personal unconscious, asserting that beneath this is a primeval, collective unconscious shared in the psychic inheritance of all members of the human family. As Jung himself explains in *The Structure and Dynamics of the Psyche* (*Collected Works*, vol. 8 [Princeton, NJ: Princeton UP, 1960]):

> If it were possible to personify the unconscious, we might think of it as a collective human being combining the characteristics of both sexes, transcending youth and age, birth and death, and, from having at its command a human experience of one or two million years, practically immortal. If such a being existed, it would be exalted over all temporal change; the present would mean neither more nor less to it than any year in the

hundredth millennium before Christ; it would be a dreamer of age-old dreams and, owing to its immeasurable experience, an incomparable prognosticator. It would have lived countless times over again the life of the individual, the family, the tribe, and the nation, and it would possess a living sense of the rhythm of growth, flowering, and decay. (349–50)

Just as certain instincts are inherited by the lower animals (for example, the instinct of the baby chicken to run from a hawk's shadow), so more complex psychic predispositions are inherited by human beings. Jung believed, contrary to eighteenth-century Lockean psychology, that "Mind is not born as a *tabula rasa* [a clean slate]. Like the body, it has its pre-established individual definiteness; namely, forms of behaviour. They become manifest in the ever-recurring patterns of psychic functioning" (*Psyche and Symbol* [Garden City, NY: Doubleday, 1958]: xv). Therefore what Jung called "myth-forming" structural elements are ever present in the unconscious psyche; he refers to the manifestations of these elements as "motifs," "primordial images," or "archetypes."

Jung was also careful to explain that archetypes are not inherited ideas or patterns of thought, but rather that they are predispositions to respond in similar ways to certain stimuli: "In reality they belong to the realm of activities of the instincts and in that sense they represent inherited forms of psychic behaviour" (xvi). In *Psychological Reflections* (New York: Harper, 1961), he maintained that these psychic instincts "are older than historical man, . . . have been ingrained in him from earliest times, and, eternally living, outlasting all generations, still make up the groundwork of the human psyche. It is only possible to live the fullest life when we are in harmony with these symbols; wisdom is a return to them" (42).

In stressing that archetypes are actually "inherited forms," Jung also went further than most of the anthropologists, who tended to see these forms as social phenomena passed down from one generation to the next through various sacred rites rather than through the structure of the psyche itself. Furthermore, in *The Archetypes and the Collective Unconscious* (New York: Pantheon, 1959), he theorized that myths do not derive from external factors such as the seasonal or solar cycle but are, in truth, the projections of innate psychic phenomena:

> All the mythologized processes of nature, such as summer and winter, the phases of the moon, the rainy seasons, and so forth, are in no sense allegories of these objective occurrences; rather they are symbolic expressions of the inner, unconscious drama of the psyche which becomes accessible to man's consciousness by way of projection—that is, mirrored in the events of nature. (6)

In other words, myths are the means by which archetypes, essentially unconscious forms, become manifest and articulate to the conscious mind. Jung indicated further that archetypes reveal themselves in the dreams of individuals, so that we might say that dreams are "personalized myths" and myths are "depersonalized dreams."

Jung detected an intimate relationship between dreams, myths, and art in that all three serve as media through which archetypes become accessible to consciousness. The great artist, as Jung observes in *Modern Man in Search of a Soul* (New York: Harcourt, n.d.; first published in 1933), is a person who possesses the "primordial vision," a special sensitivity to archetypal patterns and a gift for speaking in primordial images that enable him or her to transmit experiences of the "inner world" through art. Considering the nature of the artist's raw materials, Jung suggests it is only logical that the artist "will resort to mythology in order to give his experience its most fitting expression." This is not to say that the artist gets materials secondhand: "The primordial experience is the source of his creativeness; it cannot be fathomed, and therefore requires mythological imagery to give it form" (164).

Although Jung himself wrote relatively little that could be called literary criticism, what he did write leaves no doubt that he believed literature, and art in general, to be a vital ingredient in human civilization. Most important, his theories have expanded the horizons of literary interpretation for those critics concerned to use the tools of the mythological approach and for psychological critics who have felt too tightly constricted by Freudian theory.

1. Some Special Archetypes: Shadow, Persona, and Anima

In *The Archetypes and the Collective Unconscious* (New York: Pantheon, 1959), Jung discusses at length many of the archetypal patterns that we have already examined (for example, water, colors, rebirth). In this way, although his emphasis is psychological rather than anthropological, a good deal of his work overlaps that of Frazer and the others. But, as we have already indicated, Jung is not merely a derivative or secondary figure; he is a major influence in the growth of myth criticism. For one thing, he provided some of the favorite terminology now current among myth critics. The term "archetype" itself, though not coined by Jung, enjoys its present widespread usage among the myth critics primarily because of his influence. Also, like Freud, he was a pioneer whose brilliant flashes of insight have helped to light our way in exploring the darker recesses of the human mind.

One major contribution is Jung's theory of *individuation* as related to those archetypes designated as the *shadow*, the *persona*, and the *anima*. Individuation is a psychological growing up, the process of discovering those aspects of one's self that make one an individual different from other members of the species. It is essentially a process of recognition—that is, as one matures, the individual must consciously recognize the various aspects, unfavorable as well as favorable, of one's total self. This self-recognition requires extraordinary courage and honesty but is absolutely essential if one is to become a well-balanced individual. Jung theorizes that neuroses are the results of the person's failure to confront and accept some archetypal component of the unconscious. Instead of assimilating this unconscious element to their consciousness, neurotic individuals persist in projecting it upon some other person or object. In Jung's words, projection is an "unconscious, automatic process whereby a content that is unconscious to the subject transfers itself to an object, so that it seems to belong to that object. The projection ceases the moment it becomes conscious, that is to say when it is seen as belonging to the subject" (*Archetypes*, 60). In layman's terms, the habit of projection is reflected in the attitude that "everybody is out of step but me" or "I'm the only honest person in the crowd." It is a commonplace that we can project our own unconscious faults and weaknesses on others much more easily than we can accept them as part of our own nature.

The shadow, the persona, and the anima are structural components of the psyche that human beings have inherited, just as the chicken has inherited his built-in response to the hawk. We encounter the symbolic projections of these archetypes throughout the myths and the literatures of humankind. In melodrama, such as the television or Hollywood western, the persona, the anima, and the shadow are projected respectively in the characters of the hero, the heroine, and the villain. The shadow is the darker side of our unconscious self, the inferior and less pleasing aspects of the personality, which we wish to suppress. "Taking it in its deepest sense," writes Jung in *Psychological Reflections,* "The shadow is the invisible saurian [reptilian] tail that man still drags behind him" (217). The most common variant of this archetype, when projected, is the Devil, who, in Jung's words in *Two Essays on Analytical Psychology* (New York: Pantheon, 1953), represents the "dangerous aspect of the unrecognized dark half of the personality" (94). In literature we see symbolic representations of this archetype in such figures as Shakespeare's Iago, Milton's Satan, Goethe's Mephistopheles, and Conrad's Kurtz.

The anima is perhaps the most complex of Jung's archetypes. It is the "soul-image," the spirit of a man's élan vital, his life force or vital energy. In the sense of "soul," says Jung, anima is the "living thing in man, that which lives of itself and causes life. . . . Were it not for the leaping and twinkling of the soul, man would rot away in his greatest passion, idleness" (*Archetypes,* 26–27). Jung gives the anima a feminine designation in the male psyche, pointing out that the "anima-image is usually projected upon women" (in the female psyche this archetype is called the *animus*). In this sense, anima is the contrasexual part of a man's psyche, the image of the opposite sex that he carries in both his personal and his collective unconscious. As an old German proverb puts it, "Every man has his own Eve within him"—in other words, the human psyche is bisexual, though the psychological characteristics of the opposite sex in each of us are generally unconscious, revealing themselves only in dreams or in projections on someone in our environment. The phenomenon of love, especially love at first sight, may be explained at least in part by Jung's theory of the anima: we tend to be attracted to members of the opposite sex who mirror the characteristics of our own inner selves. In literature, Jung regards such figures as Helen of Troy, Dante's Beatrice, Milton's Eve, and H. Rider Haggard's She as personifications of the anima. Following his theory, we might say that any female figure who is invested with unusual significance or power is likely to be a symbol of the anima. (Examples for the animus come less readily to Jung; like Freud, he tended to describe features of the male psyche more than those of the female, even though both analysts' patients were nearly all women.) One other function of the anima is noteworthy here. The anima is a kind of mediator between the ego (the conscious will or thinking self) and the unconscious or inner world of the male individual. This function will be somewhat clearer if we compare the anima with the persona.

The persona is the obverse of the anima in that it mediates between our ego and the external world. Speaking metaphorically, let us say that the ego is a coin. The image on one side is the anima; on the other side, the persona. The persona is the actor's mask that we show to the world—it is our social personality, a personality that is sometimes quite different from our true self. Jung, in discussing this social mask, explains that, to achieve psychological maturity, the individual must have a flexible, viable persona that can be brought into harmonious relationship with the other components of his or her psychic makeup. He states, furthermore, that a persona that is too artificial or rigid results in such symptoms of neurotic disturbance as irritability and melancholy.

2. "Young Goodman Brown": A Failure of
Individuation
The literary relevance of Jung's theory of shadow, anima, and persona may be seen in an analysis of Hawthorne's story "Young Goodman Brown." In the first place, Brown's persona is both false and inflexible. It is the social mask of a God-

fearing, prayerful, self-righteous Puritan—the persona of a good man with all its pietistic connotations. Brown considers himself both the good Christian and the good husband married to a "blessed angel on earth." In truth, however, he is much less the good man than the bad boy. His behavior from start to finish is that of the adolescent male. His desertion of his wife, for example, is motivated by his juvenile compulsion to have one last fling as a moral Peeping Tom. His failure to recognize himself (and his own base motives) when he confronts Satan—his shadow—is merely another indication of his spiritual immaturity.

Just as his persona has proved inadequate in mediating between Brown's ego and the external world, so his anima fails in relating to his inner world. It is only fitting that his soul-image or anima should be named Faith. His trouble is that he sees Faith not as a true wifely companion but as a mother (Jung points out that, during childhood, anima is usually projected on the mother), as is revealed when he thinks that he will "cling to her skirts and follow her to heaven." In other words, if a young man's Faith has the qualities of the Good Mother, then he might expect to be occasionally indulged in his juvenile escapades. But mature faith, like marriage, is a covenant that binds both parties mutually to uphold its sacred vows. If one party breaks this covenant, as Goodman Brown does, he must face the unpleasant consequences: at worst, separation and divorce; at best, suspicion (perhaps Faith herself has been unfaithful), loss of harmony, trust, and peace of mind. It is the latter consequences that Brown has to face. Even then, he still behaves like a child. Instead of admitting to his error and working maturely for a reconciliation, he sulks.

In clinical terms, young Goodman Brown suffers from a failure of personality integration. He has been stunted in his psychological growth (individuation) because he is unable to confront his shadow, recognize it as a part of his own psyche, and assimilate it to his consciousness. He persists, instead, in projecting the shadow image: first, in the form of the Devil; then on the members of his community (Goody Cloyse, Deacon Gookin, and others); and, finally on Faith herself (his anima), so that ultimately, in his eyes, the whole world is one of shadow, or gloom. As Jung explains in *Psyche and Symbol* (Garden City NY: Doubleday, 1958), the results of such projections are often disastrous for the individual:

> The effect of projection is to isolate the subject from his environment, since instead of a real relation to it there is now only an illusory one. Projections change the world into the replica of one's own unknown face. . . . The resultant [malaise is in] turn explained by projection as the malevolence of the environment, and by means of this vicious circle the isolation is intensified. The more projections interpose themselves between the subject and the environment, the harder it becomes for the ego to see through its illusions. [Note Goodman Brown's inability to distinguish between reality and his illusory dream in the forest.]
>
> It is often tragic to see how blatantly a man bungles his own life and the lives of others yet remains totally incapable of seeing how much the whole tragedy originates in himself, and how he continually feeds it and keeps it going. Not *consciously,* of course—for consciously he is engaged in bewailing and cursing a *faithless* [our italics] world that recedes further and further into the distance. Rather, it is an unconscious factor which spins the illusions that veil his world. And what is being spun is a cocoon, which in the end will completely envelop him. (9)

Jung could hardly have diagnosed Goodman Brown's malady more accurately had he been directing these comments squarely at Hawthorne's story. That he was generalizing adds impact to his theory as well as to Hawthorne's moral insight.

3. Syntheses of Jung and Anthropology

As we can see from our interpretation of "Young Goodman Brown," the application of Jungian theory to literary analysis is likely to be closer to the psychological than to the mythological approach. We should therefore realize that most of the myth critics who use Jung's insights also use the materials of anthropology. A classic example of this kind of mythological eclecticism is Maud Bodkin's *Archetypal Patterns in Poetry*, first published in 1934 and now recognized as the pioneer work of archetypal criticism. Bodkin acknowledges her debt to Gilbert Murray and the anthropological scholars, as well as to Jung. She then proceeds to trace several major archetypal patterns through the great literature of Western civilization (for example, rebirth in Coleridge's "Rime of the Ancient Mariner"; heaven-hell in Coleridge's "Kubla Khan," Dante's *Divine Comedy*, and Milton's *Paradise Lost*; the image of woman as reflected in Homer's Thetis, Euripides's Phaedra, and Milton's Eve). The same kind of critical synthesis may be found in subsequent mythological studies like Northrop Frye's *Anatomy of Criticism*, in which literary criticism, with the support of insights provided by anthropology and Jungian psychology, promises to become a new "social science."

One of the best of these myth studies is James Baird's *Ishmael: A Study of the Symbolic Mode in Primitivism* (New York: Harper, 1960). Baird's approach derives not only from Jung and the anthropologists but also from such philosophers as Susanne Langer and Mircea Eliade. Though he ranges far beyond the works of Herman Melville, Baird's primary objective is to find an archetypal key to the multilayered meanings of *Moby-Dick* (which, incidentally, Jung considered "the greatest American novel"). He finds this key in primitive mythology, specifically in the myths of Polynesia to which young Melville had been exposed during his two years of sea duty in the South Pacific. (Melville's early success as a writer was largely due to his notoriety as the man who had lived for a month among the cannibals of Taipi.) Melville's literary primitivism is authentic, unlike the sentimental primitivism of such writers as Rousseau, says Baird, because he had absorbed certain Asian archetypes or "life symbols" and then transformed these creatively into "autotypes" (that is, individualized personal symbols).

The most instructive illustration of this creative fusion of archetype and autotype is Moby-Dick, Melville's infamous white whale. Baird points out that, throughout Asian mythology, the "great fish" recurs as a symbol of divine creation and life; in Hinduism, for example, the whale is an avatar (divine incarnation) of Vishnu, the "Preserver contained in the all being of Brahma." (We might also note that Christ was associated with fish and fishermen in Christian tradition.) Furthermore, Baird explains that *whiteness* is the archetype of the all-encompassing, inscrutable deity, the "white sign of the God of all being who has borne such Oriental names as Bhagavat, Brahma—the God of endless contradiction." Melville combined these two archetypes, the great fish or whale and whiteness, in fashioning his own unique symbol (autotype), Moby-Dick. Baird's reading of this symbol is substantiated by Melville's remarks about the contrarieties of the color white (terror, mystery, purity) in his chapter "The Whiteness of the Whale," as well as by the mysterious elusiveness and awesome power with which he invests Moby-Dick. Moby-Dick is therefore, in Baird's words, a "nonambiguous ambiguity." Ahab, the monster of intellect, destroys himself and his crew because he would "strike through the mask" in his insane compulsion to understand the eternal and unfathomable mystery of creation. Ishmael alone is saved because, through the wholesome influence of Queequeg, a Polynesian prince, he has acquired the primitive mode of accepting this divine mystery without question or hostility.

C. Myth Criticism and the American Dream: Huckleberry Finn as the American Adam

In addition to anthropology and Jungian psychology, a third influence has been prominent in myth criticism, especially in the interpretation of American literature. This influence derives not only from those already mentioned but also from a historical focus upon the informing myths of our culture. It is apparent in that cluster of indigenous myths called "the American Dream" and subsequently in an intensified effort by literary scholars to analyze those elements that constitute the peculiarly American character of our literature. The results of such analysis indicate that the major works produced by American writers possess a certain distinctiveness and this distinctiveness can largely be attributed to the influence, both positive and negative, of the American Dream.

The central facet of this myth cluster is the Myth of Edenic Possibilities, which reflects the hope of creating a second paradise, not in the next world and not outside time, but in the bright New World of the American continent. From the time of its settlement by Europeans, America was seen as a land of boundless opportunity, a place where human beings, after centuries of poverty, misery, and corruption, could have a second chance to actually fulfill their mythic yearnings for a return to paradise. As early as 1654 Captain Edward Johnson announced to the Old-World-weary people of England that America was "the place":

> All you the people of Christ that are here Oppressed, Imprisoned and scurrilously derided, gather yourselves together, your Wifes and little ones, and answer to your several Names as you shall be shipped for His service, in the Westerne World, and more especially for planting the united Colonies of new England. . . . Know this is the place where the Lord will create a new Heaven, and a new Earth in new Churches, and a new Commonwealth together.

Fredric I. Carpenter, in *American Literature and the Dream* (New York: Philosophical, 1955), points out that although the Edenic dream itself was "as old as the mind of man," the idea that "this is the place" was uniquely American:

> Earlier versions had placed it in Eden or in Heaven, in Atlantis or in Utopia; but always in some country of the imagination. Then the discovery of the new world gave substance to the old myth, and suggested the realization of it on actual earth. America became "the place" where the religious prophecies of Isaiah and the Republican ideals of Plato [and even the mythic longings of primitive man, we might add] might be realized. (6)

The themes of moral regeneration and bright expectations, which derive from this Edenic myth, form a major thread in the fabric of American literature, from J. Hector St. John Crèvecoeur's *Letters from an American Farmer* through the works of Emerson, Thoreau, and Whitman to such modern writers as Hart Crane and Thomas Wolfe.

Closely related to the Myth of Edenic Possibilities is the concept of the American Adam, the mythic New World hero. In *The American Adam* (Chicago: U of Chicago P, 1955), R. W. B. Lewis describes the type: "a radically new personality, the hero of the new adventure: an individual emancipated from history, happily bereft of ancestry, untouched and undefiled by the usual inheritances of family and race; an individual standing alone, self-reliant and self-propelling, ready to confront whatever awaited him with the aid of his own unique and inherent resources" (5). One of the early literary characterizations of this Adamic hero is James Fenimore Cooper's Natty Bumppo, the central figure of the Leatherstocking saga. With his moral purity and social innocence, Natty is an explicit version of Adam before the Fall. He is a child of the wilderness, forever in flight before the corrupting influences of civilization—and from the moral compromises of Eve (Cooper never allows his hero to marry). He is also, as we might guess, the literary great-grandfather of the Western hero. Like the hero of Owen Wister's *The Virginian* and Matt Dillon of television's "Gunsmoke," he is clean-living, straight-shooting, and celibate. In his civilized version, the American Adam is the central figure of another corollary myth of the American Dream: the dream of success. The hero in the dream of success is that popular figure epitomized in Horatio Alger's stories and subsequently treated in the novels of William Dean Howells, Jack London, Theodore Dreiser, and F. Scott Fitzgerald: the self-made man who, through luck, pluck, and all the Ben Franklin virtues, rises from abject poverty to high social estate.

More complex, and therefore more interesting, than this uncorrupted Adam is the American hero during and after the Fall. It is with this aspect of the dream rather than with the adamant innocence of a Leatherstocking that our best writers have most often concerned themselves. The symbolic loss of Edenic innocence and the painful initiation into an awareness of evil constitutes a second major pattern in American literature from the works of Hawthorne and Melville through Mark Twain and Henry James to Ernest Hemingway and William Faulkner to Stephen King. This is the darker thread in our literary fabric, which, contrasting as it does with the myth of bright expectancy, lends depth and richness to the overall design; it also reminds us of the disturbing proximity of dream and nightmare. From this standpoint, then, we may recall Hawthorne's young Goodman Brown as a representative figure—the prototypal American hero haunted by the obsession with guilt and original sin that is a somber but essential part of America's Puritan heritage.

The English novelist D. H. Lawrence was first among the

modern critics to perceive the "dark suspense" latent in the American Dream. As early as 1923 he pointed out the essential paradox of the American character in his *Studies in Classic American Literature* (New York: Viking, 1964; rpt.), a book whose cantankerous brilliance has only lately come to be fully appreciated by literary scholars. "America has never been easy," he wrote, "and is not easy today. Americans have always been at a certain tension. Their liberty is a thing of sheer will, sheer tension: a liberty of THOU SHALT NOT. And it has been so from the first. The land of THOU SHALT NOT" (5). Lawrence saw Americans as a people frantically determined to slough off the old skin of European tradition and evil, but constricted even more tightly by their New World heritage of Puritan conscience and inhibition. He pointed out the evidence of this "certain tension" in the writings of such classic American authors as Cooper, Poe, Hawthorne, and Melville. Though Lawrence is certainly not the only source of such insights, much of myth criticism of American literature—notably such works as Leslie Fiedler's *End to Innocence* (Boston: Beacon, 1955), *Love and Death in the American Novel* (New York: Criterion, 1960), and *No! in Thunder* (Boston: Beacon, 1960)—reflects his brilliantly provocative influence.

Huck Finn epitomizes the archetype of the American Adam. *Huckleberry Finn* is one of the half dozen most significant works in American literature. Many critics rank it among the masterpieces of world literature, and not a few consider it to be the Great American Novel. The reasons for this high esteem may be traced directly to the mythological implications of Twain's book: more than any other novel in our literature, *Huckleberry Finn* embodies myth that is both universal and national. The extent of its mythic content is such that we cannot hope to grasp it all in this chapter; we can, however, indicate a few of those elements that have helped to give the novel its enduring appeal.

First, *Huckleberry Finn* is informed by several archetypal patterns encountered throughout world literature:

1. *The Quest:* Like Don Quixote, Huck is a wanderer, separated from his culture, idealistically in search of one more

substantial than that embraced by the hypocritical, materialistic society he has rejected.

2. *Water Symbolism:* The great Mississippi River, like the Nile and the Ganges, is invested with sacred attributes. As T. S. Eliot has written in "The Dry Salvages," the river is a "strong brown god" (line 2); it is an archetypal symbol of the mystery of life and creation—birth, the flowing of time into eternity, and rebirth. (Note, for example, Huck's several symbolic deaths, his various disguises and new identities as he returns to the shore from the river; also note the mystical lyricism with which he describes the river's majestic beauty.) The river is also a kind of paradise, the "Great Good Place," as opposed to the shore, where Huck encounters hellish corruption and cruelty. It is, finally, an agent of purification and of divine justice.

3. *Shadow Archetype:* Huck's pap, with his sinister repulsiveness, is a classic representation of the devil figure designated by Jung as the shadow.

4. *Wise Old Man:* In contrast to pap Finn, the terrible father, Jim exemplifies the Jungian concept of the wise old man who provides spiritual guidance and moral wisdom for the young hero.

5. *Archetypal Women*
   a. The Good Mother: the Widow Douglas, Mrs. Loftus, Aunt Sally Phelps.
   b. The Terrible Mother: Miss Watson, who becomes the Good Mother at the end of the novel.
   c. The Soul-Mate: Sophia Grangerford, Mary Jane Wilks.

6. *Initiation:* Huck undergoes a series of painful experiences in passing from ignorance and innocence into spiritual maturity; he comes of age—is morally reborn—when he decides to go to hell rather than turn Jim in to the authorities.

In addition to these universal archetypes, *Huckleberry Finn* contains a mythology that is distinctively American. Huck himself is the symbolic American hero; he epitomizes conglomerate paradoxes that make up the American character. He has all the glibness and practical acuity that we admire in our businesspeople and politicians; he is truly a self-made youth, free from the materialism and morality-by-formula of the Horatio Alger hero. He possesses the simple modesty, the quickness, the daring and the guts, the stamina and the physical skill that we idolize in our athletes. He is both ingenious and ingenuous. He is mentally sharp, but not intellectual. He also displays the ingratiating capacity for buffoonery that we so dearly love in our public entertainers. Yet, with all these extraverted virtues, Huck is also a sensitive, conscience-burdened loner troubled by man's inhumanity to man and by his own occasional callousness to Jim's feelings. Notwithstanding his generally realistic outlook and his practical bent, he is a moral idealist, far ahead of his age in his sense of human decency, and at times, a mystic and a daydreamer (or, more accurately, a night dreamer) who is uncommonly sensitive to the presence of a divine beauty in nature. He is, finally, the good bad boy whom Americans have always idolized in one form or another. And, though he is exposed to as much evil in human nature as young Goodman Brown had seen, Huck is saved from Brown's pessimistic gloom by his sense of humor and, what is more crucial, by his sense of humanity.

IV. LIMITATIONS OF MYTH CRITICISM

It should be apparent from the foregoing illustrations that myth criticism offers some unusual opportunities for the enhancement of our literary appreciation and understanding. No other critical approach possesses quite the same combination of breadth and depth. As we have seen, an application of myth criticism takes us far beyond the historical and aesthetic realms of literary study—back to the beginning of humankind's oldest rituals and beliefs and deep into our own individual hearts. Because of the vastness and the complexity of mythology, a field of study whose mysteries anthropologists and psychologists are still working to penetrate, our brief introduction can give the reader only a superficial and fragmentary overview. But we hope we have given interested students a glimpse of new vistas and that they will explore myth on their own.

We should point out some of the inherent limitations of the

mythological approach. As with the psychological approach, the reader must take care that enthusiasm for a new-found interpretive key does not tempt him or her to discard other valuable critical instruments or to try to open all literary doors with this single key. Just as Freudian critics sometimes lose sight of a great work's aesthetic values in their passion for sexual symbolism, so myth critics tend to forget that literature is more than a vehicle for archetypes and ritual patterns. In other words, they run the risk of being distracted from the aesthetic experience of the work itself. They forget that literature is, above all else, art. As we have indicated before, the discreet critic will apply such extrinsic perspectives as the mythological and psychological only as far as they enhance the experience of the art form, and only as far as the structure and potential meaning of the work consistently support such approaches.

· 5 ·

Feminist Approaches

I. FEMINISM AND FEMINIST LITERARY CRITICISMS: DEFINITIONS

"I myself have never been able to find out precisely what feminism is," author and critic Rebecca West once remarked; "I only know that other people call me a feminist whenever I express sentiments that differentiate me from a doormat or prostitute" (*The Young Rebecca*, ed. Jane Marcus [London: Virago, 1982]: 219). Indeed, feminism and feminist literary criticism are often defined as a matter of what is absent rather than what is present. Unlike the other approaches we have examined, feminist literary criticism is a political attack upon other modes of criticism and theory, and because of its social orientation it moves beyond traditional literary criticism. In its diverse manifestations (and the reader will note that we refer to "feminist approaches" rather than "the" feminist approach) feminism is concerned with difference and marginalization of women. Feminists believe that our culture is a patriarchal culture, that is, one organized in favor of the interests of men. Feminist literary critics try to explain how what they term engendered power imbalances in a given culture are reflected, supported, or challenged by literary texts. Feminist critics focus on absence of women from discourse as well as meaningful spaces opened by women's discourse.

Adrienne Rich describes feminism as "the place where in the most natural, organic way subjectivity and politics have to come together" (*Adrienne Rich's Poetry*, ed. Barbara C. Gelpi and Albert Gelpi [New York: Norton, 1975]: 114). This critical stance allows feminism to connect such diverse concerns as a readdressing of the personal (as in contemporary attention to diary literature), a powerful political orientation (as the work of feminist-Marxists especially shows), and a redefinition of literary theory itself (particularly in feminist work on the psychosexual aspects of language). Feminist literary criticism is not, as Toril Moi observes, "just another interesting critical approach on a line with a concern for sea-imagery or metaphors of war in medieval poetry" ("Feminist Literary Criticism," in *Modern Literary Theory: A Comparative Introduction*, 2nd ed., ed. Ann Jefferson [Lanham, MD: Barnes Imports, 1987]: 204). The exclusion of women from the literary canon as a political as well as aesthetic act is being addressed and remedied, but impatience with such patriarchal prejudice is not enough for most feminists. The very language of literary criticism is being changed. In this radical reorientation, feminism represents the single most important social, economic, and aesthetic revolution of modern times.

It thus comes as no surprise that feminism exists in a rich diversity of forms, reflecting a complex historical development. This has been especially important as feminists try more and more to examine the experiences of women from all races and classes and cultures, including, for example, black, Hispanic, Asian, lesbian, handicapped, elderly, and Third World women. Annette Kolodny aptly describes this richness as a "playful pluralism," for it exhibits liberal tolerance, interdisciplinary links, and an insistence on connecting art to the diversities of life ("Dancing through the Minefield: Some Observations on the Theory, Practice, and Politics of a Feminist Literary Criticism," in *The New Feminist Criticism: Essays on Women, Literature and Theory*, ed. Elaine Showalter [New York: Pantheon, 1985]: 161).

Feminist critics have consciously resisted the systematizing of literary theory, particularly the structuralist, poststructuralist, and deconstructionist schools of the 1970s, which they saw as arid patriarchal methodologies. As these approaches attempted to purge themselves of subjectivity,

Elaine Showalter observes, "feminist criticism reasserted the authority of experience." Thus, rather than a promised land of sameness, feminist critics need to seek "the tumultuous and intriguing wilderness of difference itself" (Showalter, "Feminist Criticism in the Wilderness," Critical Inquiry, Special Issue on Writing and Sexual Difference, 8 [1981]: 181, 205).

Despite their diversity, feminist critics largely agree on a threefold purpose: to expose patriarchal premises and resulting prejudices, to promote discovery and reevaluation of literature by women, and to examine social, cultural, and psychosexual contexts of literature and criticism. As feminist critics reread male texts, they describe how women in those texts are constrained in culture and society; the second and third purposes thus follow naturally from the first. The male tradition can be supplemented (or replaced) with a new female tradition. With new methodologies, feminist literary critics quickly find themselves moving toward study of sexual, social, and political issues once thought to be "outside" the study of literature. Sandra Gilbert defines feminist criticism at its most ambitious: it seeks "to decode and demystify all the disguised questions and answers that have always shadowed the connections between textuality and sexuality, genre and gender, psychosexual identity and cultural authority" ("What Do Feminist Critics Want? Or, A Postcard from the Volcano," ADE Bulletin 18 [Winter 1980]: 19). As Maggie Humm notes, such critics wish to make us act as feminist readers; that is, to create "new communities of writers and readers supported by a language spoken for and by women." Literature and criticism are ideological, she argues, "since writing manipulates gender for symbolic purposes" and style is an articulation of ideology. Inevitably women's ideology will "encompass more and more contradictions than the ideology of men since women are provided with many more confusing images of themselves than are men." In psychological terms the "instability of the feminine" can thus be a source of women's power (Feminist Criticism: Women as Contemporary Critics [Brighton, Eng.: Harvester, 1986]: 14–15, 7). Among other things such analysis suggests that gender is conceived of as a complex cultural idea and a psychological component rather than as strictly tied to biological sex. Obviously, being female does not necessarily mean one holds a feminist view; nor does being male have to prohibit one from adopting a feminist critical or social approach, at least in the minds of some feminists.

Feminist criticism is *always* political and *always* revisionist, no matter what the emphasis, and most feminists now agree that despite their diverse approaches there is a distinctive feminine sensibility, even a primordial female part of the psyche reflected in style and in language itself. Annette Kolodny and Elaine Showalter, as well as Adrienne Rich, have argued powerfully for this latter notion, and have used it as a foundation for literary and political awareness. Kolodny thus argues for the feminist's right to "liberate" new and different significances from texts and events, to ask different *questions* of texts and events. In the process the feminist critic is "conscientiously decoding woman-as-sign" in literature as well as in the rest of life (Kolodny, "Dancing through the Minefield," 19–20). As Vincent Leitch aptly observes, this kind of thinking implies that English seminar students should not glory in theoretical discussions of feminism and then ignore the fact that a department secretary has been fired for attempting to start a clerical workers' union (*American Literary Criticism from the Thirties to the Eighties* [New York: Columbia UP, 1988]: 315). One of the most important mottoes adopted by feminism has been and continues to be "the personal is political."

II. HISTORICAL OVERVIEW AND MAJOR THEMES IN FEMINIST CRITICISM

Elaine Showalter has identified three historical phases of women's literary development: the "feminine" phase (1840–1880), during which women writers imitated the dominant tradition; the "feminist" phase (1880–1920), during which women protested and advocated minority rights; and the "female" phase (1920 to the present), during which dependency on opposition—that is, the focus on uncovering misogyny in male texts—is being replaced by a turn inward for identity and a resulting rediscovery of women's texts and women.

Showalter attacks traditional literary history that reduces female writers to only a few who are "accepted." She describes a women's tradition in literature that is an "imaginative continuum . . . [of] certain patterns, themes, problems, and images from generation to generation" (*A Literature of Their Own: British Women Novelists from Brontë to Lessing* [Princeton: Princeton UP, 1977]: 11). Along with Ellen Moers (*Literary Women* [Garden City, NY: Doubleday, 1976]) and Sandra Gilbert and Susan Gubar (*The Madwoman in the Attic* [New Haven: Yale UP, 1979]), Showalter has helped define the entire project of reevaluating women's reading and writing. She has contributed the very important distinction between the feminist critique, which is focused on the female reader, and gynocriticism, which emphasizes difference in the female writer.

Notwithstanding the earlier contributions of writers such as George Eliot, Mary Wollstonecraft, Virginia Woolf, Rebecca West, and Charlotte Perkins Gilman, in Showalter's context feminist literary criticism has mostly developed since the women's movement beginning in the early 1960s. Early feminist critics such as Simone de Beauvoir, Kate Millett, Betty Friedan, and Germaine Greer presented cultural analysis that centered on the female "self" as a cultural *idea* promulgated by male authors; their analysis of literature and culture concentrated on the ways male fears and anxieties were portrayed through women characters. Texts were viewed as models of power, in contrast to the practice of most academic critics of the day, and literature was seen as an instrument of socialization. This emphasis on the social function of literature—as opposed to studying literature in terms of "pure" aesthetics—has continued to generate controversy in the academy, for feminist literary criticism has maintained close ties to feminist politics.

Beauvoir's *The Second Sex* (1949; rpt. Harmondsworth: Penguin, 1972) asked, What is woman? How is she constructed differently from men? Beauvoir answered that she is constructed differently *by* men. This philosophical approach was a crucial breakthrough in thinking about woman as a cultural idea; the thesis that men write about women to find out more about men has had lasting implications. Beauvoir established the fundamental issues of modern feminism by observing, as Raman Selden puts it, that "when a woman tries to define herself, she starts by saying 'I am a woman.' No man would do so. This act reveals the basic asymmetry between the terms 'masculine' and 'feminine.' Man defines the human, not woman. . . . [H]e is the One, she the Other" (*A Reader's Guide to Contemporary Literary Theory* [Brighton, Eng.: Harvester, 1985]: 129). Friedan's *The Feminine Mystique* (1963; rpt. Harmondsworth: Penguin, 1982) demystified the dominant image of the happy American suburban housewife and mother. Friedan's book appeared amidst new women's organizations, manifestos, protests, and publications that called for an end to sex discrimination and enforcement of equal rights, including the right to abortions. An author of many articles in publications such as *Good Housekeeping*, Friedan also analyzed reductive images of women in American magazines.

Millett's *Sexual Politics* (1970; rpt. London: Virago, 1977) was the first widely read work of feminist literary criticism, and it eventually led to a reevaluation of the entire enterprise of studying English and American literature. Millett's focus, unlike Friedan's, is a critique of ideology. Distinguishing between sex as determined biologically and gender as a psychological concept that refers to culturally acquired sexual identity, Millett wrote that "the essence of politics is power," and that the most fundamental and pervasive concept of power in our society is male dominance (25). She saw literature as a record of the collective consciousness of patriarchy; her readings of D. H. Lawrence, Norman Mailer, Henry Miller, and Jean Genet offered a powerful challenge to archetypal social values of capitalism, violence against women, crude sexuality, and male power in general, while it also assaulted the reigning formalism in literary criticism of her day. As a "resisting reader" who focused on patterns of dominance and submission, Millett found that these writers distort female characters by associating deviance with femininity. Unlike Beauvoir and Friedan, Millett advocated revolution instead of reform in her attack on misogyny in literature. In *The Female Eunuch* (London: Paladin, 1971) Germaine Greer documented images of women in popular culture and literature in order to free women from their mental dependency on such images. Greer advocated plurality rather than any dominant monology.

Today perhaps the most obvious evidence of the changing status of women in American higher education is the proliferation of Women's Studies programs on college and university campuses. Accompanying scholarly work of women in many fields, these programs focus exclusively on women writers and female critical issues. As new women's scholarly journals such as *Women's Studies* (1972), *Feminist Studies* (1972), and *Signs: Journal of Women in Culture and Society* (1975) and new anthologies of women's writing appeared, in the 1970s and 1980s Women's Studies programs grew exponentially, forming their courses on the model of small discussion groups. Women's Studies programs are usually interdisciplinary, designed for undergraduates, and focused on history. They tend to challenge the prevailing curricula at their institutions and support outreach programs such as counseling and rape crisis centers. They frequently study class oppression, homophobia, and racism. Florence Howe, an important figure in developing such programs, has noted that feminists have "helped to build a body of knowledge about women and gender significant enough to suggest an epistemological shift qualitatively and quantitatively comparable to the nineteenth century's shift from theology to science. Like our nineteenth-century forefathers, we are also shifting the object of study, even as we are making the lenses of study a new issue" (*Myths of Coeducation: Selected Essays 1964–1983* [Bloomington: Indiana UP, 1984]: 236–37).

The political and methodological diversity of feminist literary criticism clearly reflects that of the larger women's movement. As more and more critics came on the scene, they worked in a greater variety of areas. Elaine Showalter's divisions of types of feminist criticism has been influential. She identifies four models of difference used in theories of women's writing: the biological, linguistic, psychoanalytic, and cultural. The biological model is the most extreme; she finds the argument that the text is marked by the body dangerous in that it can lead to essentialism in defining women writers' value, particularly on the part of male critics. Yet she praises female poets' frankness with regard to the body, and she finds in the intimate and confessional tone of many biological critics a rebuke to those women who continue to write "outside" the female body. Literary creation, she finds, can be modeled by birth rather than by insemination.

Showalter's linguistic model of difference describes women speaking men's language as a foreign tongue; purging language of sexism is not going far enough. Women must use their own language, for if women continue to speak as men do when they enter discourse, whatever they say will be subdued and alienated. Though Showalter is aware of ethnographic and mythic evidence of past "women's languages," she recognizes that today there is no separate female language and no evidence to suggest that the sexes are programmed to develop structurally different languages. Showalter thus calls for a female discourse working within male discourse that can deconstruct it and for efforts in style, strategy, and context. As she notes, "the appropriate task for feminist criticism . . . is to concentrate on women's access to language, on the available lexical range from which words can be selected, on the ideological and cultural determinants of expression."

Her psychoanalytic model locates gender difference in an author's psyche, focusing on the relation of gender to the creative process. Incorporating biology and linguistics, such a model of sexual identity stresses feminine difference as absence of the phallus, and though such an absence allows free play of meaning, since femaleness means remaining outside the closures of maleness, she describes the female artist as lonely, anxious, lacking an audience ("Feminist Criticism in the Wilderness," 186–88, 193, 196–97).

But Showalter's most important contribution has been to describe the cultural model for feminist theorists, a model that offers a more complete way of talking about the difference of women's writing because it places feminist concerns in social contexts. The female psyche as a construction of cultural forces acknowledges class, racial, national, and historical differences and determinants among women but offers a collective experience that unites women over time and space—a "binding force." What if, Showalter asks, we were to look at history through the values of a woman-centered culture? What would change from the ways we normally look at things? Some feminist critics feel they must make the effort to discover and work out of a female-only sphere, a "no man's land" or female "wild zone" in order to be authentic. But because no intellectual act is independent of the economic and political pressures of the male-dominated society, this idea of a wild zone is a "playful abstraction." Feminist critics work inside two traditions in a "double-voiced discourse" ("Feminist Criticism in the Wilderness," 197–202).

Within such a cultural context as Showalter's, K. K. Ruthven has identified seven types of contemporary feminists: *sociofeminists*, who study the social roles of women in literature; *semiofeminists*, who look at how women are coded and classified as women (semiotics being the study of signs or codes); *psychofeminists*, who focus on psychoanalytic and mythic theories of the feminine; *Marxist feminists*, who view women first as members of the oppressed working class; *socio-semio-psycho-feminists*, who combine the above approaches; *lesbian feminists*, who offer what they believe to be a distinctly feminine theory of writing metaphorically based on the female body rather than the male; and *black feminists*, who believe themselves triply oppressed as women, blacks, and workers and who often attack the other feminist theorists and critics for centering their work on upper-middle-class white women only. Ruthven also mentions poststructuralist antifeminist feminists who resist patriarchy by treating "the feminine" as an excluded "sign" rather than as something necessarily connected to women (*Feminist Literary Studies: An Introduction* [Cambridge: Cambridge UP, 1984]: 19). Vincent Leitch adds to this list existential, speech-act, reader-response, and Third World anticolonialist feminist critics (Leitch 311).

Finally, deconstruction has been very important for feminists. Indeed, even when ostensibly uninterested in deconstruction as a theory, feminists seem drawn to deconstruct. The close reading offered by deconstruction has helped feminism dispel accusations of dogmatism and reductiveness; as Elizabeth Abel has noted, its "acutely literary perspective" has also "regenerated interest in male texts, examined less as documents of sexism than as artful renditions of sexual difference." In addition, in deconstruction women are often seen as "powerful figures that elicit texts crafted to appropriate or mute their difference" ("Editor's Introduction," *Critical Inquiry*, special issue on Writing and Sexual Difference, 8 [1981]: 174). Raman Selden has recognized that the unique appeal of deconstruction for feminists lies in its refusal to assert a "masculine" authority or truth; this is important because of feminists' rejection of what they perceive as macho in theory itself. Theory's "rigor, thrusting purpose, and rampant ambition" seems to them opposed to "the often tender art of critical interpretation." Indeed, "much feminist criticism wishes to escape the 'fixities and definities' of theory and to develop a female discourse which cannot be tied down conceptually as belonging to a recognised (and therefore probably male-produced) theoretical tradition" (Selden 129). (For a discussion of deconstruction, see chapter 6.)

With these general categories in mind, in the following pages we preface our analysis of literary works by looking at the most significant movements in feminist criticism of the past two decades, combining the diverse approaches listed above into four main types currently most pervasive in feminist criticism: gender studies, Marxist studies, psychoanalytic studies, and minority studies. In all these areas, there has clearly been a general shift from a negative attack on the male writing about women, towards a positive delineation of women's redefinition of their identity in their own writing. The latter, gynocentric criticism, concentrates on female creativity, stylistics, themes, images, careers, and literary traditions. This new emphasis began with the rediscovery of neglected or forgotten women writers and has grown into the attempt to determine psychosexual and political determinants of gender in a quite broadly defined field of language and literature studies.

III. FOUR SIGNIFICANT CURRENT PRACTICES

A. Gender Studies

There are currently two major types of gender studies. First, feminists argue that gender determines everything, including value systems and language structures; as Elizabeth Abel has said, "sexuality and textuality both depend on difference" (173). Yet while some feminists stress gender differences, others argue that the entire concept of female difference is what has caused female oppression; they wish to move beyond "difference" altogether.

Because of this second emphasis, gender studies is beginning to replace feminist studies in some areas. Believing that all language, indeed all activity, is conditioned by gender, gender critics have broadened the definitions of gender by feminists. Male critics who wish to pursue feminist studies often do so under the umbrella term of gender studies, and in gay studies critics often approach their subject through the topic of gender. Both of these groups are less interested in a writer's or reader's biological sex than in certain qualities of masculinity and femininity.

Nevertheless, feminists by and large regard gender as indicating female difference and place their focus on femininity. Important to feminist distinctions concerning gender is the contrast between masculine and feminine writing. While some would argue that neither exists, that writing is writing and cannot be categorized as masculine or feminine, feminist critics disagree. Following the ideas of the French philosopher Michel Foucault, some feminists argue that the entire notion of authorship is a patriarchal notion, that ownership of a text and identifying normative ideas within a text or choosing among legitimate and illegitimate readings of a text are equally problematic. In moving away from formalism and its notion of the unified self and in embracing the disunity (termed "distortions," perhaps, by male critics) typical of feminine writing, feminist critics alert their readers to underlying patriarchal assumptions. Maggie Humm responds to arguments against this position by noting that in literary studies male critics are seen to be "unaligned," while "a feminist is seen as a case of special pleading." Male criticism, not feminism, she claims, is ideologically blind to the implications of gender (12–13).

In criticism and in literature, feminist critics identify sex-related writing strategies, including matters of subject, vocabulary, syntax, style, imagery, narrative structure, characterization, and genre preference. For example, the novel is often described as a female genre; feminists debate whether the female preference for the novel is based on its realism or on its subjectivity, and whether there is a distinction to be made between these notions. In general women writers tend to be more holistic than men. While male writers seem more interested in closure, women writers often respond with open endings. Feminine logic in writing is often associational; male logic sequential. Male objectivity is challenged by feminine subjectivity. This list of contrasts could go on, but of course exceptions are everywhere. The general difference, however, is a constant subject of study.

By studying women's writing as a gender issue, we are led to ask the general question, What is to be valued? Is diary literature or the Gothic romance genre automatically less worthy than the "realistic" novel or the "high" modernist poem? Do female writers value diversity merely for its own sake? Do they attack men and valorize women excessively? Seemingly the male tradition would have it so. The last few years have seen an unprecedented challenge to traditional thinking with greater attention paid to such suppressed or devalued artistic genres as women's letters and journals.

In the past, descriptions of prose in masculine terms (i.e., praising someone's prose as "virile") were taken as the norm; today, applying a term like *virile* might be intended to describe the *limitations* of a work. As Myra Jehlen puts it, "if literature speaks gender along with class and race, the critic has to read culture and ideology. It turns out that all the time writers and critics thought they were just creating and explicating transcendingly in a separate artistic language, willy nilly they were speaking the contemporary cultural wisdom." Jehlen is aware that many traditional critics regard such talk of gender, class, and race as threatening to diminish literature. But she counters that such reading actually complicates meaning by refusing to reduce the complexities of sexual and other interactions to a false common denominator. Jehlen believes that particularly with authors who seem unconscious of gender as an issue in their work we must make an effort to read for gender: "literary criticism involves action as much as reflection, and reading for gender makes the deed explicit." She emphasizes that as women escape the masculine "norms" of society, men also benefit: "men, . . . upon ceasing to be mankind, become, precisely, men" ("Gender," in *Critical Terms for Literary Study*, ed. Frank Lentricchia and Thomas McLaughlin [Chicago: U of Chicago P, 1990]: 263–65, 273).

B. Marxist Feminism

Marxist feminist criticism focuses on the relation of reading to social realities, refusing to accept the separation of art from life. Certainly the establishment of Women's Studies programs, bookstores, libraries, political action committees, film boards, and community groups attests to the crucial connection between theory and reality offered by feminism. Feminism constantly criticizes the prevailing ideologies of our culture, and, unlike most intellectual strategies, it *acts* on ideas. This is nowhere more evident than in Marxist feminism. Arguing from the ideas of Karl Marx (1818–1883) and Friedrich Engels (1820–1895), who argued that all historical and social developments are determined by forms of economic production, Marxist feminists attack the prevailing capitalistic system of the West, which they view as sexually as well as economically exploitative. Marxist feminists thus combine study of class with that of gender. In Marxist feminism the everyday tends to be valued more than in traditional approaches, and personal identity is not seen as separate from cultural identity. As they emphasize historical and economic contexts of literary discourse, they often direct attention toward the conditions of production of literary texts; that is, the economics of publishing and distributing texts.

Marxist feminists, like other Marxist critics, are often attacked for undervaluing or misunderstanding the nature of quality in art. For them literary value is not a transcendent property, but something conditioned by social beliefs and needs. What is "good" art for many Marxists is simply what people in a given society agree upon as good. This view has been criticized as failing to account for aesthetics or for artistic genius.

Lillian Robinson responds to such questions with a counterattack on formalism: "we have forgotten not so much that art has content but that *content has content.*" Form, style, and history are not independent of content, ideology, or politics. And formalism, she says, serves ruling-class interests; this connects it to the systematic exclusion of women, non-whites, and the working class. Feminist criticism, she contents, "is criticism with a cause, engaged criticism. . . . It must be ideological and moral criticism; it must be revolutionary" (*Sex, Class, and Culture* [Bloomington: Indiana UP, 1978]: 17, 3). Such analyses as Robinson's characterize feminism as angry, tough, and committed criticism, combining the unsentimental with the unapologetically personal and insisting upon the matter rather than the manner of a text.

In contrast to Robinson's all-out attack upon notions of formalism, Rita Felski in *Beyond Feminist Aesthetics: Feminist Literature and Social Change* (Cambridge, MA: Harvard UP, 1990) sidesteps the question of literary value by concentrating not on interpretation but on theories of production. Who writes books? Who publishes them? Who buys them? Who profits from them? Instead of scanning texts for political correctness, Felski examines how women's literature is produced, in not so much an aesthetics as a sociology of literature. She defends confessional, realistic writing by women; she is not troubled by writing that can be described as uncomplicated, a charge often leveled at women's writing. Because she defines writing as "an endless chain of signifiers," a reader is able to extend subtle ironies infinitely. Felski is interested in the *effects* of art rather than in an inherent value; that is, she works with instrumental rather than transcendent value. Felski does not claim that an ideologically correct work of art is somehow better than one that is not, but she implies that we may excuse or ignore artistic blunders in favor of an overriding social good.

This issue of quality has been a painful one for feminists of all persuasions, since they have often been attacked for admitting into discussion what are described by other critics as third-rate works of literature. Helen Vendler has been especially direct in her attacks upon the dismissal of aesthetics by Marxist feminists. She writes, for example, that "Felski's conception of art ends up where political criticism usually ends up—valuing a naive and didactic 'articulation of oppression' because it has awakened certain members of an oppressed class. It seems of no use to tell such critics that propaganda is not valuable except as you value propaganda. If you value art, you cannot value propaganda as art" ("Feminism and Literature," *New York Review of Books*, May 31, 1990, 20–21).

Michele Barrett attempts to resolve this issue by demonstrating how the ideology of gender affects the ways in which male and female works are received and canons established. But she emphasizes the fictional nature of texts, arguing that instead of naively condemning male authors for sexism in their works and approving the ways women writers challenge sexism, we should remember that texts have no fixed meanings apart from the context and ideology of the reader. This does not mean that women writers and readers should stop trying to confront gender issues, only that they must remember the simultaneity of the fictional and political aspects of perception (*Women's Oppression Today: Problems in Marxist Feminist Analysis* [London: Verso Editions, 1980]).

Some socialist feminist critics question the privileged status of gender in Marxist analysis altogether. The situation of most women in the world is described as dismal, but oppressed minorities are of both sexes. Class is the primary division, not sex. There is no essential notion of woman transcending history and culture, and women do not have a shared identity (Karen Hansen and Ilene J. Philipson, eds., *Women, Class and the Feminist Imagination: A Socialist-Feminist Reader* [Philadelphia: Temple UP, 1990]).

C. Psychoanalytic Feminism

Feminists have been especially attracted to the psychoanalytic approach, particularly that of Sigmund Freud and the Freudian revisionist Jacques Lacan. Feminists locate the feminine at the juncture of language and subjectivity, of discourse and identity. These critics define an alternative women's time and space, but they also seek to do away with notions of otherness, to dispense with the idea of woman as secondary to a male norm; indeed, some would locate the idea of otherness itself in the Freudian notion of women as those humans without penises, humans thus to be defined negatively. Psychoanalytic critics wish to readdress such descriptions and to broaden our overall sense of identity.

In America, psychoanalytic criticism has tended to be practical and not particularly terminology-ridden. The most obvious case in point is Sandra Gilbert and Susan Gubar, who have done more than any other feminist critics to make feminism central to literary studies. In *The Madwoman in the Attic*, which became the most influential feminist literary critical text of its time, Gilbert and Gubar examine female images in the works of Jane Austen, Mary Shelley, Charlotte and Emily Brontë, and George Eliot. They address mothering, enclosures, doubling, disease, and landscape, and they make the interesting argument that women writers often identify themselves with the literary characters they detest. Gilbert and Gubar point out how the monster/madwoman figure represents aspects of the author's self-image—like the angel/heroine figure—as well as elements of the author's antipatriarchal strategies. They effectively argue for a women's tradition in writing, what Elaine Showalter calls "a literature of their own." They urge women writers to find the archetypal sibyl figure within, a female Ur-poet, as a source of female creativity, and they believe that women artists must kill the angel within them as well as the monster. They describe a feminine utopia in which *wholeness* rather than *otherness* would prevail as a psychological definition of identity. Gilbert and Gubar's more recent work, the two volumes of *No Man's Land: The Place of the Woman Writer in the Twentieth Century*, volume 1: *The War of the Words*, and volume 2: *Sexchanges* (New Haven: Yale UP, 1988), carry forward their literary history of women's writing.

Other American feminists vary their psychological emphases, for example, Ellen Moers's focus on mothering in novels and narrative poems in *Literary Women*; Carolyn Heilbrun's work on androgyny in *Toward a Recognition of Androgyny* (New York: Harper Colophon, 1973) and more recently in *Hamlet's Mother and Other Women* (New York: Columbia UP, 1990); the linguistic approach of Mary Daly in

Gyn/Ecology (Boston: Beacon, 1978); and the deconstructive readings of Gayatri Chakravorty Spivak. Moers anticipated Gilbert and Gubar when she identified specific female modes, myths, and symbols. Especially important was her description of women writers' respect for each other, such as Emily Dickinson felt for Elizabeth Barrett Browning. Heilbrun asks us to accept androgyny as a means of attaining fluidity in sex roles in her discussions of Virginia Woolf, Gertrude Stein, Djuna Barnes, Carson McCullers, and Ursula Le Guin. Daly and Spivak seek to describe what is absent in a text, what is not said or what is marginalized or erased. Daly's project is to develop a new female syntax, a common language of women, and she goes about this by splintering words such as "de-partment" or "re-presentation." She argues that by splitting off prefixes, which usually intensify nouns, she is politicizing etymology by undercutting the male discourse of patriarchal nouns. But the most innovative and far-reaching use of psychoanalytic theories for feminist criticism is not to be found among the American feminist critics, but among the French.

Elaine Showalter has observed that "English feminist criticism, essentially Marxist, stresses oppression; French feminist criticism, essentially psychoanalytic, stresses repression; American feminist criticism, essentially textual, stresses expression"; yet all three have become gynocentric, searching for terminology to rescue the feminine from being a synonym for inferiority ("Feminist Criticism in the Wilderness" 186). These primary differences involve a poststructuralist influence: while French critics who practice what they call *l'écriture féminine* uphold the power of the psychological category of the feminine, they dismiss the actual sex of an author as unimportant (following their deconstructive attack upon the author or self as a meaningful term in discussion). They also reject the idea that art is mimetic or representational, for images in art are merely tropes, or effects of language; they thus refuse to think of literary works as having an ascertainable truth behind them.

Because the feminine in language, philosophy, and psychoanalysis subverts notions such as literary truth, *écriture féminine* disrupts the unities of Western discourse, pointing to its silences and urging us to rethink its assumptions in radically new ways. While the male tradition often presents the transmission of ideas as modeled by a father-son relationship (as in Harold Bloom's *Anxiety of Influence* [New York: Oxford UP, 1973]), the feminine tradition would have us focus on mother-daughter relationships, a very different model that questions authorship as ownership of texts or ideas. These thinkers conceive of writing as an activity or practice rather than as an institution with rules and regulations that define concepts such as plot or genre. Literary language can carry out feminism's goals by helping women deconstruct the institution of literature's historical and social values. French feminists thus speak of exploding the sign rather than interpreting signs. As Hélène Cixous has stated: "there has not yet been any writing that inscribes femininity" ("The Laugh of the Medusa," *Signs* 1.4 [1976]: 878). French feminism is accordingly often more interested in the manner of what is written rather than the matter, it seems, for its aim is to describe the dichotomy between what is expressed and what is repressed in language. (It is important to remember that by *female psyche* or *the feminine*, these critics do not necessarily mean persons of the female sex; indeed, James Joyce is often cited by them as a practitioner of *écriture féminine*.) The French feminists might even be described as antifeminist because they see feminism in its binary oppositions as a leftover male cultural notion from the past. Like Carolyn Heilbrun, they urge *androgyny* as an antidote to the male-female opposition. Nancy K. Miller has thus described *écriture féminine* as "fundamentally a hope" for the future ("Emphasis Added: Plots and Plausibilities in Women's Fiction," *PMLA* 96 [1981]: 37).

In Sigmund Freud's denigration of female identity French feminists find him actually *centering* everything on female identity. From Jacques Lacan comes the notion of the Imaginary, or a pre-Oedipal, stage in which the child has not yet differentiated himself or herself from the mother and has accordingly not yet learned language. The Oedipal crisis marks the entrance of the child into a world of symbolic order (language) in which everything is separate, including conscious and unconscious, self and other, words and actions. This transition also marks entry into a world ruled by the Law of the Father; Lacan calls it the phallocentric or phallogocentric universe. This law helps explain why French feminists reject any "ism" as confining and hence dangerous.

Freud and Lacan's relevance for French feminism thus arises from their treatment of language. French feminists recognize the significance of Freud's making psychoanalysis a matter of *hearing* words rather than *seeing* symptoms, as was more the case with traditional medicine; Freud's making women the center of his psychology has resulted in making the feminine the center of Western discourse. Lacan similarly finds that the unconscious is structured like a language, and like language its power arises from the sense of openness and play of meaning. When we "read" language we may identify slippages of what is signified and what signifies as a sign of the presence of the unconscious, for language is a mixture of fixed meanings and metaphors. (This process is not unrelated to the now-clichéd idea of the Freudian slip.) And because language is what identifies us as gendered subjects, identity truly occurs only when we enter into speech. Femininity is a language in which the unconscious reveals the fictional nature of sexual categories, and it is always open to redefinition. Thus, just as the notion of a coherent self is itself defined as a fiction, sexual identity is always unstable and always vital because of the disruptions of the unconscious.

French feminists who follow Lacan, particularly Hélène Cixous, often propose a utopian place, a primeval female space free of symbolic order, sex roles, otherness, and the Law of the Father in which the self is still linked with what Cixous calls the Voice of the Mother. This place, with its Voice, is the source of all feminine writing, Cixous contends; to gain access to it is to find a source of immeasurable feminine power. Luce Irigaray also describes this utopian feminine space, but Julia Kristeva is most explicit about the distinction between it and the "real" world. Kristeva calls this Mother-centered feminine realm the semiotic as opposed to the symbolic.

In such works as "The Character of 'Character'" (*New Lit-*

erary History 5.2 [1974]: 383–402) and "The Laugh of the Medusa," Cixous argues that women's language arises from the female body, which she sees as nonunitary, undefinable, unknowable. She defines a woman's language as inherently languages; her notion of *jouissance* evokes sexual orgasm in describing the pleasure of the text as arising from the pleasure of the body. Abolishing repressions of meaning abolishes the notion of meaning itself, releasing us from the laws of rational discourse. Following Jacques Derrida's deconstructive analysis, Cixous has sought to undo what she describes as binary thinking. In Western culture, such oppositions as culture/Nature, intellect/emotion, or activity/passivity, she argues, always come down to male/female, a positive/negative evaluation. In every case the female term is either passive or erased, for victory is equated with activity. Cixous denounces such a way of thinking because it leaves no positive space for the woman. She calls instead for a recognition of woman's life-giving force through a new feminine language that would subvert binarism and patriarchy. If a feminist sees feminism as only pitted against patriarchy, then feminism only shores up the system it seeks to attack. Cixous hopes that by readdressing language entirely women can come to a new understanding of themselves and their world. For example, the French language furnishes an especially compelling instance of the silencing of women in language; its mute *e*, a feminine construction, mutes women. Restoring the feminine semiotic state will allow women to retake the linguistic world, and hence the world, from the men who have devalued feminine sensuality in favor of symbolism. Cixous urges women to read for themselves and to read woman-centered texts.

Luce Irigaray seeks a new language by deconstructing the idea of realism and arguing that the feminine is a rhetorical category rather than a "natural" one. Two of her most significant works are "When Our Lips Speak Together" (*Signs* 6.1 [1980]: 69–79) and *Speculum of the Other Woman* (trans. Gillian C. Gill [Ithaca, NY: Cornell UP, 1985]). Arguing that the feminine has been excluded from Western philosophical discourse, she feels that because a woman critic finds herself speaking *as* a man, her femininity can thus only be traced through blanks and silences in her text. Irigaray describes the ways women speak to each other—shared confidences, unfinished sentences, exclamations, what Irigaray calls babble—in order to define feminine language as outside the order of the symbolic, based as it is on referential meaning and orderly syntax. She develops an alternate discourse that is multiple, fluid, and heterogeneous, basing her theory on the anatomy of female genitalia, whose shape, as Irigaray puts it, is that of two constantly touching and retouching lips. Irigaray has been criticized as biologically essentialist by Toril Moi and others.

Julia Kristeva furnishes the most psychoanalytically based version of French feminism in *Desire in Language: A Semiotic Approach to Literature and Art* (New York: Columbia UP, 1980) and other works. She believes patriarchy has defined femininity as a margin or border of the patriarchal order, and this position has caused women both to be oppressed and feared as well as giving them their unique position of power within Western civilization. The prior semiotic realm is present in symbolic discourse, she finds, as absence or contradiction, and great writers are those who offer the reader the greatest amount of disruption of the nameable. (For a discussion of the "semiotic realm," see chapter 6.) Like Cixous and Irigaray, Kristeva opposes phallocentrism with images derived from women's corporeal experiences as analyzed by psychoanalytical theory. Such theory thus attempts, as does Marxist feminist theory, to connect the personal with the social. Psychology is a fruitful source of methodology for such a project because it addresses both equally, and it is enacted not solely through biology or social constructions but through language.

One can easily see why Freud's accounts of his female patients' first-person narratives of their fantasies and diseases figure so strongly in Kristeva's work; before such maladies as Freud addresses could be treated medically, they first had to be voiced subjectively. Freud and Carl Jung stressed the textual nature of psychoanalytic cases, and they read their patients much as readers read literary texts. Jung in particular developed the artistic aspects of his psychoanalytic thought. Thus maneuvers such as bringing a subtext to light are similar in literary criticism and psychoanalysis, for the goal of both is understanding rather than repression. What Kristeva and the other French psychoanalytic feminists contribute to this comparison is the notion that such therapy is not locked into the individual psyche but is a quality of all language and experience: the feminine is the silence of the unconscious that precedes discourse; its utterance is a flow or rhythm instead of an ordered statement; expression is fluid like the free-floating sea of a womb or the milk of the breast. The semiotic for Kristeva and the other French feminists is an unrepressed flow of words, of the word—of poetry itself.

Yet this attempt to connect the individual experience with cultural realities such as language (or, more accurately, to dissolve such distinctions) has not been effective in the minds of other feminists critical of écriture féminine. Several issues are problematic. First, if feminist critics simply celebrate their notion of the feminine through the Voice of the Mother or the child's pre-Oedipal babble in language, how does this really reject patriarchy, which also values the maternal—indeed, which represses women by doing so? And what about the French feminists' dependence upon male authority figures like Freud and Lacan—does that pose a problem? Next, the semiotic realm is obviously unattainable: what would a nonverbal existence minus self-identity be like? Perhaps it would be psychosis, if it were consciousness at all. Consigning women to such a realm casts them in a primitive and in fact voiceless role. Other feminists have wondered how one actually does speak from a position outside patriarchy, or how one manages to arrive at a theory of femininity that does not at least unconsciously contain influences of patriarchal thinking. Indeed, why do Cixous, Irigaray, and Kristeva keep writing essays and publishing them when they argue in those essays that women cannot "write themselves" in language as we know it at all?

Arguing for an essentially feminine style comes dangerously close to the kind of ghettoizing of women's literature that occurs in much of men's writing about women's works, *especially* the location of meaning and style in female physiology. Many feminists, including Beauvoir, angrily respond to French feminism that they want to live their lives as they please, without being defined as different, which is in itself a male construct. Others, such as Elaine Showalter and Carolyn Heilbrun, resist any theory or practice that limits them to female-only modes.

Other attacks on French feminism come from Marxist feminists, who wonder that anyone should prefer to concern himself or herself with the libidinal when social issues are so pressing. Much if not most of the strengths of feminist criticism have come from its relationship to the larger, political women's movement; feminist deconstruction's and French feminism's movement away from this social commitment seems regressive to Toril Moi, for example, when it loses touch with "the political reality of feminism." In Kristeva's case, Moi insists, we must keep Kristeva's three categories of difference in place simultaneously if we are to avoid such a loss of contact: "1) Women demand equal access to the symbolic order. Liberal feminism. Equality. 2) Women reject the male symbolic order in the name of difference. Radical feminism. Femininity extolled. 3) Women reject the dichotomy between masculine and feminine as metaphysical." If we adopt the deconstructive third category, and forget the first two, we lose political awareness, but of course without the third category we run the risk of an inverted sexism. The third category can *transform* our awareness of the struggle taking place (214). Moi's emphasis on this third category defines the essential strength of *écriture féminine* while it also solves some of its problems.

Nancy Chodorow in *The Reproduction of Mothering: Psychoanalysis and the Sociology of Gender* (Berkeley: U of California P, 1978) offers a psychoanalytically based feminism that avoids some of the foregoing problems. Her non-Lacanian Freudian analysis describes the pre-Oedipal stage in slightly different terms than *écriture féminine,* as she points out that this stage is experienced differently by little boys and little girls. Boys are urged to seek separation and autonomy, while girls are more likely to remain bonded with the mother for the rest of their lives. As a result, Chodorow points out, women's erotic relationships may be described as secondary to this primary relationship of love and dependence, and women as a result experience richer and more diverse psychic lives. These differences can mean frustration in heterosexual relationships. Chodorow offers no theoretically complete answers as the French feminists do, but more realistically offers some practical solutions, which, though partial, are demonstrably workable. She urges equal parenting to help balance girls' dependence and boys' independence, for she sees these traits as furthering the values of a patriarchal capitalist society. But for her the bisexuality or polysexuality of the French feminists is regressive.

Before we end this section, we must mention one other type of psychological feminism, myth criticism. Though myth criticism has its own history and methodology (see chapter 4), several feminist writers have adopted its perspectives and transformed them for the purposes of feminist criticism. Notable among these is Annis Pratt, who, although she criticizes Jung for his lack of treatment of the female developing psyche, offers intriguing connections between feminism and Jungian archetypal criticism. Pratt attempts to construct archetypes of power that are useful to practicing women critics as a means of avoiding the patriarchal tradition (Pratt, *Archetypal Patterns in Women's Fiction* [Brighton: Harvester, 1982], and "The New Feminist Criticisms: Exploring the History of the New Space," in *Beyond Intellectual Sexism: A New Woman, A New Reality,* ed. Joan I. Roberts [New York: David McKay, 1976]: 175–95). Feminist myth critics tend to center their discussions on the Great Mother and other early female images and goddesses, viewing these figures as the radical others that can offer hope and wholeness as against the patriarchal repression of women. Especially popular are figures of the Medusa, Cassandra, Arachne, and Isis.

In *The Lost Tradition: Mothers and Daughters in Literature* (ed. Cathy M. Davidson and E. M. Broner [New York: Ungar, 1980]), prominent feminist myth critics, including Annis Pratt and Adrienne Rich, define myth as the key critical genre for women. Criticizing male myth critics of the 1950s and 1960s, such as Northrop Frye, for ignoring gender in their scientific classifications of myths and archetypes, these writers direct our attention to gender as well as to the actual practices of diverse ethnic groups. Since most myths are constructed and studied by men, there are some very difficult issues concerning women's representation in myths; thus the need is even greater for women's creation of their own myths. Many of these new feminist myth critics reject the Greco-Roman tradition as hegemonous and instead seek pre-Greek myths, such as those of Isis, and diverse, lesser known cultural myths in different parts of the world, such as those of Native American legend. Rich conforms to these general strategies, but focuses on the ways mothers are portrayed in mythology and literature. Although some early feminists seem to have felt that motherhood and feminism do not go comfortably together, Rich argues through myth that motherhood is the feminine status. She distinguishes between the fact of motherhood and the institution a patriarchal culture makes of it, finding that society's oppression of women comes precisely from its need to romanticize (and in a certain sense avoid facing) the terrible and wonderful powers of the mother.

Myth can teach women how to live, and it can help ethnic groups, especially oppressed minorities, reorganize and reorient themselves within a dominant culture. Myth manages to bring together private and public experiences in forms that can be as direct or as masked as the situation demands. It especially appeals to women in their identification with nature, as in the vegetation-goddess archetypes such as Ceres, and it can connect the individual woman with the totality of the cosmos, as with a goddess such as the three-faced goddess of the crossways, Diana-Selene-Hecate. Even the most destructive women in mythology, such as Medea, can be analyzed to show their attraction for modern women; it is well-documented that in many cultures, when matriarchal societies were replaced with patriarchal ones, the previously venerated goddesses are turned by the new culture into witches, seductresses, or fools. Studying these transformations reveals the powers of the goddess all over again, enriching the lives of men as well as women. Yet myth criticism in general and feminist myth criticism in particular have been attacked as too homogenizing, promoting a false universality of identity. In the next section, we will see how such an issue is central to all contemporary feminism.

D. Minority Feminist Criticism

Within the feminist minority there are still other significant minorities, most prominently black and lesbian feminists. While it is true that many black and lesbian feminists include each other in analyses of the problems of either group, it may seem to violate their most fundamental ideas to address them in a single section, since they have strongly protested both their marginalization in society and their often unwanted groupings with other minorities. But they have become a widespread pairing among feminist critics, and they are the most vocal and successful of feminist minorities within the minority. Our treatment of their concerns is meant to suggest issues that confront other minorities, including Hispanic, Asian American, Native American, Jewish, Third World, handicapped, elderly, and other groups of women. Certainly feminism has created a space in which such groups can be freer to move and work, and feminism in general has certainly allied itself with diverse arguments against racism, xenophobia, and homophobia.

Blacks and lesbians are violently attacked in all manner of ways in Western literature and culture in general; thus for them, the personal is even more political than for other women. Their work, both artistic and critical, tends to use irony as a primary literary device to focus on their self-definitions—their coming out of silence. They reject classic literary tradition as oppressive. Not only do they find most other critics racists and misogynists, but they accuse other feminist critics of developing their ideas only in reference to white, upper-middle-class women who oftentimes practice feminism only in order to become part of the patriarchal power structure they criticize for excluding them. That is, the majority of feminists want to be counted as men and share in the bounties of the dominant society, whether it be equal wages, child care, or other accepted social rights. Black and lesbian feminists thus argue that most women have more in common with men than with each other. In literature, black and lesbian women more than white heterosexual women have been written out or ignored or treated as alien. Thus the need to create a new set of traditions is more urgent for black and lesbian critics, and this helps explain the strong lesbian contingent in the civil rights movement, as well as vigorous publishing of black lesbians. Maggie Humm has suggested that "the central motifs of Black and lesbian criticism need to become pivotal to feminist criticism rather than the other way around" (106). This statement has many implications, not the least of which is that reading black or lesbian authors is a political act that must be made a self-conscious political act.

During the 1960s interest in black culture grew, as the inclusion of black writers in syllabi and anthologies attests. In the 1970s and 1980s this was followed by the emergence of black feminist critics. Criticism and theory have only been barely able to keep up with the explosion of interest in black writers. Things are happening so quickly that even the term *black* feminist is problematic. When referring to black feminists in the United States in particular or in the New World in general, the preferred usage is *African-American* feminist; we use *black* here to include black feminists elsewhere as well (though it must be noted that the most prominent work, such as that we cite here, is being done by African-American women critics). Additionally, however, the entire concept of race ("black") is being challenged as meaningless by anthropologists—so even that term is far from agreed-upon. And then there are those black women writers who reject the term *feminist*; among them is Alice Walker, one of the most successful black women authors and critics. Speaking as a woman of color, Walker writes in her collection of criticism, *In Search of Our Mothers' Gardens* (London: Women's Press, 1984), that she has replaced *feminist* with *womanist*, remarking that the former is to the latter as lavender is to purple.

Walker's focus in the well-known title essay of her collection is to identify black female creativity in earlier generations through folk art, including quilting, music, and gardening. Approaches by other black women critics are quite varied. Seeking out the autobiographies of black women writers, especially in slave narratives, has been important to many. Audre Lorde has concentrated on the intuitive language of black women and used this as a critical basis in place of traditional techniques in works such as *A Sister Outsider* (New York: Crossing Press, 1984). Barbara Smith's *Toward a Black Feminist Literary Criticism* (New York: Out and Out Books, 1977) combines a political orientation with a new aesthetics to express black women's experiences. Challenges to traditional criticism from these critics and others have included new bibliographies of neglected or suppressed works, political attacks, and a new interdisciplinary approach that organizes music, painting, literature, autobiography, and literary criticism in radically new ways. How language operates is a constant concern of black critics; they prefer holistic rather than linear notions of meaning, such as the call-and-response dialogics of the black church (for a discussion of "dialogics," see our "Additional Approaches"). Lorde asks us to seek "the Black mother in each of us": that is, to rely on intuition rather than analysis, to place private needs over others, and to see African culture's emphasis upon the mother-bond as an alternative way of thinking in a white patriarchal culture. These critics seem to unite in their attacks on white culture's preference for black male protest writing (such as the works of Richard Wright and Ralph Ellison) to the exclusion of black women's less strident and more holistically conceived works.

Black feminists believe the issues that concern black women writers and characters should be expanded and given a greater place in literary criticism in general. In this sense they tend to be engaged in a variant of the feminist critique described by Showalter, namely, the attack on male-centered literary values, but they are also celebrating the Feminine as the "black mother" in ourselves. Black feminists are interested in exploring texts for motifs of interlocking racist, sexist, and classist oppression; the portrayal of black women as complex selves; spiritual journeys of black women from victimization to the realization of personal autonomy or creativity; the centrality of female bonding; personal relationships in the family and community; reclaiming such figures as the tragic mulatta or the black mother image; validation of the epistemological power of the emotions; the iconography of women's clothing; and black female language. (See Bernard W. Bell, *The Afro-American Novel and Its Tradition* [Amherst: U of Mass. P, 1987]: 242–43, for details on many of these subjects.)

Of concern as well are the issues that interest all specialists in black literature: folklore; oral culture; what W. E. B. Du-Bois called the "double consciousness" of the person of color in a white society; "socialized ambivalence," defined by Bernard Bell as a back-and-forth movement between integration and separation (19); black survival strategies such as "signifying" (an Afro-American vernacular term that describes ritualistic verbal jousting); rejection of the notion of white cultural superiority; an alternating movement between African and Western influences; a heightened appreciation of the deconstructive category of *differance* (see ch. 6). As Bell has noted in regard to the "difference" of blacks in America,

> Traditional white American values emanate from a providential vision of history and of Euro-Americans as a chosen people, a vision that sanctions their individual and collective freedom in the pursuit of property, profit, and happiness. Radical Protestantism, Constitutional democracy, and industrial capitalism are the white American trinity of values. In contrast, black American values emanate from a cyclical, Judeo-Christian vision of history and of African-Americans as a disinherited, colonized people, a vision that sanctions their resilience of spirit and pursuit of social justice. A tragicomic vision of life, a tough-minded grip on reality, an extraordinary faith in the redemptive power of suffering and patience, a highly developed talent for dissimulation, a vigorous zest for life, a wry sense of humor, and an acute sense of timing are basic black American values.

These values, products of the resistance of African-American culture to class, color, and gender domination, are major sources of themes, characters, and forms of the African-American novel (Bell 20). Furthermore, they all contain even more relevance specifically for African-American women, who might be said to possess a "triple consciousness" of being black and female within a white male society. Literature—both its creation and its analysis—furnishes a means of overcoming the fragmentation of consciousness induced by a racist society.

As Houston Baker has pointed out, reliance on the word is particularly necessary for black women because they need to escape their interiority at the same time they wish to share it.

> [I]nteriority and the frontier of violation coalesce in the accessible body of the African woman. . . . The body's male owner and violator (the European slave trader) is in an altogether different relationship to it than its intimate occupant. And the nature of the intimacy achieved by the occupant as domesticator [servant in other's domestic spaces] is quintessential to definitions of Afro-American community. The occupant marks the "boundary case" of ownership. (*Workings of the Spirit: The Poetics of Afro-American Women's Writing* [Chicago: U of Chicago P, 1991]: 133)

The word is thus a way both of escaping the prison of the body as object and of sharing the reality of this interiority.

But until fairly recently, the written word was something denied to black women in the West. Though she had many predecessors, Harlem Renaissance writer and folklorist Zora Neale Hurston (1901–60) should be credited as the first to make black folklore and literature accessible to general readers. In her unusual career there were some interesting crossovers between the magic of the folktales recorded in *Mules and Men* and the magic of the storyteller's power of "lying" Hurston enjoyed in telling her own tales, such as *Their Eyes Were Watching God* or the short stories of *Spunk;* her reliance on "the word" thus took on varying forms. Black feminists such as Walker look to Hurston as a godmother figure because of her mingling of fact and fiction, her effort to combine community with expression of the individual self.

The dual urge to portray the self among black feminist writers and to make the black community whole are not competing interests. Black feminists accuse white feminists of furthering fragmentation of the self and of the community; black feminists are particularly anxious to evolve a world in which their men will be included, in contrast to what they perceive among white feminists, who appear to them to be separatists. As Alexis DeVeau notes:

> I see a greater and greater commitment among black women writers to understand self, multiplied in terms of the community, the community multiplied in terms of the nation, and the nation multiplied in terms of the world. You have to understand what your place as an individual is and the place of the person who is close to you. You have to understand the space between you before you can understand more complex or larger groups. (*Black Women Writers at Work*, ed. Claudia Tate [New York: Continuum, 1983]: 55)

Similarly, Andree Nicola McLaughlin observes,

> Black feminine consciousness, having its counterpart among women from every region the world over, is the basis of an expanding challenge to the postulates of domination in all forms. It protests and takes to task the economic and cultural superstructures that exploit earth's life by pollution, militarism, human oppression, and other means. This awareness, at its root, is anti-imperalist and antipatriarchal. ("A Renaissance of the Spirit: Black Women Remaking the Universe," in *Wild Women in the Whirlwind: Afra-American Culture and the Contemporary Literary Renaissance*, ed. Joanne M. Braxton and McLaughlin [New Brunswick: Rutgers UP, 1990]: xlv)

In short, "black feminine consciousness extols 'community' —independent of any single ideology" as it seeks to create a "new reality rooted in diversity and equality" (McLaughlin xlvi).

In *Inspiriting Influences: Tradition, Revision, and Afro-American Women's Novels* (New York: Columbia UP, 1989), Michael Awkward makes an important distinction between the general ways black female writers influence each other and the way males do. Awkward points out that black women writers carry out their relationships as mothers, daughters, sisters, and aunts rather than as sons vying with fathers. Harold Bloom's famous description of the "anxiety of influence" among male authors seems inappropriate to women writers in general and to black women writers in particular, where "inspiriting" is the appropriate metaphor for influence. Indeed, as Gilbert and Gubar have argued for all women writers, black women writers tend to discover their own creativity by seeking out a female precursor. Thus, the nature of intertextuality among women writers is different than among men; they are less concerned than males with establishing their own "non-repetitive" authority. This places them in contrast to Western literary tradition in general, in which Oedipal battles among men—to say things not another way but a *better* way—characterize influence. Instead, among women we find a "positive symbiotic merger" of identities (Awkward 7). Even when black women writers do deconstruct each other, the perspectives are offered as supplemental to the predecessor's courageous example. Alice Walker has been especially conscious of her debts to her "mothers," or predecessors, including Hurston, Emily Dickinson, Flannery O'Connor, Virginia Woolf, Jean Toomer, and Martin Luther King, Jr., as well as the thousands of unnamed black women of the nineteenth century whose stories have been passed down in their families (Walker, *In Search of Our Mothers' Gardens* 142–43).

Like black feminists, lesbian feminists attempt to show how criticism can be redefined to work in a positive manner for all feminists, but especially for lesbians. Lesbian critics sometimes counter their marginalization by considering lesbianism a privileged stance and a testament to the primacy of women. Terms such as "alterity," "woman-centeredness," and "difference" take on new and more sharply defined meanings when used by lesbian critics. Lesbianism has been a stumbling block for many other feminists, and lesbian feminists have at times attempted to exclude heterosexual women. Some lesbians define lesbianism as the norm of female experience, seeing heterosexuality as abnormal for women. Others go even further and argue that *only* lesbians can offer an adequate feminist analysis. Such views can lead other feminists to reject lesbian feminism. But for the most part lesbian feminists have tried to be inclusive and have offered other feminists new techniques, such as, for example, the rejection of the traditional critical essay form in favor of a more creative and unbounded style.

Lesbian feminists tend to focus on writers such as Mary Wollstonecraft, Virginia Woolf, Gertrude Stein, Ivy Compton-Burnett, and May Sarton. Important figures in lesbian criticism include Lillian Faderman, who offers a historical perspective on lesbianism in *Surpassing the Love of Men:*

Romantic Friendship and Love between Women from the Renaissance to the Present (New York: Morrow, 1981), and Margaret Cruikshank, who argues in *Lesbian Studies: Present and Future* (New York: Feminist Press, 1982) that the notion of the otherness of lesbians will encourage other feminists to see woman as other in more powerful and rewarding ways. Lesbian critics reject the notion of a unified text, much as do the French feminists; accordingly, they investigate mirror images, secret codes, dreams, and stories of identity while rejecting stereotypes of women and especially lesbians. Again, like most other feminists, ambiguity and open endings of stories are stressed, and double meanings are sought after. Lesbian critics offer new views of such accepted genres as the *Bildungsroman*, the Gothic tale, or the utopian tale, and far from tying everything into sexual practice, lesbian authors and critics identify themselves first as breakers of any tradition, reformers of traditionally restrictive textual practices, and linkers of literature to everyday life. In general, literature and life are seen in terms of polarities, and literature and literary criticism as institutions are called into serious question.

IV. FOUR FEMINIST APPROACHES

A. The Marble Vault: The Mistress in "To His Coy Mistress"

Addressing himself to a coy or putatively unwilling woman, the speaker in Andrew Marvell's poem pleads for sex using the logical argument that since they have not "world enough, and time" to delay pleasure, the couple should proceed with haste. But the poem's supposed logic and its borrowing from traditional love poetry only thinly veil darker psychosexual matters. What is most arresting about the address is not its John Donne-like argumentation, nor certainly its loving qualities, but its shocking images of the female body.

As in much of Donne's poetry, the address to a woman causes us to create that woman in ourselves. Like Donne's addressee, the woman in "To His Coy Mistress" is not only unwilling to accept the speaker, but is obviously quite intelligent; otherwise, he would not bother with such vaulting metaphysics. Yet the speaker seeks to frighten her into sexual compliance. This is most clear in his violent and grotesque descriptions of her body.

Her body is indeed the focus, not his nor theirs together. Following a series of exotic settings and references to times past and present, the speaker offers the traditional adoration of the various parts of her body, effectively dismembering her identity into discrete sexual objects, including her eyes, her forehead, her breasts, "the rest," and "every part," culminating in a wish for her to "show" her heart. This last image moves in the direction of more invasive maneuvers toward her body.

The woman is next compared to a "marble vault," and this important image occurs in the center of the poem. The speaker's problem is that despite her charms, her vault is coldly closed to him. He deftly uses this closure as an assault on the woman, however, since the word *vault* means tomb and points toward her death. He clinches the attack with the next image, the most horrifying one in the poem. If she refuses him, "then worms shall try / That long preserved virginity."

Returning to more traditional overtures, the speaker praises her "youthful hue" and dewy skin, from which, through "every pore," he urges her "willing soul" to catch fire. These pores are more openings into her body, and the connection with the earlier ones undercuts the innocence of this set of images. The violence directed at the woman is expanded and redirected toward the couple as "amorous birds of prey" who may "devour" time instead of "languish in his slow-chapped power." The closing vision of how they will "tear our pleasures with rough strife / Thorough the iron gates of life" returns to the language of assault on the female body. All in all, the woman stands to "devour" and to be "devoured" by her amorous admirer and by time itself.

Written in the context of the Renaissance poetic tradition of equating the female body with a fortress that must be assaulted, and phrased in terms of the seventeenth century's fondness for morbid and grotesque metaphysical imagery, the poem's address of the feminine powerfully depicts the marginal status of the female body. Indeed, when the speaker notes that she will find herself in her grave one day, though he does imagine his lust turned to ashes, he does not describe the moldering away of his body in the vault as he does hers. He fails to note where he will be, almost as though he will not pay the penalties she will.

But it is a mistake to see "To His Coy Mistress" as belittling women in some simplistic way; indeed, just as today, the marginality of their status in a male-dominated world indicates not their helplessness, but their awful power. The woman is goddesslike; capricious and cruel, she is one who must be complained to and served. The speaker's flattery as well as his verbal attacks belie his fear of her. To him the feminine is enclosed, unattainable, tomblike as well as womblike. The speaker's gracefulness of proposition, through the courtly love tradition, gives way to crude imagery as the woman's power is excercised in continued refusal (it is clear that she has *already* said no to him). The feminine is thus portrayed as a *negative* state. That is, she does not assent; she is not in the poem; her final decision is not stated. It is a poem about power, and the power is with the absent female, with the vault or womb, the negative space of the feminine.

As distinct from his speaker, Marvell offers a portrayal of male and female roles of his day that celebrates their various positions while sharply indicating their limitations. It is a positive and negative evaluation. On the one hand, the poem gives us a picture of the lives of sophisticated people during the time, people who enjoy sex for pleasure and who are not above making jokes and having fun arguing about it. No mention is made of procreation in the poem, nor of marriage, nor really of love. It is about sex. The poem is so "sophisticated" that instead of merely restating the courtly love tradition, it parodies it. Furthermore it is a poem to youth and passion for life, both intellectual and physical, both male and female. Yet on the other hand, as the male speaker satirizes the lady's coyness, he is even more satirizing himself and his peevish fear of the feminine as expressed in his imagistic attempts to scare her into sex with him. The repellent quality of his images of women remains with one long after his own artful invention and his own coy sense of humor fade.

B. Frailty, Thy Name Is Hamlet:
Hamlet and Women

The hero of *The Tragedy of Hamlet, Prince of Denmark* is afflicted with the world's most famous Oedipus complex, surpassing that even of the complex's namesake. The death of his father and the hasty remarriage of his mother to his uncle so threaten Hamlet's ego that he finds himself splintered, driven to action even as he resists action with doubts and delays. He is a son who must act against his "parents," the Queen and Claudius, in order to avenge his image of his real father as well as to alleviate his own psychic injury, which could be described as a symbolic castration. But because his conflict is driven by two irreconcilable father-images, Hamlet directs his fury toward his mother—and, to a lesser degree, toward his beloved Ophelia—even as he attempts to act against the father(s). Hamlet's irresolvable tension between his two fathers becomes a male-female conflict that is likewise unannealed. A feminist reading of the play has much in common with what we discussed in chapter 3 in our Freudian reading, but the viewpoint is different—very different.

In the play's father-son conflicts as well as in its depiction of a clash between genders, one finds that the role of women in the play generates its various dualities. What has been acted upon by the forces with which Hamlet contends is a woman's body, his mother's, and he in turn has been acted upon by that body. That Hamlet sees sexuality as evil is manifested in his revulsion at Claudius and his disgust at his mother as well as in his cruel denunciation of Ophelia. But his most thoroughgoing hatred is for himself, for it is his conflicting sexual feelings that threaten him from his own unconscious.

The world of *Hamlet* is riven by power struggles within and without Elsinore Castle, and the play's psychological themes are made more powerful by their contact with the other major thematic pattern in the play, that of politics. In both these arenas, it is fruitful to examine how women are represented in inner and outer worlds.

As Shakespeare was writing *Hamlet*, perhaps the advancing age of Queen Elizabeth I and the precariousness of the succession—always with the accompanying danger of war at home and abroad—was an element in the dramatist's conjoining a man's relations with women with his relation to political power. The play gives us a picture of the role of women in Elizabethan society, from the way Ophelia must obey her father without question, to the dangers maidens might expect to encounter from young men, to the inappropriateness of Queen Gertrude's sexual desire. Admittedly, cultural roles of the women of the court as in this play are not applicable to women of all classes in Elizabethan times nor our own, but what women stand for psychologically in *Hamlet* has universal significance for society.

Act I begins with descriptions of the ghost of the recently dead King Hamlet and of the just-ended war over land between him and the King of Norway, Fortinbras. Later the conflict will continue between his son, the younger Hamlet, and Fortinbras, nephew of the King of Norway. Such emphasis upon family relationships from the beginning of the play is accompanied by emphasis upon political matters. Next, the night from which the ghost emerges is identified in female terms, compounding the fear of unrest in general with fear of the feminine: the ghost lies in the "womb of earth" (I.i.137) and walks in an unwholesome night in which a "witch has power to charm" (I.i.163), banished only by the crowing "cock." The cause of the as yet unnamed unrest is demonstrated by Claudius's speech on how the Danes should cease mourning the late king and put the kingdom in order so that young Fortinbras will not attack them in their disorder. He tells the courtiers that "so far hath discretion fought with nature / That we with wisest sorrow think on him / Together with remembrance of ourselves." He has thus taken as his wife "our sometime sister, now our queen" albeit with "a defeated joy, / With an auspicious, and a dropping eye, / With mirth in funeral, and with dirge in marriage" (I.ii.5–14). Thus do *Hamlet*'s dualities multiply rapidly and even exponentially.

The father-son images in Claudius's description of matters between Denmark and Norway are followed by Claudius's fatherly behavior to young Laertes and then by the first appearance of Hamlet in the play. Hamlet's first words are directed to his mother in response to an address by Claudius; when Claudius gets as far as calling Hamlet "my son," Hamlet's aside, "A little more than kin, and less than kind" (I.ii.64), illustrates the personal and political conflicts that have arisen. Gertrude pleads with Hamlet to stop mourning his father—"Thou know'st 'tis common" to lose a father—to which he responds, "Ay, madam, it is common" (I.ii.74), an implied attack on her. Claudius asks Hamlet to think of him as a father, but Hamlet's emotions in this scene are directed at his mother, as his images of darkness attest (his "inky cloak" and "suits of solemn black") (I.ii.76–77).

What follows is the first of his many soul-searching monologues. When Hamlet thinks of himself he thinks first of "this too too sallied flesh" (for which alternate readings have offered "sullied" and "solid" for "sallied"), which he would destroy had "the Everlasting not fix'd / His canon 'gainst self-slaughter." If his flesh is sullied, his mother's is polluted: in the monologue he blames his mother's "frailty" for exchanging "Hyperion" for a "satyr." She is "unrighteous" and beastlike in her lust, even incestuous (I.ii.129–59). He is interrupted from his thoughts by the news that guards have sighted his father, and he is eager for the night to come so that the foul deeds of the past may rise.

Laertes warns his sister Ophelia against Hamlet, believing that Hamlet intends only to seduce and not marry her, but she warns him not to be a hypocrite about his own pleasures. Polonius, their father, delivers a lecture to Laertes, who is about to embark on a journey back to school, about how to be a man in the world of men. He then scoffs at Ophelia's report of Hamlet's sincerity and redoubles Laertes' warning, and she obeys (I.iii). The tensions are thus all between fathers and sons concerning their relations with women, or between men and women themselves.

When Hamlet and his friends emerge to see the ghost, the news is of the king's drunkenness that night, evidence of his "vicious mole of nature" (I.iv.26). Hamlet's attentions to the ghost of his father are thus emphasized by contrast; when the ghost appears, like Hamlet, he addresses Gertrude's sin—"O Hamlet, what a falling-off was there / From me" to the lewd wretch she has married, driven, he believes by her lust (I.v.47–57). The ghost's narrative of his death identifies his body with the temple and the city ("And in the porches of my ears did pour / The leprous distillment") while his wife's body is described as thorny vegetation. But the ghost warns Hamlet against taking revenge on his mother: "Leave her to Heaven" (I.v.63–88). Hamlet responds that he will act only for his father.

Act II begins with Polonius paying Reynaldo to spy on Laertes, especially with regard to whoring, making Ophelia's words to Laertes seem more plausible, and subtly establishing her as reliable and morally upright girl against the examples of her hypocritical brother and father. Again, by "indirections," the directions that are sought have to do with sexuality. The troubled Ophelia narrates Hamlet's mad intrusion on her sewing in her closet, and Polonius puts Hamlet's frenzy down to lovesickness. But from what the audience has seen of Hamlet's feelings, his treatment of her could also evince his disgust with women. Again the two lines of meaning are joined: Hamlet's madness may be feigned for political reasons, or it may be genuine for personal ones.

In the next scene Claudius and Gertrude ask Rosencrantz and Guildenstern to spy on Hamlet to find out what troubles him, compounding the pattern of deception. But Gertrude knows what is wrong with her son: "I doubt it is no other but the main, / His father's death and our o'erhasty marriage" (II.ii.56–57). Polonius offers another plot to the royal pair; he will loose his daughter upon Hamlet and then they will watch what transpires between her and Hamlet: "I have a daughter—have while she is mine— / . . . in her duty and obedience" (II.ii.106–7). When Polonius encounters Hamlet, Hamlet seems to have overheard or at least unconsciously intuited Ophelia's victimization by Polonius: "Have you a daughter?" he asks Polonius, advising "Let her not walk i' th' sun," punning on sun/son (II.ii.182–84). Hamlet next encounters Rosencrantz and Guildenstern, whose crude sex jokes characterize first the earth and then Fortune as whores. Again negative images of women are associated with unwholesome characters, even as Ophelia's innocence serves as an ironic contrast.

When a troupe of players comes to the castle, Hamlet asks one of them to repeat Aeneas's speech to Dido, an appropriate male-versus-female scene in that Aeneas abandons Dido in order to pursue his political destiny (one also recognizes the relevance of Aeneas's famous role as model son to his aged father when the two escaped from Troy). The "strumpet" Fortune is referred to again, but the most important image of a woman offered in Hamlet's conversation with the players is the mention of Hecuba, wife of King Priam, who, unlike Hamlet's mother, mourns for her children rather than they for her. Hamlet thinks of his own genuine grief in response to the players' delineation of Hecuba's, and calls himself a "whore" and "drab" who must only "unpack" his heart with words—acting, that is, instead of taking action (II.ii.585–86).

Act III begins with an increasingly weak Queen, whose half-hearted questions and empty statements indicate both her powerlessness and her sharp sense of guilt. Her voiced hopes for Hamlet are wishful thinking, for she knows the truth. Continuing the whore image, Claudius in an aside compares his own pious playacting to a "harlot's cheek, beautied with plast'ring art" (III.i.50).

These doubts about truth in love and in kingship are followed by Hamlet's famous musing on suicide in his "To be, or not to be" speech (III.i.55–87), a speech ended by the entrance of Ophelia, whom Hamlet addresses as "nymph" and whom he asks to pray for him. But as they speak he begins to berate her violently; this is the first time he has fully revealed his emotions to another person. In a way he now pours poison into someone's ears. His demands of her as to whether she is "honest and fair" escalate into his shouting at her, "Get thee to a nunnery" (nunnery being Elizabethan slang for brothel), and his cruel rehearsal of the advice against young men she has heard from Polonius and Laertes. He ends by accusing her and all women of making monsters of men. It is noteworthy that Hamlet never voices his anger to Claudius, only to Gertrude and Ophelia.

Ophelia's response is "Heavenly powers, restore him!"; she sadly tells herself after he has fled the stage, "O, what a noble mind is here o'erthrown!" Hamlet was the model for manhood, "Th' expectation and rose of the fair state, / The glass of fashion and the mould of form." Ophelia's punning on "the fair state" to mean women as well as his polity effectively yokes together the feminine and the male political power that denies it. Hamlet is falling prey to the denial of his own "feminine" traits—gentleness, a forgiving heart, stability—caught as he is in the personal and political throes of male ego struggles. Ophelia too is coopted by destructive male definitions of women and what women represent when she characterizes herself as "of ladies most deject and wretched, / That suck'd the honey of his music vows" (III.i.150–61). But her reason, her love, and her faith outshine anyone else's in the play, and from here until her death she demonstrates her value more and more.

The players and Hamlet converse about appearance and reality as well, as Hamlet bids them to "hold as 'twere the mirror up to nature" in their portrayal of the play (part of which he has written for them), the murder of a king by his rival. Hamlet next tells Horatio of his plan, and while his speech characterizes his action as masculine it describes his soul as feminine, Nature as feminine, and Fortune as feminine, thus blurring the lines between the feminine's positive and negative aspects for him.

But on the night of the play he is all performance and in no doubt about what he is doing or feeling. He falsely jests with the Queen and Claudius, fittingly, about "country matters," or sex. He rests his head in Ophelia's lap. His pretense of being jolly is designed to make the guilty pair think he is cured of his melancholy by using the very issue, sex, that is so dangerously at hand. The prefatory "dumb show" of "The Murder of Gonzago" follows, upsetting Gertrude and Claudius, and causing Ophelia to utter: "Belike this show imports the argument of the play" (III.ii.140). Her reaction is important; since here she seems to span the distance between the play-within-the-play and *Hamlet* itself, Ophelia places herself in the realm of the author and reader, adding to the self-reflexivity or metafiction of this scene (its concern with itself as artifice). Ophelia is the one who after the dumb show of the King's murder asks Hamlet, "What means this, my lord?" (II.ii.136). At this moment Hamlet and Ophelia are very close, although they do not know it. That the audience is here able to put together their masculine and feminine selves into a mental and spiritual whole is of utmost importance in the reception of the play's artistic forms and moral arguments.

Hamlet continues to interject comments, with most of them directed at women's fickleness. Perhaps in response to these comments, the queen senses something false in the play's portrayal of women. She says, "The lady doth protest too much, methinks" (III.ii.230), as though she knows personally how the scene should be played. Hamlet continues to interject sex jokes directed at Ophelia, so much so that she comments, "You are as good as a chorus, my lord" (III.ii.245). But his interpretations climax with the actor's pouring poison in the ear of the play-king; the real king and queen, having had enough, exit. In the play-within-the-play, the line that precedes this abrupt ending is the description of the poison as "With Hecat's ban thrice blasted, thrice infected" (III.ii.258). Both "blasted" and "infected" are words that refer to venereal disease, like the "sullied" with which Hamlet earlier described himself and like the terms used to describe whores. This is what has been driving everyone: the "sin" of sexuality.

The queen sends Rosencrantz, Guildenstern, and Polonius to bring Hamlet to her following this debacle. Hamlet will not be played by them, however (III.ii.364). Instead, he thinks of the hour as "the very witching time of night," and when he stumbles upon Claudius at his prayers, he thinks of murdering him, but desists because his soul would fly to heaven cleansed. That it is Hamlet's midnight and Claudius's noon in this scene highlights the dramatic climax that is occurring: this is Hamlet's time to act, from now until the end of the play. But he resolves that in seeing his mother, he will "speak daggers" to her but "use none"; he will make his "tongue and soul . . . hypocrites" (III.ii.396–97). It is indeed the "witching time."

We sense that the scene between Hamlet and his mother is that which has been put off so long, but in typically misdirected anger Hamlet murders the foolish Polonius instead of taking direct action and confronting himself. Polonius said that he would hide behind the arras to judge Hamlet's remarks better than a mother could, and he instructs Gertrude prior to her audience with Hamlet. But he ends up with a closer identification with Gertrude than he imagined. Polonius springs out because Gertrude asks Hamlet if he will murder her. In a sense all Hamlet's words of anger directed at her were directed at Polonius's deception too.

After Polonius's death, Hamlet denounces his mother with renewed fury, seeing her as the agent of evil and the dead Polonius as only a "wretched rash intruding fool" (III.iv.31). He damns her for her violations of her marriage in a lengthy speech filled with frightening, grotesque imagery. Her response is: "O Hamlet, speak no more! / Thou turn'st my eyes into my very soul, / And there I see such black and grained spots" (III.iv.88–90). Interestingly, after killing a father-image, no matter how unmanly an example, Hamlet is free to vent his full fury. The ghost of Hamlet's father appears and begs him to look upon Gertrude's "amazement" and "fighting soul," to speak to her, for "Conceit in weakest bodies strongest grows" (III.iv.112–14). Though the injunctions against blaming his mother make him blame her more, Hamlet relents as much as he is able. The ghost represents Hamlet's best version of a father, merciful even, and it seems that this vision arises out of his confusion about his mother and his doubt about his severity towards her. Hamlet accordingly addresses his mother in a new tone, urging her out of her hypocrisy: "Mother, for love of grace, / Lay not that flattering unction to your soul. . . . Confess yourself to heaven, / Repent what's past, avoid what is to come, / And do not spread the compost on the weeds / To make them ranker" (III.iv.114–52). In other words, stop sleeping with Claudius. Throughout the scene, Hamlet's obsession with his mother is so great that his emotional attention is directed at her rather than at Polonius or his father.

Act IV begins with the king showing a similar disregard for the murdered old man; he first says that it could have been he, and then that he will be blamed for not locking Hamlet up. He plans to control the story by sending Hamlet away. The disarray of the world around them is made clearer by Hamlet's retaliation against his former friends, Rosencrantz and Guildenstern, and by Fortinbras's plans against Poland.

The king and the queen next turn their attention to silencing Ophelia, who, in her madness over her father's death, might name Hamlet as the murderer. Again Hamlet and Ophelia are similar; they are both made mad over the death of a father. Once both their hearts were innocent; now, apparently only Ophelia's still is. Ophelia is genuinely mad. Yet even in her affliction she still seems to see truth, as when she sings to Gertrude of a dead love and shares with her the herb rue for sorrow and repentance. Her songs are rather like Hamlet's rambling jests, and they also focus on jilted love. Her parting lines are a mock dirge for all women: "Good night, ladies, good night. Sweet ladies, good night, good night" (IV.v.72–73).

When Laertes learns of his father's murder, the images in which he expresses his frustrated desire to revenge him are those of whoredom: "That drop of blood that's calm proclaims me bastard, / Cries cuckold to my father, brands the harlot / Even here between the chaste unsmirched brow / Of my true mother" (IV.v.118–21). Following Hamlet's failure the feminine is again perceived as negative by the males, and this in contrast to Ophelia's pitiful grief. Who comforts her? Even Hamlet, who perhaps longs to, is incapable of helping. Claudius tells Laertes he protected Hamlet only for his mother's sake, and the two of them plan the poisoned sword fight that they hope will kill Hamlet. But their plans are interrupted by Gertrude's news of Ophelia's death, to which Laertes responds that after his tears over her he will cry no more: "when these are gone, / The woman will be out" (IV.vii.188–89), as though it were a shame to be so feminine as to cry.

Act V begins with Hamlet and Laertes fighting in Ophelia's newly dug grave, after which Hamlet confesses his love for her, a question that has been left hanging until now. It is important that he confesses it to Gertrude; it may be that Ophelia's death has awakened Hamlet to his true nature as a lover of women instead of a victim of them. As Hamlet remarks to Horatio in the next scene, "When our deep plots do pall, and that should learn us / There's a divinity that shapes our ends, / Rough-hew them how we will—" (V.ii.11–13). Laertes, now described as the gentleman's model instead of

Hamlet, has taken Hamlet's masculine, provoking, revenge-seeking place. They fight, and each is wounded with the poisoned sword of the father-king, Claudius. The queen drinks a poisoned cup. She says she "carouses" to Hamlet's "fortune" and urges him against his opponent: "He's fat, and scant of breath. / Here, Hamlet, take my napkin, and rub thy brows," just as a proud and loving mother would do—perhaps she has been healed as well (V.ii.288–90). Dying, Hamlet forces Claudius to drink from the cup. But it is all too late, even for revenge, and it is left to Horatio to tell Hamlet's story. Hamlet and the two women he loved join his two fathers and Laertes in death. Political stability is restored by Fortinbras with a manly flourish. The crisis of sons and fathers is over, and the world of male political power is renewed, but the problem of the feminine is obliterated instead of solved.

C. Men, Women, and the Loss of Faith in "Young Goodman Brown"

From a feminist point of view, Nathaniel Hawthorne is unusual for his time, the mid-nineteenth century, in that his portraits of women go against the prevailing literary sexism of his day. Despite his comment that he was tired of competing with the "mob of scribbling women" writing romance novels, he generally used women not just as symbols of wholeness and goodness, as they are often traditionally pictured, but as empowered with knowledge that approaches that of the author and narrator. In many of the stories and novels by Hawthorne in which women appear, they may be thought of as creative and moral spokeswomen for the author, rather like Ophelia in *Hamlet*. Hawthorne also treated women with more realism and more depth than did most other writers, especially male writers. This treatment of the other or the feminine by Hawthorne became an important legacy to such modern American writers as Henry James and William Faulkner, who also portray women as powerful moral agents.

Hawthorne's most interesting women characters are Hester Prynne and Pearl of *The Scarlet Letter*, Zenobia of *The*

Blithedale Romance, Hepzibah Pyncheon of *The House of the Seven Gables,* Miriam of *The Marble Faun,* Georgiana Aylmer of "The Birthmark," and Beatrice Rappaccini of "Rappaccini's Daughter." All these women engage in conflict with the men in their lives and all of them have the sympathy—to varying degrees—of the author. Hester is Hawthorne's greatest creation of a character, male or female, and from the lips of the magnificent Zenobia, Hawthorne gives us as eloquent a speech on women's rights as was ever penned.

Faith Brown of "Young Goodman Brown" is not a three-dimensional character, however. Like Phoebe Pyncheon of *The House of the Seven Gables* or Priscilla of *The Blithedale Romance,* Faith is a symbolic character. With her allegorical name and small role in the story, readers might be likely to overlook her significance. In fact, the story centers on her, more specifically, on her husband's rejection of her. The tale is a psychosexual parable of the rejection of the feminine in favor of the father-figure symbolized by the Devil. Good and evil are thus engendered qualities in the story. Hawthorne's sympathies are with the woman and not with the misconstrued masculinity of the rigid Brown.

"Young Goodman Brown" portrays the spiritual and moral failure of the title character, a young newlywed whose rejection of his wife's sexuality leads him to seek fulfillment in a fling he wishes to pursue in the wilderness. It is not stated exactly what Brown is seeking, but if he knows there is a witch meeting to which he might go, then the implication of sexual activity is certainly there. But what he finds is a frightening masculine figure who actually resembles his own father. In this Brown gives up his adult sexuality in favor of a regression to a powerless state at the mercy of the Terrible Father.

Wiser than Brown, Faith begs him not to go, whispering "softly and sadly." She pleads that a lone woman is troubled by bad dreams, forecasting Brown's own coming troubles. Yet in the end, when he returns to town a broken and bitter man, Faith has not gone mad as he has. She runs to kiss him. He rejects her, sentencing her to a lifetime with a physically cold and spiritually empty man.

When Faith protests Brown's departure, he asks her, "dost

thou doubt me already?," turning the tables on her. But she responds, "Then God bless you!," adding the warning, "and may you find all well when you come back." As he strolls along, he thinks "Poor little Faith!" He worries that his deed will "kill her," but his worries do not turn him homeward. He instead decides that after this night he will "cling to her skirts and follow her to heaven." Yet when he later has the chance to do this he does not; he instead sees Faith as an obstacle throughout the story, as when he tells the man he meets on the path, the Devil, that he is sorry for being late, for "Faith kept me back awhile."

The sexual aspect of Brown's mission is indicated in the repeated mentions of the women who will be present at the witch meeting, from Goody Cloyse to the governor's wife to the most spent of prostitutes. It is important that Brown learned his religion from Goody Cloyse, and that when he wonders whether he is hurting Faith by continuing his wilderness journey, the Devil produces Goody Cloyse to make Brown suspect women. Goody Cloyse says she is attending the black mass to see a man, while the men on horseback Brown overhears seem to be drawn there to see the women whom they expect. Sex is certainly at the heart of things: the Devil's snaky staff is an appropriate phallic symbol, and the ritual itself turns out to be a satanic wedding. But the sexuality and the general tone are far from celebrating male and female sexuality; women are victimized in this story. Significantly, the Devil mentions his having helped Brown's grandfather to persecute Quaker women in Salem.

Much has been written about Faith's pink ribbons; when Brown hears her voice from a cloud overhead as he proceeds, the ribbons, symbol of her faith as well as her sexual allure, float down. With this flimsy symbol of her *instead* of her, he can proceed. Later, when Brown has the chance to be saved from the unholy rite into which he stumbles, he calls out to Faith too late; she disappears. Brown has the ribbons but loses the woman's body they adorn, and he rejects the mock wedding in the forest just as he rejected his real marriage in the town.

Women are seen by Brown to be temptresses, from Martha Carrier, a "rampant hag" who "had received the devil's promise to be queen of hell" to the seemingly unspotted maidens who crowd around the evil altar. To Brown, sex with women is alluring but deadly, and this is reinforced in the juxtaposition of the bloody basin and Brown's paleness at the altar of unholy matrimony.

In contrast, the Devil is a man among men in a world much larger than Brown's. He is very attractive to Brown. He tells Brown: "I have a very general acquaintance here in New England. The deacons of many a church have drunk the communion wine with me; the selectmen of divers towns make me their chairman; and a majority of the Great and General Court are firm supporters of my interest. The governor and I, too—But these are state secrets." The Devil also has the only really good lines in the story. Though he symbolizes evil, Hawthorne also speaks through him in a double discourse. For example, in his speech to "my children," he wisely seems to see humans as mixtures of good and evil, which is of course what Brown will not be able to do. Brown takes the Devil's lesson only partway: he sees only the evil in those around him, missing the point about the "sympathy of your human hearts" implied in the Devil's words. When he sees his wife at the altar, it is only as a "polluted wretch."

Of course, in the end, Brown is the one polluted; as the narrator tells us, he is the most "frightful" figure in the forest, and he returns home to rebuke and frighten his wife. We are told that his "hoary corpse" is carried to his grave followed by his wife and his children and grandchildren, and instead of shuddering at his gloomy death, one shudders instead to think that Faith and her children had lived all those empty years with his blighted self, a failed husband and father.

D. Women and "Sivilization" in *Huckleberry Finn*

When Huck Finn dresses as a girl, "Sarah Mary Williams," and tries to get information from the clever housewife, Mrs. Judith Loftus, in chapter 11 of *Huckleberry Finn*, is Twain reminding female readers that they are not persons who could enjoy Huck's male adventuring, or is he illustrating the close connection between Huck's willingness to cross gender lines and his later freedom to cross racial lines? The novel's overall theme of the individual versus the community uses the feminine in very important ways. In one sense, the feminine represents what the novel's satire so often suggests, what is *not* there in community that ought to be. If the feminine is shown as a negative quality in the world of *Huckleberry Finn*, it is possible to judge that negativity both as an attack on women and as an upholding of their neglected value to the community and to the hero Huck. The most positive figure in the story, the slave Jim, is a happily married man who in the end is to be reunited with his wife and family; indeed, the most poignant speech of Jim's is his detailing of his unjustified punishment of his deaf and dumb daughter. This sort of tenderness is the book's most positive feminine quality, and with the exception of Jim, it is almost entirely absent in the men Huck encounters. It might even be argued that the entire novel is about the Death of the Father leading to a quest for contact with the Feminine. In a broader sense, it challenges all designations of gender identities, as it does those of race.

The Judith Loftus scene casts all categorization of male and female roles into question. It accomplishes this in several ways. Huck mistakenly misjudges Mrs. Loftus's keen intelligence and her kind of wisdom, a highly positive feminine value. Twain himself seems confused in the episode about the lines between maleness and femaleness. Mrs. Loftus is quite schooled in the *appearance* of gender; after Huck is found out, her list of how women do things as opposed to how men do them demonstrates not the absoluteness of gender but its existence as merely interpretable behavior. For example, Mrs. Loftus tells Huck that "when you set out to thread a needle, don't hold the thread still and fetch the needle up to it; hold the needle still and poke the thread at it—that's the way a woman most always does; but a man always does it t'other way." Yet in chapter 13 of *The Prince and the Pauper*, Twain has Miles Hendon do it exactly opposite: "He did as men have always done, and probably always will do, to the end of time—held the needle still, and tried to thrust the thread through the eye, which is the opposite of a woman's way." Twain is obviously unsure of true women's or men's ways.

One gets the feeling that Mrs. Loftus is bored and that she likes the diversion of Huck. She is anything but the arid disciplinarian Miss Watson; she is more akin to the kind Widow Douglas. But Mrs. Loftus is more. In this scene, the combination of her sagacity, kindness, and willingness to playact seem to set the stage for Huck's own lies and performances throughout the novel in the various disguises he dons to protect himself and Jim. The notion that femininity is a role is important to Huck's learning about role-playing by people in society in general, and it points toward the role assigned to blacks in the South of this novel as well. What is important, the scene teaches, is the reality within, not the appearance of a person on the outside.

The Judith Loftus episode is the most important use of the feminine in the novel, but the role of women is important throughout. Though on the surface the novel is a classic rendition of the American male hero and a compelling portrait of male friendship, and though its themes of love and community between people transcend gender, women figure in almost all major episodes. Their representation is paradoxical: on the one hand, they represent the conventionality of society, the very conventionality from which Huck and Jim are attempting to escape. One recalls that the book's climax, the letter Huck tears up giving Jim's location away, is written to Miss Watson, a mother figure. Most women in the book are reduced to titles: "Widow," "Miss," "Aunt," and "Sister," instead of full names. On the other hand, however, buried in many of these somewhat reductive characterizations is another message of the feminine, if not all individual women, as representing a principle of union, nurture, and moral stability—a very positive image indeed.

A brief look at the major women characters bears out both these assertions. The Widow Douglas, kind but ineffectual, is both the one from whom Huck runs away and the one whom he reveres. Miss Watson, the old maid, represents a balance to the Widow: she is the worst of "sivilization," hypocritical, self-righteous, repressed, even cruel. And yet she frees Jim in the end. Huck generally includes women in his fictional family and friends, who are usually said by him to be in some sort of danger. A motherless child, Huck has been the victim of his father, as frightening an image of the Father principle as any child could experience. Jim's softer traits balance pap Finn's cruel masculinity. In Huck's lies about his family, there is always a dead mother and a threatened sister or family friend (such as "Miss Hooker"—a name Twain may have gotten from his wife's close friend, Alice Hooker Day, niece of Henry Ward Beecher). Aunt Sally Phelps seems at first to be a flibbertigibbet, but when she mothers Huck he says he feels "mean," and the comedy of the last chapters depends almost entirely on her exaggerated reactions to the boys' pranks. One wonders just how much she really does know is going on. Yet when she wants to adopt and "sivilize" Huck, he heads for the Territory. Aunt Sally is another alternate mother, both in a mother's good and bad aspects. She is accompanied by her know-it-all friend, Sister Hotchkiss, who in chapter 41 provides comic relief from the suspenseful last events of the story. Expressing the outrage of the conventional community, Sister Hotchkiss blames the "niggers" for all the problems caused by Tom and Huck. A final mother figure is Aunt Polly, who appears like a dea ex machina—she is severe on the boys, but tender to a fault in her concern for Tom, who has caused her to journey several hundred miles downriver.

Younger women figure importantly as well. Emmeline Grangerford represents the shallow, anemic romanticism of the literature of the time; her "art" of poetry is heavily satirized. Hypocrisy is again an issue because all her sorrow is wasted on her insubstantial self. Yet she also points interestingly toward important social characterizations of women at the time, particularly the cult of hysteria and the cult of idleness, reminding one of Edgar Allan Poe's female characters or of the heroine of Charlotte Perkins Gilman's "The Yellow Wallpaper." Sophia Grangerford is a welcome contrast. She is a beautiful and courageous young woman, and in her romantic elopement she is the eternal bride. It is important that through her friendship with Huck she manages to escape her fear-ridden upper-class world. The Harelip Huck meets at the Wilks house is a humorous interlude between dangers: she wryly sees through his ridiculous lies—made to impress her—concerning life in England. Sober, serious-minded, and shrewd, despite her knowledge of Huck she does not attack him. Mary Jane Wilks is like Sophia, only wiser and calmer, and like the Harelip, only more attractive. Her friendship with Huck allows him to save her and her family from the evil manipulations of the King and the Duke. Huck seems to respect this more verbal and down-to-earth version of Sophia more than any other woman in the book, or any other character except Jim, for that matter.

Huckleberry Finn does seem to want to turn women into mothers, it does keep sex out of the picture, and it does use women for the most part as symbols of undesirable cultural conventionality. But in its variety of types and in its constant return to the nurturing side of the feminine, it offers an admiring and enriching view of women. In Huck and Jim's journey downriver one may even see a journey on to the Eternal Feminine.

V. THE FUTURE OF FEMINIST LITERARY STUDIES: SOME PROBLEMS AND LIMITATIONS

Feminism has caused a major reorientation of values in literary studies and elsewhere in Western culture, and it will continue to challenge long-held beliefs and practices. It is a vigorous, growing, diverse school of thought, and though it is prone to internal strife, it has thus far managed to use its internal conflicts to further its dynamic growth. But it has obviously not been popular in all quarters. It has been powerfully attacked by those who are suspicious of its social values and who fear its politicizing of artistic value. Though it is no longer acceptable to display sexist values in the academy, hostility to feminists has by no means lessened. One still hears males in business or in universities refer to female colleagues as "girls" or, worse, "gals," and everything from unfair hiring practices to sexual harrassment continues. But aside from the lingering male knee-jerk rejection of feminism, there are some major problems that will continue to raise important questions about feminist theory and practice.

First, many women who think of themselves as feminists are somehow not considered feminist enough by more radical feminists, and this often leads women in the first group to reject feminism as a field of study altogether. Older women in particular often feel that they are being judged and found wanting by their younger and more theoretically inclined sisters. These women complain that to be allowed to do as one pleases and think as one pleases, not adopting a rigid party line, should be the goal of feminism (for an example of this thinking, see Felicia Bonaparte's letter to the editor of *The New York Review of Books*, August 16, 1990, 59).

Many women critics also feel that feminist literary criticism has become too theoretical and too radical and has lost sight of both its application in reading texts and its social roots. Myra Jehlen has sharply criticized the separatism of feminist criticism and theory, which for her results in a certain ineffectiveness. She reaffirms the autonomy of the work of art and urges us to remember that art can contain good ideas as well as bad ones, but that this does not determine literary *value*. She believes that we should ask the questions of a work of art that it asks us to ask, and not others ("Archimedes and the Paradox of Feminist Criticism," *Signs*, 6.2 [1981]: 575–601). Other feminist critics such as Elaine Showalter and Annette Kolodny have criticized feminism's growing bifurcation of literary scholarship and political action.

Jehlen's call to formalism is echoed by women critics who do not think of themselves primarily as feminists, as well as by male critics of many different types. When one has proclaimed that "all art is political," one ought to wonder exactly what one has said. Is it like saying, "all human activity has to do with sex" or "all human endeavors are based on the opposable thumb"? The reductiveness of some feminist theory, like that of much of Marxist theory, is a nagging problem that has not been adequately addressed. As has often been noted, radical movements do not like compromise or being reasonable about issues, and this tendency has both strengthened feminism and cost it supporters. Also, some feel that the issue of art as a matter of aesthetics versus art as a political statement is not a question about women's rights, but about the entire nature of literary criticism. Is there something in art that transcends the particulars of a historical time and place, or of place?

Helen Vendler has been extremely willing to engage this question, and her criticisms of feminism's political biases have caused a storm of protests from feminist critics, particularly Sandra Gilbert and Susan Gubar. In two numbers of *The New York Review of Books* in 1990 (May 31, and August 16), Vendler takes on prominent feminist critics and in turn is attacked by them. The debate centers on formalism versus political meaning. Vendler seems to have little to complain of in feminist political theory in general, but in feminist literary criticism she finds much that is lacking. She criticizes early feminist critics for looking at women in male novels in a naive fashion, treating the characters as real people and predictably not finding them portrayed correctly—as sympathetically as they would like. She also finds fruitless the attempt of later feminist critics to discover a distinctively female way of writing or a women's language. She finds that "feminism's unacknowledged problem, visible from its inception, has been its ascription of special virtue to women. In its most sentimental form, feminism assumes that men, as a class, are base and women are moral; in its angry version, that men are oppressors and women the oppressed." But she argues that it is the possession of power that determines who is oppressed, and she gives examples of women as oppressors: "there are choleric, sadistic, indifferent, and cold women just as there are such men." If feminism is to succeed, she says, it must de-idealize women "to the extent that truth is preferable to cant" (19–22).

A persistent problem for feminism is the question of what to do with male feminist critics. Unlike the other approaches described in this handbook, males who perform feminist criticism can have a hard time of it. Many feminists believe that no man can possibly read or write or teach as a feminist; some even feel that men should be barred from teaching as feminists. Maggie Humm argues that no man can read as a feminist because at any time he can escape into patriarchy; the extent of "difference," she feels, is "infinite" (13–14). Toril Moi first distinguishes between men and women feminists as like white and black antiracists—men can be feminists but will always speak from a different position from women, and their political strategies must take this into account. But she goes on to aver that "in practice, therefore, the would-be male feminist critic ought to ask himself whether he as a male is really doing feminism a service in our present situation by muscling in on the one cultural and intellectual space women have created for themselves within 'his' male-dominated discipline" (208).

But many well-known male critics have read and taught as feminist critics, including Houston A. Baker, Paul Lauter, Wayne C. Booth, Jonathan Culler, Terry Eagleton, Lawrence Lipking, and Robert Scholes, and there will undoubtedly be more male students and critics who find the issues raised by feminism to be the issues they want to study. Feminism continues to flourish in its many forms, and it will continue to offer society and literary studies a fruitful and exciting set of intellectual problems. Most feminists do not want to abolish maleness or male values; instead, they wish to do away with such gender-typed categories altogether.

· 6 ·

Structuralism and Poststructuralism

I. STRUCTURALISM

A. Context and Definition

Structuralism and its derivative theories, especially deconstruction, have proposed to alter drastically the direction of literary studies during the last twenty-five or thirty years. By no means confined to the study of literature, structuralism has been applied to linguistics, psychology, sociology, anthropology, folklore, mythology, and Biblical studies—in fact, to all social and cultural phenomena. Its attractions are considerable: structuralism is, at least seemingly, scientific and objective. It identifies *structures*, systems of relationships, which endow signs (e.g., words) or items (e.g., clothes, cars, table manners, rituals) with identities and meanings, and shows us *the ways in which we think*.

Structuralism claims intellectual linkage to the prestigious line of French rationalism stretching from Voltaire to Jean-Paul Sartre. Its representatives in Britain and the United States tend to retain French terminology and, as some say, to sound French. Structuralists emphasize that novelty is only apparent and that any essential description of any phenomenon or artifact without placement in the broader systems which generate it is misleading if not impossible. Accordingly, they have developed analytical, systematic ap-

proaches to literary texts that free us from piecemeal traditional categories like plot, character, setting, theme, tone,
and the like. Even more significantly, however, structuralists
tend to deny the text any inherent privilege, meaning, or
authority; to them the text is simply a system that poses the
simple question of *how* such a construct of language can
contain meaning for us.

Such a view denies any claim of privilege for any author,
any school, any period, and any "correct" explication (which
is familiar in the reductive acceptance that the teacher
is somehow always "right" and the student nearly always
"wrong"). The structuralists have encouraged us to reread,
rethink, and restudy all literary works and to equate them
with all other cultural and social phenomena—for example,
language, landscaping, architecture, kinship, marriage customs, fashion, menus, furniture, politics, and so on.

B. The Linguistic Model

Structuralism emerged from the structural linguistics developed, mainly in his lectures at the University of Geneva between 1906 and 1911, by Ferdinand de Saussure. Not available in English until 1959, Saussure's *Course in General
Linguistics* in French (1916) attracted thinkers far beyond
Switzerland, linguistics, and universities: it became the
model for Russian formalism, semiology or semiotics, French
structuralism, and deconstruction. Saussure's model is acceptable as an analogy for the study of many systems beyond
language.

Saussure's theory of language systems distinguishes between *la langue* (language, the system possessed and used by
all members of a particular language community—English,
French, or Urdu, etc.) and *la parole* (word; by extension,
speech-event or any specific application of *la langue* in
speech or writing). The *parole* is impossible without the support, the structural validity, generation, meaning, conferred
upon it by the *langue*, the source of grammar, phonetics,
morphology, syntax, and semantics. As Saussure explained,
paroles appear as phonetic and semantic *signs* (*phonemes*
and *semes*). A linguistic sign joins a *signifier* (a conventional

sound construction) to a *signification* (semantic value, meaning). Such a sign does not join a thing and its name, but an allowable concept to a "sound image" (Philip Pettit, *The Concept of Structuralism: A Critical Analysis* [Berkeley: U of California P, 1975]: 6). The sign thus has meaning *only within its system*—a *langue* or some other context. To a speaker of English the words "Help!" or "Murder!" or "Fire!" communicate crises, acute social disorders, particularly if spoken with feeling and urgency; they might be meaningless to a speaker of Arabic who does not know English. Similarly a red light in a traffic signal advises us emphatically to stop our vehicles; red in the context of Catholic liturgy has quite another meaning—martyrdom. A tuxedo and black tie communicate formality within the dress code of Europe and North America but would simply puzzle the inhabitants of the jungle along the Amazon. An item is meaningful only within its originating system.

Saussure stressed the importance of considering each item in relationship to all other items within the system. The syntax of English allows a meaning to "Tom takes a bath" but not to "a takes bath Tom." A purple light in a traffic signal would create perplexity because the traffic signal system we know does not accredit the color purple. Yellow and green are validated by the system as contrasts to red and are thus given the respective meanings of "caution" and "go." In a sentence we are accustomed to recognizing subjects in distinction to predicates.

The approach to analyzing sentences is *syntagmatic*— word by word in the horizontal sequence of the parts or *syntagms* of the sentence. Saussure's "structural" linguistics furnishes a functional explanation of language in process and according to its structural hierarchy—that is, structures within structures. He suggested that his system for studying language had profound implications for the study of other disciplines. In the study of a literary work, Saussure's syntagmatic approach explains our usual, instinctive approach: we read the poem from its start to its finish, we see the narrative work in terms of the sequence of events or the scenes of the play, we inventory the details from the first to the last, from their start to their finish. This approach emphasizes the *sur-*

face structures of the work, as it does for the sentence in Saussure's scheme, as opposed to the *deep structures*, those not on the surface—the latent, understood but unexpressed signs. Saussurean linguistics applies, moreover, to *synchronic* features (i.e., language as it exists at a particular time) rather than to *diachronic* features (details of language considered in their historical process of development).

C. Russian Formalism: Estrangement and Defamiliarization in "To His Coy Mistress" and Other Works

A group of scholars in Moscow during World War I perceived the dynamic possibilities of using Saussure's work as a model for their investigations of phenomena other than language. Vladimir Propp studied Russian folktales as structural units that together contained a limited number of characterizations and actions (Propp called them "functions"). The functions recur and thus constitute in their unity the grammar or rules for such tales. To recall the Saussurean model, we can say that the entire group of functions is the *langue*; the individual tale is a *parole*. A number of these *actants* (characters) and functions were introduced in our chapter on mythological approaches; for example, Propp's theory identifies hero, rival or opponent, villain, helper, king, princess, and so on, and such actions as the arrival and the departure of the hero, the unmasking of the villain, sets of adventures, and the return and reward of the hero. The possibilities for applications of such a scheme to literary works are apparent. We see clearly identified in *Hamlet* a hero and a villain—Hamlet and Claudius; in *Huckleberry Finn*, the functions of a hero's emergence and departure and of a corrupt society's opposition (a corporate villain!) to a hero's quest for freedom; in "Young Goodman Brown," a false hero. Other features of Russian Formalism will be discussed in chapter 7, in the section on "Dialogics."

Victor Shklovsky pointed out literature's constant tendency toward *estrangement* and *defamiliarization*, away from habitual responses to ordinary experience and/or ordinary language. In poetry, we see a particular drive toward the strange and away from the familiar in its lineation of words, its rhythmic patternings, its choice of language (sometimes called "poetic diction"). Its texture is typically packed with meanings and suggestions; it tends to be more or less arcane or even ritualistic, and it calls attention to itself as different. Why should Marvell go to such lengths to dramatize—in effect, make a dramatic monologue for—"To His Coy Mistress," a version of a poetic mode or convention (*carpe diem*) already old and quite familiar to all educated readers by his time? Why the comparison of the exotic (the Ganges River, a metonym for India, Asia, the Far East) and the familiar (the Humber River in England)? Why the exaggerations, the outrageously hyperbolic (a hundred years to praise a mistress's eyes; thirty thousand to praise the rest of her)? Does Marvell *defamiliarize* his dramatic situation, his language, the metrical "confinement" of language, in order just to praise seduction or to structure a context much more serious though patently witty? Is connivance in the joys of sexuality a structured way to signify an attitude toward language, perhaps poetry itself? The imagery of time's pressures contains (or does it mirror?) a concern for expressing oneself truthfully, because the language of poetry, valuable and pleasurable, like sexual intercourse, is always threatened by time's "slow-chapped power." The surface subject matter, of course, echoes other poems and even real life. But its imagery of the remote and fantastic, its touch of the graveyard—lust in the dust—takes the discourse of the poem beyond ordinary discourse. It is the *defamiliarization*, the *estrangement*, of his material that makes Marvell's poem interesting and lures the reader into the complex intellectual and emotional experience that has been the trademark of English Metaphysical poetry, at least since T. S. Eliot called our attention to it some seventy years ago. Without the defamiliarization, the poem would risk banality and, worst of all, imitativeness.

Shklovsky also emphasized that narrative has two aspects: *story*, the events or functions in normal chronological sequence, and *plot*, the artful, subversive rearrangement and thus defamiliarization of the parts of that sequence. Story is the elementary narrative that seeks relatively easy recognition as in most nursery tales, whereas plot estranges, prolongs, complicates perception as in, say, one of Henry James's fictions. Thus we may observe that in "Young Goodman Brown," a seemingly straightforward sequence of events —what happened to young Brown between sunset and the next morning—reveals to a perceptive reader many touches of defamiliarization, such as the staginess and theatricality of its four main scenes, not to mention the formality of the dialogue even when we make allowances for Puritan stiffness; the blurring of the natural and the supernatural; or the ominous tone of the work's final sentence that invites prolonged contemplation.

In general, the Russian formalists adapted Saussure's *syntagmatic*, linear approach—examining structures in the sequence of their appearance—but showed how to use Saussure's theory in disciplines far beyond linguistics. Propp and Shklovsky demonstrated that literature can be made the equivalent of *langue* and the individual literary work or *pericope* (part of a work) the equivalent of *parole*. The work of the Russian formalists, still impressive and frequently cited, reminds us to some extent of American New Criticism in its concern for linking form to its constituent devices or conventions.

D. Semiotics or Semiology

Structuralism as it attracted interest in the United States came only after the publication of Claude Lévi-Strauss's *Structural Anthropology* in the 1950s, even though an American edition did not appear until 1963. In contrast to Saussure and the Russian formalists, Levi-Strauss concentrated on the *paradigmatic* approach—that is, on the *deep* or *imbedded structures* of discourse that seem to evade a conscious arrangement by the artisan but are somehow embedded vertically, latently, within texts and can be represented only as abstractions or "algebraic" formulas or paired opposites (*binary oppositions*). Lévi-Strauss, an anthropologist who studied myths of aboriginal peoples in central Brazil, combined psychology and sociology in cross-cultural studies and found structures comparable to those discovered by Saussure in language—that is, systems reducible to structural features. He traced structural linkages of riddles, the Oedipus myth, American Indian myths, the Grail cycle, and anything else that might be found to structure codes of kinship (including codes of chastity and incest), which, he believed, reach out to embrace the most profound mysteries of human experience and which may very well remind us of the simultaneous layers of literary and mythic images in Eliot's *The Waste Land*, Joyce's *Ulysses* or *Finnegans Wake*, and Conrad's *Heart of Darkness*. Incest and the solution of a riddle have *no* relationship in their external facts (as in a *syntagmatic* approach) but reveal deep internal relationship of reason—that is, in a *paradigmatic* scheme (Lévi-Strauss, *Structural Anthropology*, vol. 2, trans. Monique Layton [New York: Basic, 1976]: 23). The theory and method of Lévi-Strauss's approach assume the investigator's command of multidisciplinary knowledge necessary to construct adequate models. For example, he warns literary critic-scholars specifically not to attempt structural or "semiotic" studies from a solely literary fund of knowledge: "The fundamental vice of literary criticism with structural pretensions stems from its too often being limited to a play of mirrors, in which it is impossible to distinguish the object from its subjective image in the subject's consciousness." Instead of identifying cross-fertilizing systems, the critic is apt to get merely a "reciprocal relativism"—a "structural analysis with raw material rather than a finished contribution" (Lévi-Strauss 275). In short, as in almost any critical approach to a literary work or pericope, we sometimes find ourselves unable to move anywhere beyond our literary facts because we lack extraliterary (and sometimes, alas, even intraliterary) insights. Some so-called structuralist criticism strikes us as simply a summary of the plot or an inventory of the surface features rather than genuine analysis, explanation of *how* meaning appears from discourse.

The myth studies of Lévi-Strauss suggest the kind of links we infer between, say, *Oedipus Rex* and *Hamlet* (the emphases implied on kinship, marriage, incest, gross sexuality; the sacrificial scapegoat: the health of society vis-à-vis the throne; the uses of reason), or between *King Lear* and *Moby-Dick* (structures of sight, the tyranny of pride, or "reason in madness"), or between the *Divine Comedy* and *Leaves of Grass* (the poet's quest, the scope of vision, and the sequence of confrontations). Lévi-Strauss recommended the semiotic approach (semiotics or semiology being the study of *signs*) because it links *messages* (in individual works or discourses) "of a most disheartening complexity . . . which previously appeared to defeat all attempts to decipher them" to their respective *codes*, the larger system which permits individual expression—from *langue* to *parole*, that is (Lévi-Strauss, *The Raw and the Cooked* [London: Jonathan Cape, 1970]: 147). Like studies by the Russian formalists, Lévi-Strauss's semiotics is important for us because it prepares for mainstream structuralism so significantly that most explanations of structuralism identify Lévi-Strauss as its major founding father.

Lévi-Strauss and his disciples determined that the adaptation of Saussure's linguistic model to problems of human science was sound because Saussure had followed a rigorous, objective scientific method, which identifies and defines constituent parts, studies relationships within a system, and accepts mathematical analysis. Language and culture are alike because they are composed of "oppositions, correlations, and logical relations" (Claire Jackson, "Translator's Preface," in Lévi-Strauss, *Structural Anthropology*, Vol. 1 [New York: Basic, 1963]: xii). To Lévi-Strauss, the structures of myth point to the structures of the human mind common to all people—that is, to the way all human beings *think* (cf. our discussion in chapter 4 of the universality of myth). Myth thus becomes a language—a universal narrative mode that transcends cultural or temporal barriers and speaks to all people, in the process tapping deep reservoirs of feeling and experience and often invested with divine origins. To Lévi-Strauss, even though we have no knowledge of any *entire* mythology, such myths as we do uncover reveal the existence within any culture of a system of abstractions by which that culture structures its life. In his study of the Oedipus myth, Lévi-Strauss found a set of *mythemes*—units of myth analogous to linguistic terms like morphemes, phonemes, or tagmemes, and like those linguistic counterparts based in *binary oppositions*—whose structural patterns invest the myth with meaning. For example, Oedipus kills his father (a sign of the undervaluation of kinship) and marries his mother, Jocasta (an overvaluation of kinship). In either case, Oedipus has choices, although a pitying reader may not think so: what he does plus what he does not do are significant binary oppositions within the myth (as they are in Sophocles's tragedy). Although Lévi-Strauss was not interested in the *literariness* of myths, some of his contemporaries saw in his work promising implications for purely literary studies, particularly studies of narrative.

E. *Nouvelle Critique* (French Structuralism): Decoding "Young Goodman Brown" and Other Works

Admittedly a response to Lévi-Strauss, the school of Paris, as it is often called, produced a French new criticism in the 1950s and 1960s in the work of Roland Barthes, Jacques Derrida, Michel Foucault, and Tzvetan Todorov. Mainly these writers have been interested in relatively sophisticated narrative—the fiction of Bernanos, Proust, Racine, Balzac—and some popular modes like mystery novels and humor, rather than in folk or naive art. Yet they too accept the Saussurean linguistic model and thus an essentially syntagmatic approach to texts. They have viewed narrative as a kind of analogy to the sentence: the text, like the entire sentence, expresses the writer's mind and is a whole composed of distinguishable parts. Instead of the Russian formalists' distinction between *story* and *plot*, the French structuralists use the terms *histoire* (essentially the sequence of events from their beginning to the end) and *discours* (discourse; the narrative rearranged and reconstructed for its own purposes and aesthetic effects—one thinks of the artful, intricate rearrangements of time and events in Faulkner's "Rose for Emily," which conceals or withholds an essential fact until the very last sentence in the work). Such manipulation of the raw data of the *histoire* is another instance of the *defamiliarization* we associate with and expect in literary art; other kinds of estrangement are flashbacks, shifts of viewpoint or speaker, the absence of viewpoint (as in the French *nouveau roman* of Alain Robbe-Grillet and others; see Jonathan Culler, *Struc-*

turalist Poetics: Structuralism, Linguistics, and the Study of Literature [Ithaca, NY: Cornell UP, 1975]: 190–92), unequal treatment of time, alternation of dramatic and expository passages, and so on. The text is a *message* (the particular, concrete, individualized telling of a story) which can be understood only by references to the *code* (the internalized formal structure consisting of certain semantic possibilities, which explain and validate the content of a *message*). The reader gets the message (*parole*) only by knowing the code (*langue*) that lies behind it. Structuralist reading is essentially the quest for the code, as life itself is the quest for signs (cf. *Moby-Dick, The Waste Land, My Ántonia, Roderick Random, Lord Jim, Leaves of Grass, The Bridge*), in each of which the reader, like the protagonist, is challenged to read a message about the criteria for, meaning of, and urgent summons to a quest equivalent to that for a holy grail, a portentous sign for a system read as regenerated life.

Todorov has assured us that structuralism cannot interpret any literary work; it can only show us how to identify a work's characteristic features and perhaps how to perceive their likenesses to or differences from structures in other works. Barthes, usually considered the preeminent structuralist (and later, a major French deconstructionist), and Todorov declared their indifference to authors, who after all cannot claim any originality since authorship is merely the rearrangement of structures already present in the code. Any literary criticism inevitably will be totally subjective; even if a critic claims to be a Freudian or a Marxist, the artifact is irreducible to any such semiological, psychological, or political systems.

From the linguistic model comes the structural concept of *binary opposition*; indeed, sets of differences underlie all human acts and practices—for example, the writing of poems or fictions. Choices constantly must be made between opposites: young Goodman Brown must stay at home with Faith or enter the darkening forest; the ghost's charge to Hamlet is either noble or diabolical. In Racine's tragedies, Barthes identified binary oppositions leading either to exclusions or to inclusions of the characters and composing the center of the plays' systems (Culler, *Structuralist Poetics* 99–100).

Similar structures can be found in *Hamlet:* Claudius schemes, along with almost everyone else, to seal Hamlet's exclusion from authority, truth, Denmark, and finally life itself; Hamlet schemes, almost totally alone, to achieve inclusion of his rights to the throne, justice, the order of nature, a cleansed Denmark, and, finally, rest for the ghost.

Sometimes the opposition with functions or discourse can be organized as distinctions between structural figurative principles. *Metonymy* identifies parts by their membership in a whole or context; for example, Huck Finn's catalogue of bad art in the Grangerford household metonymically reveals the quality of the family's values, the Southern code, and the society that engenders and defends those values by a despicable blend of "sentimentering" with violence, exploitation, and denigration of life.

Metaphor, on the other hand, reveals similarities and thus union—as, for example, in surrealistic painting, whose images reconstitute the psychological or dream state; in Ezra Pound's "In a Station of the Metro," the vision of the crowd and the state of mind coalesces in the image of "Petals on a wet, black bough." Allusions in Eliot's *Waste Land* create a metaphoric structure in which bits and pieces of history, myth, and literature create a dismal contemporary world.

The parties in Fitzgerald's *Great Gatsby* potentially reflect both figurative principles: they are, at once, telling parts of the Jazz Age, and wholes that reflect the tempo and means of quest for the age. Roman Jakobson suggests, however, that each period of literary history tends to favor one or the other principle: the age of realism favors the metonymic system; romanticism and modernism, the metaphoric. Henry Fleming in Crane's *Red Badge of Courage* is constantly figured forth as a "piece" of war, army, confused humanity, and nature. Tom Outland in Willa Cather's *Professor's House* embodies and incarnates an ideal, a generosity formed by the love of beauty and order—an idealism that the professor-protagonist seeks to defend and preserve.

Barthes also clarifies the roles of *code* and *message* by showing their presence in our reading of menus, selection of clothes, arrangement of furniture, response to architecture, and so on. Such operations dominate our lives, and all the contexts we ever know come to us as codes of language, whether in the episodes of the *Odyssey* or in the texts and pictures of *Vogue* or *Better Homes and Gardens*. Because the code allows it, messages can come to us syntagmatically or paradigmatically—syntagmatically, if we read the menu to put together a suitable, conventional dinner; paradigmatically, if we read the menu only for vegetables or for desserts. We can read *Heart of Darkness* for the recurring images of the inscrutable jungle or the inscrutable human heart; or we can read it for the sequences by which Marlow acquires vision.

It is, or should be, the object of a structuralist poetics "to specify the codes and conventions [i.e., the codes of art] which make . . . meanings possible" (Jonathan Culler, "Foreword," in Tzvetan Todorov, *The Poetics of Prose*, trans. Richard Howard [Ithaca, NY: Cornell UP, 1977]: 8). We can learn those codes and conventions, of course, only by experience.

The author *encodes* a work; the reader must try to *decode* it. For example, Todorov points out, Henry James's fictions typically encode an essential secret in the narrative machinery, so subtly that it can entrap the unwary or inexperienced reader. Other notable specialists in the encoding of essential secrets are Coleridge, Eliot, Pound, Conrad, Hawthorne, Melville, Faulkner, and Joyce. Hawthorne never tells us exactly what young Brown seeks in the forest, why Faith tries so eloquently to keep her impetuous husband at home, how the assignation with the soberly dressed and gravely speaking gentleman has been arranged, whether or not the staff really wriggles, how the gentleman resembles the males of the Brown family, the level of reality to which the experience is assigned, and so on. *Hamlet* too poses secrets: Is Hamlet, at least sometimes, really mad? Is the ghost in part malevolent? Is Gertrude an accomplice in the murder of her first husband and a conniver in her "o'erhasty" marriage to Claudius? Is nastiness integral to the human condition? Such messages seem to expect of the reader (and the theater audience of Shakespeare's time) considerable expertise in decoding them. The complexities in decoding the messages of *Hamlet* arise from the characters' baring their inner selves only in soliloquy or aside; how then may they hope to judge each other adequately? The body politic is somehow affected by the health or sickness of the throne; however, the Danish folk are scarcely present. Tragedy's decorum dictates that nobility or royalty speak in relatively lofty verse and that comic characters (e.g., the gravediggers in act V, scene i) use a relatively earthy prose—who is the more honest? Such problems help us to see the structures by which Shakespeare, his contemporaries, his audience, and we as readers think.

In his often cited analysis of a story by Balzac, Barthes, in *S/Z* (Paris: Seuil, 1970), classifies literary codes. The *proairetic* code (code of actions) asks the reader to find meaning in the sequence of events. The *hermeneutic* code (code of puzzles) raises the questions to be answered. The *cultural* code refers to all the systems of "knowledge and values invoked by a text." The *connotative* code expresses themes developed around the characters. The *symbolic* code refers to the theme as we have generally considered it—the meaning of the work. (See Robert Scholes, *Structuralism in Literature: An Introduction* [New Haven: Yale UP, 1975]: 153–55, for a full explanation of Barthes' scheme of fiction's five codes. Scholes suggests that in practice we need not identify all five codes and that we may choose to blend two or more in an analysis.) We may note that the *action* code appears in the straightforward sequence of events taking young Goodman Brown from village into the forest and darkness back to the village and daylight; but just the mention of darkness and light or the poles of village and forest suggests links between action on the one hand, and hermeneutic, cultural, connotative, and symbolic considerations on the other, so that all five codes can be seen as coalescing as one structure. The linkage of all the codes is often particularly striking in Romantic or modernist fiction—for example, "The Fall of the House of Usher," in which house, family, protagonist, and ordeal insist upon unity out of complexity; Virginia Woolf's *To the Lighthouse*, in which a seemingly trivial excursion becomes a nexus of past, present, and future and a code of action that generates a rich symbolic structure uniting life and art; or Cather's *My Ántonia*, in which the narrator reviews the rich past to see what hermeneutic and cultural codes may be salvaged in an unpromising present. Realistic fiction, on the other hand, especially fiction like O. Henry's "The Gift of the Magi," Maupassant's "The Necklace," or Howells's "Editha," may reveal only thin structures beyond the code of action. The awareness of the tendency of the codes to coalesce, to appear and disappear, may remind us of formalism's concentration on the theory of organic form.

In S/Z, Barthes also defined the *lexie* as the basic unit of a narrative text—"the minimal unit of reading, a stretch of text which is isolated as having a specific effect or function different from that of the neighboring stretches of text" (Culler, *Structuralist Poetics* 202). In size, the *lexie* can be anything from a single word to a nexus of several sentences which will fit into and support one of the five narrative codes. An early *lexie* in *Hamlet* expresses the sentinels' malaise and foreboding. We recognize the roles of the *lexies* by our experience of the world and other literary works. For instance, young Goodman Brown's departure from wife and village clearly is a recognizable stretch of text different from that which follows with his entry into the forest; it is a *lexie*, or function, we can assign to the code of action, but it also belongs to a paradigm formed from worldly experience and other literary texts about men forsaking safe homes and loving wives for dark ends, which in old melodramas used to be drink, gambling, and loose women. Likewise, the satanic congregation in the forest not only mirrors the Puritan church in the village but also fits into popular superstitions about devil worship, black masses, and the like. Each character's introduction is a new *lexie*—a speech-event, a syntagm, taking Brown closer to the climactic revelation to him of evil's totality and the paradigm of lost innocence. It is likewise a *lexie* that Brown is late for his appointment with the gentleman, but its importance structurally may not be entirely clear until we perceive that Brown's irresolution and delaying tactics in his movement toward the depths of the forest form a pattern including his feeble guilt feelings about abandoning faith (as well as Faith) and innocence. Whereas the lexies of action follow a straightforward, horizontal, syntagmatic line, those that function to establish the *cultural code* (Puritanism, Protestantism, American) or *hermeneutic* code (theme) are apt to be vertical and *paradigmatic*, as would be intertextual linkages (Faustian, Biblical, Romantic) that relate "Young Good-man Brown" to, say, Melville's "Benito Cereno" and *Moby-Dick* or Conrad's "Heart of Darkness."

F. Anglo-American Interpreters

Jonathan Culler is usually credited with the greatest success in mediating European structuralism to students of critical theory in Britain and the United States, mainly through his *Structuralist Poetics;* however, Robert Scholes's *Structuralism in Literature* may have done more to simplify and clarify the issues and the practical possibilities of structuralism for nonprofessional students of literature. Although both Culler and Scholes pass along the pervasive structuralist caveat that favors theory of literature in general over analysis of particular texts, in fact they repeatedly express regret that texts are neglected in structuralist studies. Probably the atmosphere that favors interpretation or explication established by the New Criticism also steered Anglo-American scholar-critics and teachers of literature toward textual studies—or what the New Critics had called practical criticism.

However, Culler, by specifying a structuralist poetics based on the model of Saussurean linguistic theory, invites intelligent, unprejudiced, genial readers to contribute to the expansion of that poetics, which he defines simply as the "procedures of reading" that ought to be found in any discourse *about* literature. Literature, Culler insists, can have no existence beyond a display of literary conventions to which able readers respond in identifying in a text the sign-system that they already know and that is analogous to the way we read sentences by recognizing internalized phonetic, semantic, and grammatical structures in them. Through experience, ·readers acquire degrees of literary competence (just as children gradually acquire degrees of syntactical and grammatical complexity) to permit degrees of textual penetration. Culler stresses that it is the reader's business to *find* contexts that make a text intelligible and to reduce the "strangeness" or defamiliarization achieved by the text. Learning literary conventions (the equal of Saussure's *langue*) and resisting any inclination to grant the text autonomy (to privilege the text) dispose the structuralist reader to search out and identify structures within the system of the text and, if possible, expand poetics rather than to explicate the organic form of a privileged text.

II. STRUCTURALIST APPROACHES TO *HUCKLEBERRY FINN*

With narrative, as we have seen, any structuralist reading may start or even end with a syntagmatic notation of each event within a sequence of actions. Such a notation would give us the order of events or *lexies* by which Huck's longing for freedom generates his journey down the Mississippi from St. Petersburg to the Phelps farm (and beyond?). Each risky visit to shore accentuates Huck's need for freedom and society's parallel determination to restrain individualism—that is, the personal freedom promised to Americans. The journey itself as structure—a means of getting from somewhere to somewhere else, from one condition to another, from slavery to slavery, for Huck and Jim—strikes us as familiar from our previous literary experience, which after all began with nursery tales about Red Riding Hood, Alice in Wonderland, Pinocchio, and other travelers. "Young Goodman Brown," as we have noticed in every critical context, establishes a structure of journey, which we may note is both a *metaphor* for a chain of events, or functions, and the operative term for the *code of action*. As terms like "freedom" and "restraint" enter the context of a reading, we add to the syntagmatic tracing of action hermeneutic, cultural, connotative, and symbolic codes, as we associate Huck with one set of terms and restrictive society with their opposites. At every point the action code of fictional events suggests cultural and thematic associations, all of which are involved in the total signification of the text. The characters, most of all Huck himself, are identified largely by the action; all American culture is probed and questioned by codes. Codes connect with other codes, all of which open up to readers who bring competence and their knowledge of other American social and cultural codes.

III. *HAMLET:* A PARABLE OF AMBIGUITY

Hamlet, it seems, has always eluded attempts to capture and articulate its meaning, despite continuing and intensive textual scrutiny. Moralistic critics have delivered grave pronouncements about the dangers of delaying action, of excessive deliberation, or of ardent attachments of sons to mothers; historicists have linked the play's concern with monarchy and the state to Elizabethan political dialectic; formalists have traced patterns of imagery; Jungians have excavated archetypal parallels; and feminists have identified the problems associated with engagement with the Feminine. Yet the essential mysteries of *Hamlet* remain to tease us into and out of thought. Is this play, unanimously considered great, another instance of a text at war with itself, a Penelope's tapestry that seems to unravel as fast as it is hermeneutically stitched together? Do arguments for the play's organic form, for the unity of its structure, simply wrench the text into a unity that only partial if not partisan scrutiny supports?

The play begins in the vague but troubling uncertainty of predawn with anxious sentinels questioning Norway's intentions, the state of Denmark, and the provenance of a ghost; it ends with simultaneous deference to old and new kings and questions about the meaning and endurance of human deeds and loyalties. In between, the play generates endless riddles about language and its consequences, the reading of signs (like the guilt of a murderer and usurper), heavenly versus earthly judgments, the motives or meaning of love, the idealism of the spirit that accommodates the bestiality of the flesh, the difficulty of assigning meanings in a life riddled with ambiguity and paradox. Hamlet cannot trust anyone except the rare nobleman of nature, Horatio, while the hallowed, sacramental ties of marriage, the sacred link between parent and child, the vows of lovers, the bonds of friendship prove to be maddeningly unstable and unpredictable. Are they perhaps only blinders that permit treachery and violation to flourish unchecked? Is not the world's touted morality merely a few tiresome bromides like Polonius's parting instructions to Laertes? Is not idealism dangerous to a hero's health?

The state's peace is disturbed by the "presence of feared events" (I.i.121): "A mote it is to trouble the mind's eye" (I.i.112). In the gravedigger's scene, riddles that beset the human ordeal are reducible to the unspeaking but eternally teasing skull of the dead Yorick. Is Hamlet mad, or is he the only sane person among the dramatis personae? The text of the play supports either proposition: Hamlet himself admits that his being "mad in craft" may be interpreted as "madness" (III.iv.88–89). Is there the implication that any effort to cleanse "rotten Denmark" inevitably leads to madness? Or that the idealist will always seem mad in a world geared to duplicity and ignoble sexuality and lust for power?

The play within the play tries to mirror life, and, to the extent that it does, Hamlet's madness "in craft" succeeds beautifully. But questions still abound: Is life merely a series of plotted gestures and regal splendors? Is all human conduct carefully directed and staged? How can we tell the difference between artifice and "reality"? Is prayer essentially a yoking of rhetoric and stage directions—that is, rituals? The play's text seems to be at war with itself even at the end, where we expect complete resolution for plot and theme, for do not Hamlet's cunning inventions and dogged travail lead only to his own undoing and back to the riddle of the grinning skull? Melodrama compromises tragic structure; grim buffoonery checkmates heroic grandeur. Questions abound and answers dissolve: for, as has been said of the equally enigmatic *King Lear*, "If we trace the exposed loose strands that lead into the rupture [say, between the vision of goodness and the omnipresence of vice], we follow them back and forth forever" (Jasper Neel, "Plot, Character, or Theme? *Lear* and the Teacher," in Atkins and Johnson, *Writing and Reading Differently: Deconstruction and the Teaching of Composition and Literature* [Lawrence, KS: UP of Kansas, 1985]: 186).

IV. POSTSTRUCTURALISM: DECONSTRUCTION

Deconstruction is the most significant of all poststructuralist developments in literary critical thought. Indeed, poststructuralism is virtually synonymous with deconstruction. Deconstruction arises out of the structuralism of Roland Barthes as a reaction against the certainties of structuralism. Like structuralism, deconstruction identifies textual features, but concentrates on the *rhetorical* rather than following structuralism's emphasis upon the *grammatical*. As Paul de Man points out: "The grammatical model of the question becomes rhetorical not when we have, on the one hand, a literal meaning and on the other hand a figural meaning, but when it is impossible to decide by grammatical or other linguistic devices which of the two meanings (that can be entirely incompatible) prevails" (*Allegories of Reading: Figural Language in Rousseau, Nietzsche, Rilke, and Proust* [New Haven: Yale UP, 1979]: 10).

Deconstruction accepts the analogy of text to syntax as discovered by Ferdinand de Saussure and adapted by the structuralists. But whereas structuralism finds order and meaning in the text as in the sentence, deconstruction finds disorder and a constant tendency of the language to refute its apparent sense. Hence the approach's name: texts are found to deconstruct themselves rather than to provide a stable, identifiable meaning.

How does deconstruction envision and celebrate a work's self-destruction, or self-emptying? It views texts as subversively undermining an apparent or surface meaning, and it denies any final explication or statement of meaning as it questions the presence of any objective structure or content in a text. Instead of alarm or dismay at their discoveries, the practitioners of deconstruction celebrate the text's self-destruction, that inevitable seed of its own internal contradiction, as a never-ending free play of language. Instead of discovering one ultimate meaning for the text, as formalism seems to promise, deconstruction describes the text as always in a dynamic state of change, furnishing only provisional meanings. All texts are thus open-ended constructs, and sign and signification are only arbitrary relationships. Meaning can only point to infinite other meanings.

Methodology in deconstruction thus involves taking apart any philosophical meaning to reveal contradictory structures hidden within. Neither meaning nor the text that seeks to express it has any privilege over the other, and this extends to critical statements about the text, making literary theory itself a form of literature.

The break with structuralism is profound. Since structuralism claims kinship between systems of meaning in a text and structuralist theory itself—both would reveal the way human intelligence works—when deconstruction denies connections of mind, textual meaning, and methodological approach it represents nihilism and anarchy to structuralists. It has appeared this way to many other critics as well. This reaction against deconstruction not only has come about because of its denial of meaning to any work of literature or any literary approach, but also has been heightened, undoubtedly, by the method's set of descriptive terms themselves, including the "self-destruction" of texts and the "overturning" or "attacking" or "undermining" of texts, as well as the other generally militaristic and negative ways in which it describes the reader's relationship to the text. One may contrast it, for example, with the emphasis upon community and relationship in an approach such as dialogics (see chapter 7).

Other important terms are less negative in tone, but also point to the attack on any form of perceived authority. Deconstruction opposes logocentrism, the notion that written language contains a self-evident meaning that points to an unchanging meaning authenticated by the whole of Western tradition. It would demythologize literature and thus remove the privilege it has enjoyed in academe. In deconstruction, knowledge is viewed as embedded in texts, not authenticated within some intellectual discipline. Since meaning in language shifts and remains indeterminate, deconstructionists argue that so do all forms of institutional authority. Since there is no possibility of absolute truth, deconstructionists seek to undermine all pretentions to authority, or power systems, in language. Some adherents of deconstruction, notably Marxist and feminist deconstructionists, see it as more than a literary theory and have extended its attack on meaning to other institutions in society beyond literature.

The most important figure in deconstruction has been the French philosopher Jacques Derrida, whose philosophical skepticism became widely adopted when his work was translated in the early 1970s. Derrida offered a decisive critique of structuralist and phenomenological thought and oriented himself toward certain new psychoanalytic and literary topics. His most significant work, including *Speech and Phenomena: And Other Essays on Husserl's Theory of Signs* (trans. David B. Allison [Evanston: Northwestern UP, 1973]), *Of Grammatology* (trans. Gayatri Chakravorty Spivak [Baltimore: Johns Hopkins UP, 1974, 1976]), and *Writing and Difference* (trans. Alan Bass [Chicago: U of Chicago P, 1978]), have been important sources to other deconstructionists, including Harold Bloom, Eugenio Donato, Geoffrey H. Hartman, J. Hillis Miller, Joseph Riddel, Barbara Johnson, Shoshana Felman, Jeffrey Mehlman, and Gayatri Chakravorty Spivak. Because of the academic location of many of these critics, deconstruction came to be known as the Yale school of criticism.

Derrida claims that the dominant Western tradition of thought, in its drive to establish meaning, actually represses meaning by repressing the limitless vitality of language. He rejects the "transcendental signified" that defines a center at the expense of the margin by demonstrating that the marginal term has already contaminated the center or privileged term. Thus there can be no stable self-identity: the Western "metaphysics of presence" is faulty. His most influential term, "differance," combines the French words for "difference" and "deferral" to create a definition of how language ceaselessly postpones meaning. Thus language may be described as an endless chain of the "play of difference" rather than a fixed set of determinate meanings. Yet while Derrida argues to subvert the dominant Western mindset, he also recognizes that there is no privileged position outside the instabilities of language from which to attack. Thus deconstruction deconstructs itself; in a self-contradictory effort, it manages to leave things the way they were, the only difference (differance) being our expanded consciousness of the inherent play of language-as-thought. Indeed, it accomplishes its readings precisely by playing with words; this has been a problem to those interested in deconstruction who do not read French, since much of Derrida's own philosophical word-play is lost even in the most excellent of translations. Deconstruction's wordplay and strange neologisms have mystified traditional critics as they have entertained and expanded the thinking of other critics.

Numerous critics have found themselves using parts of deconstruction for their own particular projects while not adopting the entire approach; these include prominently Wayne C. Booth, Jonathan Culler, Fredric Jameson, and Edward Said. Still others have attempted to extirpate deconstruction, including M. H. Abrams, Robert Alter, W. Jackson Bate, Denis Donoghue, Gerald Graff, E. D. Hirsch, and Robert Scholes.

The major attacks on deconstruction have responded to its seeming lack of seriousness about reading literature, and, more seriously, its refusal to privilege such reading as an act at all. Its opponents feel that it threatens the stability of the literary academy, that it promotes philosophical and professional nihilism, that it is too dogmatic, that it is wilfully obscure and clique-ridden, that it is mostly responsible for the heavy emphasis on theory over practical criticism in recent years. Various critiques of deconstruction have pointed out that deconstructive readings all sound oddly similar, that it does not seem to matter if the author under study is Nietzsche or Wordsworth. Furthermore, deconstructive readings have a tendency always to arrive at the same point, and always seem to start out with a set conclusion, lacking any sense of suspense as to the outcome of the reading. Yet for others, it is the sense of pursuing the theory's own play itself, and not attending to a particular author, that makes it an attractive literary theory. Over the last twenty years, deconstruction has certainly declined in importance—it once held sway without contest—and it has, like hermeneutics, found itself absorbed and adopted by other approaches and not so much practiced on its own as it once was.

For examples of deconstructive criticism in practice, see the essays of *Deconstruction and Criticism* (New York: Seabury, 1979) by Harold Bloom, Paul de Man, Jacques Derrida, Geoffrey Hartman, and J. Hillis Miller. De Man, Hartman, and Miller furnish particularly good applications of deconstruction to authors and texts. Their discussions of Shelley and Wordsworth are especially exhilarating and rewarding.

· 7 ·

Additional Approaches

In mature interpretation of a piece of literature, the reader may respond at any given moment from one particular orientation—perhaps a biographical, historical, formalistic, or psychological approach. Ideally, however, the reader's ultimate response should be multiple and eclectic. Because a work of literary art embodies a potential human experience and because human experience is multidimensional, the reader needs a variety of ways to approach and realize (make real) that experience.

In this chapter we therefore indicate other contemporary approaches to literature, either because they are different from those already detailed or because they place different emphases upon some of the aspects of those approaches. In addition, we depart in this chapter from the established pattern in that we do not offer explications of the four works analyzed in each of the first six chapters. Rather, we suggest a number of approaches that may profitably be studied further.

Beside the eclectic emphasis of this chapter, we may note several other characteristics. First, after the great and salutary success of the New Critics in teaching us to return to the printed text and to see it as artifact, it is not surprising—indeed, it was to be expected—that a reaction should take us somewhat away from formalism. A number of schools or approaches have suggested that in emphasizing the artifact the formalists neglected the humanness of the experience of literature, neglected the social milieu of literature, or falsified the vision of a true eclectic. Several of the approaches we survey here reflect this swing away from the allegedly objective approach of the New Critics. Second, some of the approaches in this chapter have European origins. The great movement of the New Critics was centered in America, with some ties to or even some origins in England. But more recent movements—Marxism, stylistics, the phenomenological approach—that have exerted significant influence upon American criticism have predominantly Continental origins. Third, a number of the approaches in this chapter are in some respects quite traditional and are partly suggested in chapter 1. However, not all approaches could be discussed in the context of chapter 1, and some have new dimensions. Consequently, although their roots are traditional enough, their contemporary manifestations may seem quite different. We have in mind, for example, the neo-Aristotelians, the rhetorical approach, the genre approach, the history of ideas, source and influence studies (including some aspects of "genetic" criticism), and the new historicism.

Still another characteristic of this chapter is its special kind of utility. As already indicated, the following sections do not offer readings of *Huckleberry Finn, Hamlet*, "Young Goodman Brown," or "To His Coy Mistress." Rather, the thirteen sections are as many stepping-stones to that eclectic reading we have stressed, whatever work the reader is interested in. Each of the sections provides orientation to the approach under discussion, including perhaps something of its history, sometimes an indication of how it overlaps with other approaches, frequently the names of critics and their works associated with the approach, and occasionally some very brief examples of the approach in practice. There is usually a range within a section, so that neophytes may at the very least become aware of the existence of the approach and more advanced readers can see where they might go to learn more. It is for this last reason that the sections are consciously allusive and sometimes resemble annotated bibliographies. Not all the books and articles cited explicitly deal with the approach under consideration, but they are germane or at least suggestive and frequently contain annotations and bibliographies of their own that will lead the reader to additional materials.

In sum, this chapter describing additional approaches continues the notion that has informed this handbook throughout: because literature is the verbal artistic expression of humankind qua humankind, with all the richness, depth, and complexity that concept suggests, literary criticism is necessarily the composite of many ways of reaching to that experience. It is necessarily eclectic because no one way of reading a piece of literature can capture all that is in it, just as no simplistic notion can embody all that the human being is. For that we need many approaches; in brief form we now offer thirteen more.

I. ARISTOTELIAN CRITICISM (INCLUDING THE CHICAGO SCHOOL)

Few works of literary criticism can hope to wear so well, or so long, as Aristotle's *Poetics* (fourth century B.C.). Our theories of drama and of the epic, the recognition of genres as a way of studying a piece of literature, and our methodology of studying a work or group of works and then inducing theory from practice can all find beginnings in the *Poetics*. More specifically, from the *Poetics* we have such basic notions as catharsis, the characteristics of the tragic hero (the noble figure; tragic pride, or hubris; the tragic flaw), the formative elements of drama (action or plot, character, thought, diction, melody, and spectacle), the necessary unity of plot, and perhaps most significantly, the basic concept of mimesis, or imitation, the idea that works of literature are imitations of actions, the differences among them coming by means, by objects, and by manner.

In practice, readers may be Aristotelian when they distinguish one genre from another, when they question whether Arthur Miller's Willy Loman can be tragic or affirm that Melville's Ahab is, when they stress plot rather than character or diction, or when they stress the mimetic role of literature. In formal criticism readers will do well to study Matthew Arnold's 1853 preface to his poems as a notable example of Aristotelian criticism in the nineteenth century.

In the twentieth century one critic has said that the "ideal critic" will be neo-Aristotelian if he or she "scrupulously [induces] from practice" (Stanley Edgar Hyman, *The Armed Vision* [New York: Random (Vintage), 1955]: 387).

The most concerted use in this century of Aristotelian principles (and the reason for placing Aristotelianism in this chapter rather than in chapter 1 as a traditional approach, though the principles are operative there) is that associated with a group of critics who were colleagues at the University of Chicago during the 1940s. Though stressing their humanistic concern and their pervading hope for a broadly based literary criticism, the members of the Chicago school were in part reacting against what seemed to them to be an inadequacy in the work of the New Critics: for Ronald S. Crane, "bankruptcy" and "critical monism." Consequently, they called for an openness to critical perspectives, a "plurality" of methods, and advocated using Aristotle's principles as comprehensive and systematic enough to be developed beyond what Aristotle himself had set down.

That call was made by Crane, in his introduction to *Critics and Criticism: Ancient and Modern* (Chicago: U of Chicago, 1952: 18); in that book Crane and his colleagues—W. R. Keast, Richard McKeon, Norman Maclean, Elder Olson, and Bernard Weinberg—gathered papers that they had published as early as 1936. Arranged in three sections, the essays deal with "representative critics of the present day," "figures and episodes in the history of criticism from the Greeks through the eighteenth century," and "theoretical questions relating to the criticism of criticism and of poetic forms" (1). Not all the essays are explicitly "Aristotelian," though Crane makes it clear that a "major concern of the essays . . . is with the capacities for modern development and use of . . . the poetic method of Aristotle" (12). (Given the humanistic and traditional background, the rhetorical bent of some of the essays is understandable and points up the necessity of our understanding that labeling any approach to literature should not mislead us into thinking that critical approaches are mutually exclusive. Compare for example, both the rhetorical and the generic approaches below.)

Another work of importance produced by this movement

is Crane's *The Languages of Criticism and the Structure of Poetry* (Toronto: U of Toronto, 1953). But although this book and *Critics and Criticism* are frequently alluded to by other critics, the Chicago neo-Aristotelians made less of an impact than the New Critics against whom they took up arms. Ten years after *The Languages of Criticism*, for example, Walter Sutton wrote:

> The neo-Aristotelian position developed by Crane and his colleagues is valuable for its emphasis upon theory, historical perspective, and scholarly discipline. It has, however, stimulated very little practical criticism. The active movement that the leaders hoped for never developed, although their work undoubtedly encouraged a renewal of interest in genre criticism, shared by critics like Kenneth Burke and Northrop Frye in modified forms and by many other scholar critics within the universities. (*Modern American Criticism* [Englewood Cliffs, NJ: Prentice, 1963]: 173)

The reader will find Sutton's chapter on the neo-Aristotelians helpful because it provides not only interpretive summaries of some of their work but also comments on possible deficiencies of the neo-Aristotelians and on their relationships to other schools of criticism.

Although Sutton may have been correct in saying that the neo-Aristotelian efforts of the Chicago school stimulated very little practical criticism, renewed efforts in rhetorical criticism, and possibly in genre criticism and in some of the later stresses on fiction as a serial form, may be mutations of Aristotelian criticism. This seems to be the thought of Donald Pizer when he writes that neo-Aristotelianism can be seen in "new and influential forms." Pizer cites as examples two books, Charles C. Walcutt, *Man's Changing Mask* (Minneapolis: U of Minnesota P, 1966), and Robert Scholes and Robert Kellogg, *The Nature of Narrative* (New York: Oxford UP, 1966). These books, according to Pizer, deal with the similarities between "all serial art" and fiction, which is seen as "a late example of . . . permanent elements" of narrative and characterization. Because of these similarities, the books by Walcutt and by Scholes and Kellogg, among others, "reflect what can be called an Aristotelian temper" ("A Primer of Fictional Aesthetics," *College English* 30 [April 1969]: 574).

II. GENRE CRITICISM

Genre criticism, criticism of kinds or types, like several of the approaches described in this chapter (Aristotelian, rhetorical, history of ideas, source study), is a traditional way of approaching a piece of literature, having been used in this case at least as far back as Aristotle's *Poetics*. But, like some other traditional approaches, genre criticism has been given revitalized attention in this century, modifying what was accepted as genre criticism for some two thousand years.

Since the time of the classical Greeks and especially during the neoclassical period, it was assumed that if readers knew into what genre a piece of literature fell, they knew much about the work itself. Put simply, Athenian citizens going to see a play by Sophocles knew in advance that the story would be acted out by a small group of actors, that they would be seeing and hearing a chorus as part of the production, and that a certain kind of music would accompany the chorus. When Virgil set out to write an epic for Augustan Rome, he chose to work within the genre that he knew already from Homer. According to the conventions of epic, he announced his theme in his opening line, he set his hero out on journeys and placed him in combat situations, he saw to it that the gods were involved as they had been in the *Iliad* and the *Odyssey*, and in the two halves of his *Aeneid* he even provided actions that were roughly parallel to the actions of the *Odyssey* (journey) and the *Iliad* (warfare). Because Alexander Pope and his readers were schooled in the classics, and the genres of classic literature, his parody of the epic, his turning it inside out, was easily recognizable in his mock-epic, *The Rape of the Lock*. Pope took the conventions of the epic genre and deliberately reversed them: the epic theme is "mighty contests" arising from "trivial things"; the hero is a flirtatious woman with her appropriate "arms"; the journey is to Hampton Court, a place of socializing and gossip; the battle is joined over a card table, with the cards as troops; the epic weapon is a pair of scissors; the epic boast is about cutting off a lock of hair. The same use of a genre with deliberate twisting of its conventions can be found in Thomas Gray's "Ode on the Death of a Favourite Cat, Drowned in a Tub of Gold Fishes," where again high style and low matter join. Here odes, death, cats, and goldfish come together in such fashion that one genre becomes its mirror image. Instead of a serious ode (or elegy) on a serious matter, we have a humorous, even a bathetic poem.

Such are the kinds of observations that traditional genre criticism could provide. It held sway through the eighteenth century, when it was even dominant. It was less vital as a form of criticism in the nineteenth century, although the conventional types, such as drama, lyric, and romance, were still recognized and useful for terminology, as they still are. In this century, however, new interest has been developed in genre criticism, especially in theoretical matters.

Of major significance is Northrop Frye's *Anatomy of Criticism: Four Essays* (Princeton, NJ: Princeton UP, 1957). In his introduction Frye points to our debt to the Greeks for our terminology for and our distinctions among some genres, and he also notes that we have not gone much beyond what the Greeks gave us (13). This he proposes to correct in his anatomy. Although much of his book is archetypal criticism, and hence has relevance for Chapter 4, much of it—especially the first essay, "Historical Criticism: Theory of Modes," and the fourth essay, "Rhetorical Criticism: Theory of Genres"—also bears upon genre criticism. Summarizing Frye is a challenge we shall gladly ignore, but two passages in particular will illustrate his technique of illuminating a critical problem and will provide insight especially into genre criticism. Calling attention to the "origin of the words drama, epic and lyric," Frye says that the "central principle of genre is simple enough. The basis of generic distinctions in literature appears to be the radical of presentation. Words may be acted in front of a spectator; they may be spoken in front of a listener; they may be sung or chanted; or they may be written for the reader" (246–47). Later he says, "The purpose of criticism by genres is not so much to classify as to clarify such traditions and affinities, thereby bringing out a large number of literary relationships that would not be noticed as long as there were no context established for them" (247–48). On the face of it these passages, though helpful, are not much different from what Aristotle offered in the *Poetics,* but on such bases Frye ranges far and wide (much more than we can here suggest) in his study of modes and genres, classifying, describing, dividing, subdividing.

Monumental as the work is, it provoked mixed responses, and we may cite two works that differ from Frye's, sometimes explicitly, as they offer other insights into genre criticism.

E. D. Hirsch's *Validity in Interpretation* (New Haven: Yale UP, 1967) makes only small reference to Frye, and presents (among other things) a quite different approach to genre criticism (ch. 3, "The Concept of Genre"). Less concerned with the extensive anatomizing of literature and of literary criticism (Hirsch implies that Frye's classification is "illegitimate" [110–11]), Hirsch insists on the individuality of any given work. More important, he shows again and again how the reader's understanding of meaning is dependent on the reader's accurate perception of the genre that the author intended as he wrote the work. (Hirsch is not, however, thinking simplistically of short story, for example, in contrast to masque, epic, or the like.) If the reader assumes that a work is in one genre but it is really in another, only misreading can result: "An interpreter's notion of the type of meaning he confronts will powerfully influence his understanding of details. This phenomenon will recur at every level of sophistication and is the primary reason for disagreements among qualified interpreters" (75). And again: "Understanding can occur only if the interpreter proceeds under the same system of expectations" as the speaker or writer (80). Such a statement reminds us that if a person reads *The Rape of the Lock* without any previous knowledge of the epic, we must wonder whether he or she has truly read Pope's poem. For if readers do not recognize conventions, they are reading at best at a superficial level. As Hirsch says "every shared type of meaning [every genre] can be defined as a system of conventions" (92). Elsewhere, Hirsch is helpful in showing that when we read a work with which we are not previously familiar or read a work that is creating a new genre, we operate ("triangulate") by moving back and forth from what we know to what we do not know well yet.

Still another work that qualifies Frye's treatment of genres while offering its own insights (though basically on fiction) is Robert Scholes's *Structuralism in Literature: An Introduction* (New Haven: Yale UP, 1974). Scholes's discussion (117–41) is closer to Frye's than is Hirsch's, but it brings to the treatment not only qualifications of Frye's classifications but also the influences of recent work in structuralism (see chapter 6), whereas Frye's emphasis is archetypal and rhetorical.

All three of these works—those of Frye, Hirsch, and Scholes—although they are challenging and stimulating, are sometimes difficult. Part of the difficulty when they are dealing with genres derives from the fact that pieces of literature do not simply and neatly fall into categories, or genres (even the folk ballad, seemingly obvious as a narrative form, partakes of the lyric and of the drama, the latter through its dialogue). This difficulty arises from the nature of literature itself: it is original, imaginative, creative, and hence individualistic. But regardless of literature's protean quality, our interpretation of it is easier if we can recognize a genre, if we can therefore be provided with a set of "expectations" and conventions, and if we can then recognize when the expectations are fulfilled and when they are imaginatively adapted. Perhaps one of the most beneficial aspects of engaging in genre criticism is that, in our efforts to decide into what genre a challenging piece falls, we come to experience the literature more fully: "how we finally categorize the poem becomes irrelevant, for the fact of trying to categorize—even through the crudest approach—has brought us near enough to its individual qualities for genre-criticism to give way to something more subtle" (Allan Rodway, "Generic Criticism: The Approach through Type, Mode, and Kind," in *Contemporary Criticism*, Stratford-Upon-Avon Studies 12, ed. Malcolm Bradbury and David Palmer [London: Arnold, 1970]: 91). More recent inquiries into genre have been carried out in the books and articles of Gérard Genette, Gary Saul Morson, and Wendy Steiner.

III. THE HISTORY OF IDEAS

Studying a piece of literature in the light of the history of ideas is somewhat similar to the historical approach presented in chapter 1. For example, the concept of revenge as found in certain Renaissance English plays can be tied directly to the study of Seneca in the universities of the sixteenth century and thus to the rise of interest in classical Greece and Rome during the Renaissance. Such a consideration must deal simultaneously with an idea, with historical developments, and with the contents of literature. Similarly, any consideration of the biography of an author may well deal with the history of his period and with the influence of the spirit of the times on a given work, an influence that may in turn be the result of antecedent developments. In this century the relationship between existentialism and the theater of the absurd provides an example.

However, the history of ideas may be taken to refer more precisely to an area of philosophy than of history, and to a form of study different from the traditional historical-biographical approach of chapter 1 or the more recent new historicism (see below). This subdivision of philosophy has been best described by the scholar Arthur O. Lovejoy, in his classic *The Great Chain of Being: A Study of the History of an Idea* (first published in 1936). Of the history of ideas, Stanley Edgar Hyman has said what most other readers would acknowledge: the history of ideas is a "philosophic field largely invented and pre-empted by Professor Arthur O. Lovejoy of Johns Hopkins. The history of ideas is the tracing of the unit ideas of philosophies through intellectual history, and just as it finds its chief clues in literary expression, literary criticism can draw on it for the philosophic background of literature" (*The Armed Vision*, rev. ed. [New York: Random (Vintage), 1955]: 187–88). The unit ideas mentioned by Hyman are of central importance in this field of study, as we can see from Lovejoy's introductory chapter in *The Great Chain of Being*.

There Lovejoy says, "By the history of ideas I mean something at once more specific and less restricted than the history of philosophy. It is differentiated primarily by the character of the units with which it concerns itself" (New York: Harper [Torchbook], 1960: 3), Lovejoy differentiates a unit idea from the compounds whose names usually end with *-ism:* idealism, romanticism, rationalism, transcendentalism, pragmatism, and the like. Among the principal types of ideas, that which most concerns him he describes in part as follows:

> any unit-idea which the historian thus isolates he next seeks to trace through more than one—ultimately, indeed, through all—of the provinces of history in which it figures in any important degree, whether those provinces are called philosophy, science, literature, art, religion, or politics. . . . [The history of ideas] is concerned only with a certain group of factors in history, and with these only in so far as they can be seen at work in what are commonly considered separate divisions of the intellectual world; and it is especially interested in the processes by which influences pass over from one province to another. (15–16)

Lovejoy singles out the need for seeing the relationships between philosophy and modern literature:

> Most teachers of literature would perhaps readily enough admit that it is to be *studied*—I by no means say, can solely be enjoyed—chiefly for its thought-content, and that the interest of the history of literature is largely as a record of the movement of ideas. . . . [It] is by first distinguishing and analyzing the major ideas which appear again and again [in literature], and by observing each of them as a recurrent unit in many contexts, that the philosophic background of literature can best be illuminated. (16–17)

After this introduction, Lovejoy puts into practice the concept of pursuing a unit idea by studying the notion of the great chain of being in its genesis and its later diffusion in various places, eras, philosophers, and literary figures.

At the risk of extending the concept of the unit idea to works that Lovejoy may not have accepted as examples of the history of ideas (some may be labeled simply as comparative literature, and one even antedates *The Great Chain of Being*), we shall now cite by title some books that at least partake of Lovejoy's approach and method. The titles themselves provide glosses on how the history of ideas impinges upon literary criticism:

Abrams, M. H. Natural Supernaturalism: Tradition and Revolution in Romantic Literature. New York: Norton, 1971.

Armstrong, Elizabeth. Ronsard and the Age of Gold. London: Cambridge UP, 1968.

Barkan, Leonard. Nature's Work of Art: The Human Body as Image of the World. New Haven: Yale UP, 1975.

Bate, Walter Jackson. From Classic to Romantic: Premises of Taste in Eighteenth-Century England. 1946; New York: Harper (Torchbook), 1961.

Bercovitch, Sacvan. The Puritan Origins of the American Self. New Haven: Yale UP, 1975.

Conn, Peter. The Divided Mind: Ideology and Imagination in America, 1898–1917. Cambridge: Cambridge UP, 1983.

Economou, George D. The Goddess Natura in Medieval Literature. Cambridge, MA: Harvard UP, 1972.

Elliott, Emory. Revolutionary Writers: Literature and Authority in the New Republic, 1725–1810. New York: Oxford UP, 1982.

Heninger, S. K., Jr. Touches of Sweet Harmony: Pythagorean Cosmology and Renaissance Poetics. San Marino, CA: Huntington, 1974.

Levin, Harry. The Myth of the Golden Age in the Renaissance. Bloomington: Indiana UP, 1969.

Lewis, C. S. The Allegory of Love: A Study in Medieval Tradition. 1936; New York: Oxford UP (Galaxy), 1958.

Marx, Leo. The Machine in the Garden: Technology and the Pastoral Ideal in America. New York: Oxford UP, 1964.

Patch, H. R. The Goddess Fortuna in Medieval Literature. 1927; New York: Octagon, 1967.

Perella, Nicolas James. The Kiss Sacred and Profane: An Interpretive History of Kiss Symbolism and Related Religio-Erotic Themes. Berkeley: U of California P, 1969.

Poirier, Richard. A World Elsewhere: The Place of Style in American Literature. New York: Oxford UP, 1966.

Quinones, Ricardo J. The Renaissance Discovery of Time. Cambridge, MA: Harvard UP, 1972.

Reising, Russell J. The Unusable Past. New York: Methuen, 1986.

Rousseau, George S. , ed. Organic Form: The Life of an Idea. London: Routledge, 1972.

Stewart, Stanley. The Enclosed Garden: The Tradition and the Image in Seventeenth-Century Poetry. Madison: U of Wisconsin P, 1966.

Williams, Raymond. *Keywords: A Vocabulary of Culture and Society*. New York: Oxford UP, 1979.

The list could go on, but these titles suggest the richness of the possibilities when one pursues an idea discovered in a piece of literature and traces it to its philosophic roots and its manifestations in other disciplines, such as religion and science. Finally, we may mention the *Journal of the History of Ideas*, founded by Lovejoy; *The History of Ideas: An Introduction* (New York: Scribner, 1969), by George Boas, with whom Lovejoy sometimes collaborated; and Lovejoy's *Essays in the History of Ideas* (Baltimore: Johns Hopkins UP, 1948). This last is a collection of some of Lovejoy's articles, the first of which, "The Historiography of Ideas," is helpful for showing in brief compass the interdisciplinary nature of this kind of study.

IV. LINGUISTICS AND LITERATURE

In the twentieth century, linguistics, the study of language, has become a discipline in itself (even, many would claim, a science), and in a university it is not unusual for the department of linguistics to be separate from the department of literature (or English). As the new discipline transcended traditional philological studies, a number of new questions arose: Is literature merely a part of language? Is the object of literary criticism totally different from the object of linguistic analysis? Is the language of literature susceptible to the same kind of study that can be brought to bear, for example, on language in its spoken form? If it is so susceptible, then in what manner and in what areas can a work of art be studied by a scientific or quasi-scientific discipline? Surely this kind of study—whether it derives from structural linguistics, transformational-generative grammar, or some other modulation of modern linguistics—goes beyond the historical study of language that was important to the nineteenth and early-twentieth-century philologists, who concerned themselves, for example, with the rediscovery of Chaucer's pronunciation and the meanings of the words in *Beowulf*.

Consequently, there is debate about linguistics as an approach to literature, and not the least of the difficulties in pursuing it is the overlapping quality of what in this handbook we are separating, for the sake of convenience and clarification, into several different approaches to literature—the linguistic, the stylistic, the structuralist, and in some instances the rhetorical approaches.

Both the debate and the overlap can be seen as we now turn to some items that illustrate the work being done either within the approach or because of it. Two articles intended for the general reader exemplify the cautionary view. Mark Lester begins "The Relation of Linguistics to Literature" (*College English* 30 [Feb. 1969]: 366–75) by asking, ". . . what can a student of literature learn from the discipline of linguistics?" He suggests "two main claims" that are made by those who wish to see linguistics as an approach to literature: the proposition that since "language is the medium of literature . . . the more we know about the medium, the more we will know about literature" and the proposition that the "critic may gain insight into the writer or the work or both by discovering patterns in the linguistic choices that the writer, consciously or unconsciously, has made." He concludes that one area of modern language study, structural linguistics, has provided attempts at literary analysis that were mostly "out-and-out failures" and that another, transformational grammar, overlaps the interests of the literary critic only slightly, so that even in the area where it might help, the area of metrics and stylistics, the "insight into literature [is] correspondingly small" (375). Similarly, at the conclusion of "Notes on Linguistics and Literature" (*College English* 32 [Nov. 1970]: 184–90), Elias Schwartz says flatly, "There is no such thing as a distinctive literary language. And if this is true, it means that, though linguists may tell us a great deal about language, they tell us nothing about literature" (190).

But although we might accept, at least partially, this cautionary view about the limitations of linguistics for literary analysis, there are areas that we might investigate profitably from a linguistic perspective. Such a middle view might be represented by Eugene V. Mohr, who cites these difficulties (and Lester's article) at the same time that he demonstrates briefly but cogently some areas of successful overlap ("Linguistics and the Literature Major," *CEA Forum* 3 [Apr. 1973]: 4–6). It is easy to agree with Mohr's first point—that a knowledge of historical information about language changes and about "devices which were operative at earlier stages in the development of English" is helpful in literary interpretation. Mohr cites the distinction between *thou* and *you* in an earlier day, and how Shakespeare, among others, could use the distinction. Second, he suggests that close analysis of the language of criticism itself (for example, Wordsworth's statements about language in the preface to the second edition of *Lyrical Ballads*) can help us to be more perceptive students of criticism. This leads him to comment on how the objective data provided by modern linguistics can be helpful in appreciating stylistics (see the discussion of stylistics below). Finally, he cites dialect study as a potential field for the legitimate overlap of linguistics and literary criticism, because dialects (clearly open to linguistics study) have been used for literary effects in English at least since the time of Chaucer and have been used by novelists and playwrights since then.

Another article that illustrates the qualified view of the literary critic who is willing to use linguistic techniques (while pointing out the difficulties) is by Stanley B. Greenfield ("Grammar and Meaning in Poetry," *PMLA* 82 (Oct. 1967): 377–87). Less helpful to the beginning student because of its range and depth, it is useful at least for its bibliographical citations and annotations. In addition, it shows how linguistics and stylistics overlap (Greenfield uses the term "linguistics and stylistics" [378] and implicitly suggests how linguistics analysis shades into stylistics criticism). Using 1960 as a convenient date for citing the appearance of several linguistics-oriented studies, Greenfield almost casually helps locate this approach to literature in terms of others: "[Here] is a testimony to the 'new linguistics' that gave rise to both linguistic and critical interest in elements of language other than diction, imagery, and symbolism, those staples of the New Criticism" (377).

From Mohr's and Greenfield's qualified view of the usefulness of linguistics as an approach to literature, we might move to more concerted efforts to bring the one kind of study to bear upon the other. The importance of structural linguistics (Lester's view notwithstanding) in the development of structuralism as an approach to literature (see chapter 6) is increasingly and extensively recognized. Somewhat more intensive is the study of prosody, alluded to by Lester. We may illustrate this study by beginning with some less-than-qualified comments from a review of a study of contemporary English. In 1956 Harold Whitehall reiterated (as he originally expressed in 1951) that *An Outline of English Structure*, by George L. Trager and Henry Lee Smith, was a "work that literary criticism cannot afford to ignore. . . ." "No criticism," he said, "can go beyond its linguistics. And the kind of linguistics needed by recent criticism for the solution of the pressing problems of metrics and stylistics, in fact, for all problems of the linguistic surface of letters, is not semantics, either epistemological or communicative, but down-to-the-surface linguistics, microlinguistics not metalinguistics" ("From Linguistics to Criticism," *Kenyon Review* 18 [Summer 1956]: 415; reprint of a review in the Autumn 1951 issue). When Whitehall reprinted this review in 1956, he did so as a prologue to some detailed applications to verse ("Prosodic Implications," 416–21). Whitehall's article, we should note, is part of a group of studies in that issue, entitled "English Verse and What It Sounds Like," which provide further evidence of what linguistics critics try to do in the specialized area of prosody. The group of studies is "moderated" by poet-critic John Crowe Ransom, who sees himself as a "prosodist," not as a linguist. Ransom concludes that both the prosodists and the linguists have something to tell each other.

Both the intensiveness and the extensiveness of the linguistics approach to literature can be suggested by calling attention to *Essays on the Language of Literature* (ed. Seymour Chatman and Samuel R. Levin [Boston: Houghton, 1967]). In the preface the editors admonish their readers that the causes of the "rift" between the linguists and the literary critics "are less important than its repair." "Reconciliation" between the two disciplines is in the air, a reconciliation that in some ways would take us back to the nineteenth and early twentieth centuries, when scholars would have seen, as a matter of course, that "linguistics and literary history were simply two peas in the philological pod." This return, however, would be made in the light of later investigations, and the editors cite the evidence of some "classics of criticism in the past three decades [that is, up to 1967]—*Seven Types of Ambiguity, The Structure of Complex Words, Articulate Energy: An Enquiry into the Syntax of English Poetry, The Verbal Icon, English Poetry and the English Language, Words and Poetry, The Language of Poetry. . . .*" Their anthology includes works from as early as 1900, but most are from the 1950s and 1960s. Among the essayists are some whose names can be found in bibliographies for structuralism, and among the five sections is one entitled "Style and Stylistics," so that on both counts we must be aware of the interrelationships of structuralism, stylistics, and linguistics as approaches to literature. (Their four other sections are "Sound Texture," "Metrics," "Grammar," and "Literary Form and Meaning.")

The interrelationships between linguistics and stylistics as approaches to literature can be further illustrated by an older work, Leo Spitzer's *Linguistics and Literary History*, the subtitle of which is *Essays in Stylistics* (Princeton, NJ: Princeton UP, 1948; New York: Russell, 1962). Spitzer's opening essay, which gives the book its name, describes and illustrates his "philological circle," in which he uses, among other things, etymologies, recurrence of words, and word patterns to get at the spirit not only of the work but also of the times in which it was written, and even at the "'psychogram' of the individual artist" (15).

Spitzer, whose early work dates back to 1910, serves to remind us that the linguistic approach in itself is not new. As we said earlier, however, linguistics as a discipline is rather new, and with the increasing use of the computer in linguistic and stylistic analysis, we must be alert to the possibilities that might be developed by critics who have both the gifts of literary perceptiveness and the skills of the new discipline and new technology. Particularly important practitioners of various aspects of linguistic theory in recent times include Roman Jakobson, Samuel Levin, and Michael Riffaterre.

V. STYLISTICS

Style has traditionally been a concern of rhetoric (see the next section, "The Rhetorical Approach"), but recently style has had such a development in its own right that we will treat it here not only as a division of the new rhetoric but also as an approach in its own right, as stylistics.

Stylistics, defined in a most rudimentary way, is not the study of the words and grammar an author uses, but the study of the *way* the author uses words and grammar—as well as other elements—both within the sentence (where some would see it) and within the text as a whole. Although the distinction between linguistics and stylistics as approaches to literature is difficult (and sometimes impossible) to make (see the preceding section, "Linguistics and Literature"), we offer the following parallel statements. Linguistics is a study of the materials available to users of language—the syntactic forms, the grammar—materials, in other words, available to all users by virtue of the users' ability to recognize and to duplicate sentence patterns (that is, grammar can be formulated in advance of its implementation in a given sentence of literary work). Stylistics is a study of the particular choices an author makes from the available materials, choices that are largely culture oriented and situation bound.

This distinction seems to be Roger Fowler's meaning when he says, "A text is structured in a certain way because it is a distinct use of certain distinctive materials given in advance. We need to make a fundamental division between the . . . linguistic materials available (grammatical facts) and the use made of them (stylistic facts)" ("The Structure of Criticism and the Languages of Poetry: An Approach through Language," in *Contemporary Criticism*, ed. Malcolm Bradbury and David Palmer, Stratford-upon-Avon Studies 12 [London: Edward Arnold, 1970]: 182). Fowler recognizes the assistance offered the literary critic by the linguist, but calls for a "sufficiently rich theory of linguistic *performance*" (185; our italics). Language, he asserts, has a cultural dimension so that not merely "grammatical competence" but "sociolinguistic competence" (187) is important to the "mature member of the English-speaking community"—and of course to

the author and reader therein. The performance, consequently, becomes the crucial element in a stylistic approach
to a literary work. Fowler gives brief examples from Alexander Pope's *Essay on Man*, Henry James's *Washington
Square*, and Jane Austen's *Mansfield Park* to show the principles at work and then says:

> A language is a structured repository of concepts, and every
> use of language is a particular ordering in a (partly language-
> dependent) circumscribed cultural situation. This ought to be
> a tacit principle for criticism, because it is an inevitable fact of
> all writing. The reader, whose linguistic conceptualization of
> experience answers closely to that of a poet who uses the same
> language, has his perceptions guided by the poet's perfor-
> mance in language. (193)

Because of its emphasis on choices and performance
(rather than on the "availability" of grammar), stylistics concerns the full text rather than the sentence. Although it may
move toward evaluation of texts, this last point is debated.
On the one hand, for example, David Lodge says that a certain kind of stylistics "can never become a fully comprehensive method of literary criticism" (*Language of Fiction* [New
York: Columbia UP, 1966]: 56). According to Lodge, the stylistician looks at a linguistic element in the context of the
language as a whole, whereas the critic takes as his context
the text as a whole; more importantly, the stylistician is less
concerned with questions of value than is the critic, who
"undertakes to combine analysis with evaluation": "It is the
essential characteristic of literature that it concerns values.
And values are not amenable to scientific method" (57). On
the other hand, Stanley B. Greenfield, in "Grammar and
Meaning in Poetry" (*PMLA* 82 [1967]: 377–87), takes issue
with that statement, and tries to demonstrate not only how
linguistics moves into the domain of stylistics, but also how
stylistics can be judgmental. Greenfield, whose annotations
will be of great help to anyone seeking to go further in these
related approaches, expresses his concern as being

> particularly with the voyages of *linguists* among poetic texts,
> and the values of their new methodologies, for our understand-
> ing of "how a poem means." They have staked large claims for
> the objectivity and scientific nature of their investigations as

opposed to the intuitive and impressionistic vagaries of the critical mind. And it is perhaps time that their newfoundland became as widely known and explored by readers of critical journals as it has been by those of linguistics ones. (378)

Just about the time that Greenfield was calling for a wider acceptance in the critical journals of the techniques being presented in the linguistics publications, the journal *Style* was being founded (1967). *Style* had as its expressed aims the publication of "meritorious analyses of style, particularly those which deal with literature in the English language and which provide systematic methods of description and evaluation of style," the reviewing of books "which contribute significantly to our understanding of style," and the provision of an "annual bibliography of stylistic criticism" as well as of occasional bibliographies that focus on aspects of style. The journal also aims to survey international developments in stylistics and to provide information on the teaching of style. We may note here that the reference to international matters of stylistics in an American journal that emphasizes literature in English reminds us of the European roots of stylistics. It was only a year earlier that David Lodge was saying: "The first thing that must be said about modern stylistics is that it is largely a Continental phenomenon. Stylistics as such scarcely exists as an influential force in Anglo-American criticism of literature in English. We have no Spitzer, no Auerbach, no Ullmann" (*Language of Fiction* 52).

A further appreciation of the interest in stylistics can be gained from two major collections of papers and commentaries on those papers: *Style in Language* (ed. Thomas A. Sebeok [Cambridge, MA: MIT P, 1960]); and *Literary Style: A Symposium* (ed. Seymour Chatman [New York: Oxford UP, 1971]). Each book is the result of a major conference on style. The first, held at Indiana University in spring 1958, was attended by Americans and was heavily interdisciplinary; the second, held in August 1969, was less interdisciplinary in its intent, but was very much international, with ten European nations being represented. Each book has short biographical sketches of the participants and extensive annotations; Sebeok's additionally has a list of 762 references. Taking the two volumes together, one can realize the difficulty of deciding what style is, the variety of approaches to the nature of style, the European—especially Continental—sources of this critical approach, and the variety of stylistic aspects of literary works that can be investigated.

There are almost thirty papers and "statements" in the Sebeok volume, but one essay that might be cited for its accessibility to the beginning student is John B. Carroll's "Vectors of Prose Style" (283–92). Crediting the psychologist L. L. Thurstone's *The Vectors of Mind* (1935) with introducing the technique called factor analysis, Carroll employs a "statistical procedure for identifying and measuring the fundamental dimensions ('vectors') that account for the variation to be observed in any set of phenomena" (283). Carroll describes six factors: good-bad, personal-impersonal, ornamental-plain, abstract-concrete, serious-humorous, and characterizing-narrating. His procedure is to have a team of raters evaluate selected passages of literature in terms of these factors. The result of such a close evaluation can be produced in graph form, so that, for example, the different profiles of two authors might be superimposed and compared.

Another essay in the Sebeok volume, Sol Saporta's "The Application of Linguistics to the Study of Poetic Language" (82–93), raises points we have already alluded to: Is poetry language? Can poetry, as language, be studied scientifically? Can linguistics be the scientific study of poetry? Saporta suggests that

> there seems to be an essential difference in the aims and consequently the results of linguistics and what can be called stylistics. A linguistic description is adequate to the extent that it predicts grammatical sentences beyond those in the corpus on which the description is based. Now, stylistic analysis is apparently primarily classificatory rather than predictive in this sense. . . . [the] result of a linguistic analysis is a grammar which generates unobserved (as well as observed) utterances. The aim of stylistic analysis would seem to be a typology which would indicate the features by which they may be further separated into subclasses. (86)

Saporta adds: "Such a view suggests that whereas linguistics is concerned with the description of a code, stylistics is concerned with the differences among the messages generated in accordance with the rules of that code" (87). This sounds like the nonevaluative methodology that Greenfield was seeking to transcend; nevertheless, the passages do supply another differentiation between linguistics and stylistics, and, like other commentators, Saporta points out that linguistics tends to take the sentence as its limit of concern, whereas stylistics must concern itself with a "larger unit, the text, . . . as the basis for stylistic analysis" (88).

The volume edited by Seymour Chatman, with another score of papers and contributors, is also rich and wide-ranging. The papers are arranged under six headings: "Theory of Style," "Stylistics and Related Disciplines," "Style Features," "Period Style," "Genre Style," and "Styles of Individual Authors and Texts." In his introduction Chatman says, "It would be presumptuous to attempt to summarize in a brief introduction the richness and complexity of this Symposium. But perhaps I can give the reader some sense of its unity and focus by noting certain persistent themes in the discussion. They are perhaps best formulated by [three] questions" (xi): (1) "What is style?" (2) "How do style features emerge?" and (3) "Is linguistics sufficient to describe literary style? That is, Is stylistics merely a branch of linguistics?" To this last, Chatman says, "The general opinion of the Symposium is 'No.'" He then cites from the essays a number of passages that "should relieve the fears of those (mostly Anglo-Saxon) scholars who worry about the 'encroachment' of linguistics into literary studies" (xiv). The consensus is that there are important "things in literature *beyond* language" that deserve study.

One critic cited by Chatman is René Wellek, who had also contributed to the Sebeok volume. Because Wellek, very much the literary critic, voiced a healthy corrective to the Indiana conference, it is appropriate to quote here his words at the very end of the "closing statement" of that volume:

> Interpretation, understanding, explication, analysis are not, at least in the study of literature and art, separate from evaluation and criticism. There is no collection of neutral, value-proof traits that can be analyzed by a science of stylistics. A work of literature is, by its very nature, a totality of values which do not merely adhere to the structure but constitute its very nature.

Thus criticism, a study of values, cannot be expelled from a meaningful concept of literary scholarship. This is not of course a recommendation of pure subjectivity, of "appreciation," of arbitrary opinion. It is a plea for literary scholarship as a systematic inquiry into structures, norms, and functions which contain and *are* values. Stylistics will form an important part of this inquiry, but only a part. (*Style in Language* 419)

But it is just as appropriate to close this section on stylistics by citing an article with a different bent, one that would complement—even implement—Wellek's words, particularly his last sentence. Erwin R. Steinberg, in "Stylistics as a Humanistic Discipline" (*Style* 10 [1976]: 67–68), argues that it is a human activity to seek and analyze "verbal patterns" in works of literature, for the very discovery and reporting of such patterns says, ". . . here is an ordering, an organizing principle, that has not been pointed out before. What might have been thought to have been random is in fact patterned. Here is the evidence: the tools, the procedure, and the results" (68). Steinberg would see no inconsistency between humanistic endeavor and the use of statistics or even of the computer. He suggests five ways of commenting on a discovered pattern: merely reporting the pattern for the sake of recording it; reporting the way a certain effect, already perceived, is achieved; citing evidence of how a pattern supports the meaning of a text; noting a pattern counter to the surface meaning that in turn sets up a "tension which adds an additional dimension to the meaning of the work"; and citing evidence of a pattern that weakens a work of art "by running counter to all the other patterns available in the text" (68–69). Steinberg then asks, "Is such pattern-seeking an appropriate occupation for followers of humanistic disciplines? Inasmuch as pattern-seeking itself seems to be a basic intellectual interest and inasmuch as the finding and describing of language patterns can help us to understand texts that assert 'the dignity and worth of man and his capacity for self-realization through reason,' I would assume that it is" (69).

VI. THE RHETORICAL APPROACH

Rhetorical criticism in the second half of this century, like several of the other approaches treated in this chapter, has been seen as a corrective to the New Critics' tendency to set up what John C. Gerber has called a *"cordon sanitaire between the reader and the work that distances the work almost as successfully as the historical approach"* ("Literature—Our Untamable Discipline," *College English* 28 [Feb. 1967]: 354). Gerber ascribes the rise of this "new" rhetoric (for rhetoric is among our most ancient disciplines) to the interest in the late 1940s in communication skills, out of which came the renewed awareness that in communication, a something must be communicated to a someone. Gradually, what was after World War II a pragmatic, elementary need in composition classes became (or became again) a method of literary criticism that preserved the New Critics' interest in the work but also directed attention to author and audience.

Looking back at that development in his very helpful introduction to a collection of rhetorical analyses, Edward P. J. Corbett has written that

> rhetorical criticism is that mode of internal criticism which considers the interactions between the work, the author, and the audience. As such, it is interested in the product, the process, and the *effect* of linguistic activity, whether of the imaginative kind or the utilitarian kind. When rhetorical criticism is applied to imaginative literature, it regards the work not so much as an object of aesthetic contemplation but as an artistically structured instrument for communication. It is more interested in a literary work for what it does than for what it is. (*Rhetorical Analyses of Literary Works* [New York: Oxford UP, 1969]: xxii)

While dealing with the work itself (hence, "internal"), rhetorical criticism considers external factors insofar as it "uses the text for its 'readings' about the author and the audience" (xviii). Particularly important is the effect of the work on its audience (what it *does*). This is not surprising, in that the original emphasis of rhetoric was on persuasion, and for that we go back to the classical Greeks.

As a matter of fact, literary criticism itself really had some of its beginnings in rhetorical analysis, for our first critics— Plato, Aristotle, Longinus, Horace—were devoted students, indeed formulators, of rhetoric. (Corbett, by the way, would stress the influence of Horace more than that of Aristotle in the later development of rhetorical criticism.) As late as the eighteenth century, rhetorical considerations played an important role in criticism, for learned men and women still knew and practiced formal rhetoric. Today much of the criticism of medieval, Renaissance, and neoclassical English and Continental literature can still profitably explore rhetorical strategies if only because we have and can work from the evidence of textbooks and manuals of rhetoric that were earnestly studied by the writers of those ages. Recently, however, the conscious and often impressive efforts to realize once again the advantages of rhetorical analyses of literature have not been limited to such earlier works. Even further, one area of rhetorical criticism—style—has developed so much that it now deserves its own emphasis (see the earlier section on stylistics). Today's new rhetoric may be expressed either in terms of classical rhetoric or through the insights gained in practical rhetoric without the use of Greek and Latin terms. Corbett, for example, points out that many a piece of practical criticism may be good rhetorical criticism even though the critic seems to be unaware of the long history of the mode within which he is operating, and may not at all use the terminology of the rhetorical critic. Similarly, creative authors may address themselves to the *audience*, while *arranging* their *argument* and working within a *style*, without realizing that these are four of the traditional concerns of rhetoricians. Readers who desire a convenient compilation of traditional terms of classical rhetoric should consult Richard A. Lanham, *A Handlist of Rhetorical Terms: A Guide for Students of English Literature* (Berkeley: U of California P, 1968).

As already indicated, a rhetorical approach helps us to stay inside the work, although we may go outside it for terms and naming strategies, being always aware that the original author was a person who chose between available options. In this methodology, then, rhetorical analysis, on the one hand, is similar to and supportive of the formalistic approach, but, on the other, may go beyond it. Among the questions raised by the rhetorical approach are these: What can we know of the speaker or narrator? To whom is he or she allegedly speaking? What is the nature of that addressee, that audience? What setting is established or implied? How are we asked to respond to the situation created? Are we being asked to make a distinction between the *ethos* (the ethical stance) of the author and the statements of the central character (for example, a distinction between the comprehensive view of Mark Twain and the limited view of Huckleberry Finn)?

As persuasive discourse, the rhetoric of a literary work requires or invites the reader to participate in an imagined experience. If we recognize such rhetorical devices as metaphor, irony, syllogism, and induction, so much the better. But even without such terminology and accompanying sophistication, by close reading and from the experience of even a good course in freshman composition, we can recognize that Marvell (or his persona) is skillfully using persuasive discourse. Consequently, "To His Coy Mistress" takes on the structure of argument. We could see just as easily that lyric poems can be structured on the basis of a definition, a process, an analysis, a causal relation; that they may provide examples; or that they may be arranged in a spatial or temporal pattern—rhetorical matters all. In every case, we can see that literature must be related to established forms of saying things; even syntax and diction, punch lines and sober conclusions, arrangement and emphasis are forms familiar to the writer before he or she begins his or her work, just as they are forms familiar to us before we read the work. The awareness of such special features and structures of words tells us a great deal about the author and the created voice. Our response to manipulated language tells us even more about the *meaning* of the work and quite a bit about ourselves as registers of meaning. Although lyric poetry seems to be the favorite genre for displays of rhetorical analysis, the method can be used effectively with fiction, as has been demonstrated by Ian Watt ("The First Paragraph of the Ambassadors: An Explication," *Essays in Criticism* 10 [July 1960]: 250–74; rpt. in Corbett 184–203). There Watt examines diction and syntax in six sentences for the implications of their functions within the paragraph, the effects upon the reader, the revelations of the character of Lambert Strether, James's own attitudes toward experience, and our understanding of the meaning of the style.

Though his collection includes studies of modern and traditional authors, Corbett noted that he could not find a study of the short story from a rhetorical perspective. To fill that gap, the reader might find helpful the essay by Michael Squires, "Teaching a Story Rhetorically: An Approach to a Short Story by D. H. Lawrence" (*College Composition and Communication* 24 [May 1973]: 150–56). Except perhaps for the short story, Corbett provides an extensive bibliography of works of rhetorical criticism, including both general background and specific works and authors. Not the least of the works in the first list is Wayne Booth's *The Rhetoric of Fiction* (Chicago: U of Chicago P, 1961), a work regularly cited by practitioners of rhetorical criticism. Among other essays that show what is being done in this approach are Walter J. Ong, S.J., "The Writer's Audience Is Always a Fiction" (*PMLA* 90 [Jan. 1975]: 9–21); S. M. Halloran, "On the End of Rhetoric, Classical and Modern" (*College English* 36 [Feb. 1975]: 621–31); and E. D. Hirsch, Jr., "'Intrinsic' Criticism" (*College English* 36 [Dec. 1974]: 446–57), which is especially helpful for the comments on reader-writer relationships and the ethical-social role of literature; Steven Mailloux, "Rhetorical Hermeneutics" (*Critical Inquiry* 11 [June 1985]: 620–42); and Don H. Bialostosky, "Dialogics as an Art of Discourse in Literary Criticism" (*PMLA* 101 [Oct. 1986]: 788–97).

VII. PHENOMENOLOGICAL CRITICISM (THE CRITICISM OF CONSCIOUSNESS)

Through much of Henry James's famous novel *The Ambassadors*, the reader shares with the central intelligence, Lambert Strether, a particular set of notions and beliefs, only to find at a later time along with Strether that a considerable reorientation and reinterpretation of apparent facts is necessary. In T. S. Eliot's *Waste Land*, we are deliberately deprived of transitions and explanations, forcing us to perceive juxtapositions, to grasp allusions and echoes, and to develop patterns of relationships. In F. Scott Fitzgerald's *Great Gatsby* the reader must see the world of Gatsby through the eyes of Nick Carraway, but must simultaneously evaluate and then accept or reject Nick's judgments about Gatsby and the people around him. In William Faulkner's *The Sound and the Fury* the reader must first experience the world as perceived by Benjy, the idiot from whose point of view is told the first of four sections of Faulkner's novel; then the reader moves through the other three sections, each with its own point of view, so that we must successively reorient our consciousness to live in the world (or worlds) created. For that matter, we might take all of Faulkner, or any other writer whose works form a totality, and live not only in an individual work but in the full consciousness of the author, what has been called the living unity of his work. The mind of the artist, a consciousness, has created an art object, or a number of them, with which the mind of the reader, a different consciousness, must interact in a dynamic process of perception, so dynamic that objects may cease to exist as objects, becoming subsumed in the subjective reality of the reader's consciousness.

In other words, when we place ourselves in the hands of an author, surrendering our time and attention to the author's creation, we begin to live within the world that the author has created. Conversely, the text, which has been waiting for us, begins to come alive, for the text can live only when read. The space and time dimensions of our everyday life and the facts of that life do not cease to exist, of course, but they are augmented by the space-time relationships and the facts of the fictive world that we now inhabit. In addition, the manner in which we now live, discover, and experience in that world is akin to the manner in which we live, learn, and experience in "real" life; a subjective consciousness is involved in that world, and seemingly objective data are important to us insofar as they merge into subjective consciousness. In the first half of the twentieth century the perceptions of the phenomena of "reality" became the concern of phenomenological philosophy and psychology. In the second half of the twentieth century the phenomena of the fictive world, the perceptions within that world, the very process of reading, and the understanding of consciousness (the author's and the critic's) have become the subject matter of literary criticism as well.

The development of this approach to literature is under-standable because the made object (novel, play, epic), the various occurrences and realities of the fictive world, and the reader himself are all coexistent phenomena. David Halliburton, using a concept credited to Hans-Georg Gadamer, has suggested that art "is not a means of securing pleasure, but a revelation of being. The work is a phenomenon through which we come to know the world" (*Edgar Allan Poe: A Phenomenological View* [Princeton, NJ: Princeton UP, 1973]: 32). Halliburton says:

> my chief concern is with the existential situation of the work—the way it stands against the horizon of interrelated phenom-ena that we call life. I am not speaking of some mystic spirit of *Geist*, but of everyday things: of consciousness, identity, pro-cess, body, love, fear, struggle, the material world. Phenome-nology, as a philosophical discipline, has investigated these things, and, within its powers, has described and analyzed their operations and structures. "Literary phenomenology," in its own way, must, I believe, try to do the same. (34)

The philosophical discipline alluded to by Halliburton dates especially from the works of Edmund Husserl, early in the twentieth century, and includes, at least to some extent, the work of Martin Heidegger, Jean-Paul Sartre, Maurice Merleau-Ponty, and Pierre Thévenaz. Merleau-Ponty's *Phe-nomenology of Perception* is frequently cited. Thévenaz is not as well known as some of these others, but a useful work by him is *What Is Phenomenology? And Other Essays* (ed. James M. Edie [New York: Quadrangle, 1962]). Besides "What Is Phenomenology?" the book contains other essays by Thévenaz, bibliographies, and an introduction by the editor. Near the end of the title essay, Thévenaz repeats the question, and answers: "[Phenomenology] is above all method—a method for changing our relation to the world, for becoming more acutely aware of it. But at the same time and by that very fact, it is already a certain attitude vis-à-vis the world" (90). As a method and an attitude, the philosoph-ical discipline known as phenomenology reaches out to touch such other concerns as psychology, psychiatry, social studies, and literary criticism. In this far-reaching effect, phenomenology compares with structuralism (see chapter 6) as a movement that has had European roots but that is now felt in America, including American literary criticism.

One of the more influential European phenomenological critics whose works are now available in English is the Belgian Georges Poulet, whose *Proustian Space* has been translated by Elliott Coleman (Baltimore: Johns Hopkins UP, 1977). The emphasis on the interrelationship of space and time in Proust (time transformed into space) is consonant with the concerns with time and space in other phenomenological critics. (Other works by Poulet that Coleman has translated are *The Metamorphoses of the Circle*, *The Interior Distance*, and *Studies in Human Time*.)

This spreading from Europe to America can be further illustrated by Wolfgang Iser's *Implied Reader: Patterns of Communication in Prose Fiction from Bunyan to Beckett* (German ed., Munich, 1972; Baltimore: Johns Hopkins UP, 1974). Iser's treatment of the novel is especially important in helping us to perceive the world within which the reader can live—for the reader is involved in the world of the novel in such manner that he or she better understands that world—"and ultimately his own world—more clearly" (xi). Iser's final essay (which had appeared earlier in *New Literary History* 3 [1972]: 279–99), "The Reading Process: A Phenomenological Approach," gives a helpful overview of the process of reading as seen phenomenologically, stressing "not only the actual text but also, and in equal measure, the actions involved in responding to that text" (274). Among other things, Iser deals with time and its importance in the reading process. For example, reading a work of fiction involves us in a process that has duration, and necessarily involves a changing self as the reader reads. Similarly, subsequent re-readings of a text create an interaction between text and reader that is necessarily different, because the reader is different, because he or she now knows what is to come, and reads in a different way from the initial reading, thus experiencing the phenomena in a different way.

If something of *time* can be seen in Iser's work, Cary Nelson uses *space* as his central fact and metaphor in *The Incarnate Word: Literature as Verbal Space* (Urbana: U of Illinois P, 1973). Nelson would see "literature as a unique process in which the self of the reader is transformed by an external verbal structure" (4). Individual works—or even chapters of his book—set beside one another, "are a series of alternative spaces which can be entered and energized by the imagination" (5). When we read, the "word becomes flesh," for "we evacuate a space in our bodies which we . . . encircle and fill. . . . To read is to fold the world into the body's house" (6). Nelson, unlike Iser, who concentrates on fiction, offers chapters on a range of genres—a medieval poem, Shakespeare's *The Tempest*, Milton's *Paradise Regained*, and other prose and poetry from the eighteenth, nineteenth, and twentieth centuries.

Sometimes the term *criticism of consciousness* is used to describe the kind of literary criticism represented by Poulet, Iser, and Nelson, but the term often extends beyond the study of particular works. In criticism of consciousness, some critics would pursue not only the text, but the whole range of texts of an author—his or her corpus—so that the critic's consciousness tries to identify with the author's consciousness, a "union of subject with another subject," an approach not particularly favored by W. K. Wimsatt ("Battering the Object: The Ontological Approach," in *Contemporary Criticism*, Stratford-upon-Avon Studies 12 [London: Arnold, 1970]: 66).

Associated with this approach is the so-called Geneva school, which includes, for example, Georges Poulet (mentioned earlier) and, in this country, J. Hillis Miller and Geoffrey H. Hartman (for example, *The Unmediated Vision* [New Haven: Yale UP, 1954]). Miller and Hartman have since abandoned phenomenological criticism in favor of deconstruction, but their earlier work remains influential. Poulet has written that "When reading a literary work, there is a moment when it seems to me that the subject *present* in this work disengages itself from all that surrounds it, and stands alone"; at such a time, he senses that he has "reached the common essence present in all the works of a great master," an essence that now stands out and beyond the particular manifestations in individual works ("Phenomenology of Reading," *New Literary History* 1 [1969]: 68). Two of Miller's works—*The Disappearance of God: Five Nineteenth-*

Century Writers (Cambridge, MA: Harvard UP, 1968) and *Poets of Reality: Six Twentieth-Century Writers* (Cambridge, MA: Harvard UP, 1965)—have been cited (by Halliburton [17]) as important in preparing the way for such developments in American literary thought. The following passage from Miller will help to show how criticism of consciousness drives from, or is synonymous with, the phenomenological movement:

> Literature is a form of consciousness, and literary criticism is the analysis of this form in all its varieties. Though literature is made of words, these words embody states of mind and make them available to others. The comprehension of literature is a process of what Gabriel Marcel calls "intersubjectivity." Criticism demands above all that gift of participation, that power to put oneself within the life of another person, which Keats called negative capability. If literature is a form of consciousness the task of a critic is to identify himself with the subjectivity expressed in the words, to relive that life from the inside, and to constitute it anew in his criticism. (*The Disappearance of God* ix)

Such a view compares with what Halliburton says early in his study of Poe. Calling into question the emphasis of earlier twentieth-century criticism on seeing the literary work as a "discrete object, a kind of inert and neutral 'thing,'" in the studying of which the critic need not be concerned with the author's intentions, Halliburton points out that such a view totally disconnects the text "from the consciousness that creates it and from the consciousness that interprets it." "The phenomenologist holds a different view. Without denying that the work has, in some sense, a life of its own, the phenomenologist believes that the work cannot be cut off from the intentionality that made it or from the intentionality that experiences it after it is made" (21). In stressing intention, the phenomenologist would therefore call us back to the consciousness of the author and the critic, a call that would set him or her apart from the formalistic or New Critical approach. But the phenomenologists, like the New Critics, would emphasize the text, for "The intentionality [they seek] out is not in the author but in the text" (22). In pursuing a comprehension of the work, the phenomenologist must seek out in each work "its own way of going." The interpreter must "find this way and go along with it, experiencing the process of the work *as a process*" (36).

Among the other works in the growing bibliography of phenomenological criticism, we might mention these:

Brodtkorb, Paul, Jr. *Ishmael's White World: A Phenomenological Reading of "Moby Dick."* New Haven: Yale UP, 1965.

Lawall, Sarah N. *Critics of Consciousness: The Existential Structures of Literature.* Cambridge, MA: Harvard UP, 1968.

Magliola, Robert. *Phenomenology and Literature: An Introduction.* West Lafayette: Purdue UP, 1977.

Vernon, John. *The Garden and the Map: Schizophrenia in Twentieth-Century Literature and Culture.* Urbana: U of Illinois P, 1973.

VIII. SOURCE STUDY AND RELATED APPROACHES (GENETIC CRITICISM)

The kind of approach, or the set of related approaches, discussed in this section does not have a generally accepted name. It would be pleasant but not altogether helpful if we could settle upon what Kenneth Burke called it—a "high class kind of gossip"—for Burke was describing part of what we are interested in: the "inspection of successive drafts, notebooks, the author's literary habits in general" (*Poems in the Making*, ed. Walker Gibson [Boston: Houghton, 1963]: 171).

We might call the approach *genetic*, because that is the word sometimes used when a work is considered in terms of its origins. We would find the term appropriate in studying the growth and development of the work, its genesis, as from its sources. However, the term seems effectively to have been preempted by critics for the method of criticism that, as David Daiches says, accounts for the "characteristics of the writer's work" by looking at the sociological and psychological phenomena out of which the work grew ("Criticism and Sociology," in *Critical Approaches to Literature* [Englewood Cliffs, NJ: Prentice, 1956]: 358–75). Similarly, the *Princeton Encyclopedia of Poetry and Poetics*, enlarged edition (ed. Alex Preminger [Princeton, NJ: Princeton UP, 1974]) uses the

term *genetic* in surveying the methods of criticism that treat
how the work "came into being, and what influences were at
work to give it exactly the qualities that it has. Characteristically, [genetic critics] try to suggest what is in the poem
by showing what lies behind it" (167). These phrases would
come near to what we are calling "source study and related
approaches," except for the fact that these statements tend to
have a sociological context, where the work is seen as a piece
of documentary evidence for the milieu that gave rise to it.
(This sort of criticism is now the province of the new historicists; see the section entitled "The New Historicism.")

More precisely, then, by "source study and related approaches" we mean the growth and development of a work as
seen through a study of the author's manuscripts during the
stages of composition of the work, of notebooks, of sources
and analogues, and of various other influences (not necessarily sociological or psychological) that lie in the background of the work. In such study, our assumption is that we
can derive from the background clues to a richer, more accurate appreciation of the work. It may be that such an assumption is something of a will-o'-the-wisp, for we can never be
precisely sure of how the creative process works, of the accuracy of our guesses, of the "intention" of the author (a vexed
question in modern criticism). Well suited as an introduction
to this kind of criticism and a pleasant indication of both the
advantages and the disadvantages of this approach to literature is the collection of pieces from which we took the Kenneth Burke quotation: Walker Gibson's *Poems in the Making.*
Introducing the pieces he has gathered, Gibson calls attention to the problem of the "relevance of any or all of these
accounts" in our gaining a "richer appreciation of poetry,"
but at the same time he clearly believes that this high-class
kind of gossip offers possibilities. Accordingly, he provides a
variety of specific approaches—different kinds of manuscript study, essays by the original authors (for example,
Edgar Allan Poe and Stephen Spender on their own works),
the classic study (in part) of "Kubla Khan" by John Livingston Lowes, and T. S. Eliot's devastating attack on that
kind of scholarship. Not in Gibson's compendium but of interest because of the popularity of the poem is a similar study

of Robert Frost's "Stopping by Woods on a Snowy Evening." An analysis of the manuscript of the poem shows how Frost worked out his words and his rhyme scheme, crossing out words not conducive to the experience of the poem. At the same time, Frost's own (separate) comments on the writing of the poem help us to interpret what the marks in the manuscript suggest (for this study see Charles W. Cooper and John Holmes, *Preface to Poetry* [New York: Harcourt, 1946]). An excellent example of this kind of work is Robert Gittings's *Odes of Keats and Their Earliest Known Manuscripts* (Kent, OH: Kent State UP, 1970), a handsome volume that provides an essay on how five of Keats's greatest poems were written and numerous, clear facsimile pages of the manuscripts.

These examples tend to come from poems of the nineteenth and twentieth centuries, but source and analogue study has long been a staple of traditional scholarship on literature of an earlier day, such as various works on Shakespeare's plays and *Sources and Analogues of Chaucer's Canterbury Tales* (ed. W. F. Bryan and Germaine Dempster, 1941; New York: Humanities, 1958). A work like this last, it should be noted, provides materials for the scholar or student to work with, whereas other works are applications of such materials. An example of application can be found in the study of Sir Thomas Malory's *Morte Darthur*. Study of Malory's French and English sources helps us greatly in evaluating the art of his romance and the establishment of his purposes and has contributed to the debate as to whether he intended to write one book (see *Malory's Originality: A Critical Study of Le Morte Darthur*, ed. R. M. Lumiansky [Baltimore: Johns Hopkins UP, 1964]) or a compendium of eight stories (see *The Works of Sir Thomas Malory*, 2nd ed., ed. Eugene Vinaver [Oxford: Oxford UP, 1967]). Milton's notes and manuscripts over a long period of time show us how he gradually came to write *Paradise Lost* and something of his conception of what he was working toward. This and more can be seen, aided again by facsimile pages, in Allan H. Gilbert, *On the Composition of Paradise Lost: A Study of the Ordering and Insertion of Material* (1947; New York: Octagon, 1966). More helpful to the beginning student is the somewhat broader view of a briefer work by Milton offered by Scott Elledge in *Milton's "Lycidas," Edited to Serve as an Introduction to Criticism* (New York: Harper, 1966). There Elledge provides not only manuscript facsimiles of the poem, but materials on the pastoral tradition, examples of the genre, passages on the theory of monody, and information both from Milton's life and from his times.

For an example of the application of this approach to fiction, the reader might look at Matthew J. Bruccoli, *The Composition of "Tender Is the Night": A Study of the Manuscripts* (Pittsburgh: U of Pittsburgh P, 1963). Bruccoli worked from thirty-five hundred pages of holograph manuscript and typescript, plus proof sheets, which represented seventeen drafts and three versions of the novel (xv). Perhaps this is more than the beginning student cares to have in this critical approach to literature. It may be well to mention, therefore, that, like Gibson's and Elledge's works on poetry cited earlier, there are some books on pieces of fiction that are intended for the student and offer opportunity to approach a piece of fiction by means of source and influence study. Such are, for example, some of the novels (*The Scarlet Letter*, *Adventures of Huckleberry Finn*, *The Red Badge of Courage*) in the Norton Critical Editions, where the text of the novel is accompanied by source and influence materials. Similar to these is *Bear, Man, and God: Eight Approaches to William Faulkner's "The Bear,"* 2nd ed. (ed. Francis Lee Utley, Lynn Z. Bloom, and Arthur F. Kinney [New York: Random, 1971]). In introducing the section on "Other Versions of 'The Bear,'" the editors point to some of the advantages of this kind of study:

> Criticism based on a close comparison of texts has recently come under attack; often such collation is seen as pedantic and fruitless. But a short time ago an examination of Mark Twain papers demonstrated that Twain had never composed "The Mysterious Stranger"; rather, an editor had combined selected fragments of his writing after his death to "make" the book. Perhaps in the same spirit of inquiry, critics have examined the various texts of "The Bear" in order to determine through textual changes something of Faulkner's evolving art: such an examination is the closest we can come to seeing Faulkner in his workshop. (121)

Perhaps that is a good place to engage in a high-class kind of gossip.

IX. HERMENEUTICS

Hermeneutics refers to the theory of interpretation. The term was originally used by nineteenth-century German theologians to designate a new kind of interpretation of the Bible, and it included interpretation both as the formulation of rules regarding how meaning is established in reading and as exegesis, or commentary on meanings expressed in the text. Those theologians strove to maintain some sense of meaning or truth in the Bible without necessarily always admitting a literal truth; their historical position within the context of innovations in science occasioned their struggle both to have meaning and to preserve the Bible from being disproven by science. As used in literary studies, hermeneutics denotes a theoretical and critical practice that denies the notion of a single truth expressed by a given work of art, and asks instead for critical approaches that allow multiple interpretations. Phenomenology, deconstruction, dialogics, feminism, reader-response criticism, and new historicism all emphasize hermeneutics; it is less, then, a particular approach or method than a general emphasis in many types of current literary theory and criticism.

Vincent Leitch has characterized the American intellectual scene of the 1960s and 1970s—what he calls the Vietnam era—as uniquely suited to the development of hermeneutics. Approaches arising from this era of diversification share a certain antinomianism in their attacks on New Criticism. Leitch identifies developments in American literary theory as arising from interpretations of the ideas of philosophers Martin Heidegger and Hans-Georg Gadamer, which replaced an older hermeneutic tradition rooted in the writings of Friedrich Schleiermacher (1768–1834) and Wilhelm Dilthey (1833–1911). Although E. D. Hirsch, among others, followed the older tradition, a "new hermeneutics" that completely denied objectivity evolved, "privileg[ing] the speaking voice over the dead letter," as Leitch puts it (*American Literary Criticism from the Thirties to the Eighties* [New York: Columbia UP, 1988] 182–210). Other important resources for applying hermeneutic philosophy to literary studies include:

Hirsch, E. D. *The Aims of Interpretation*. Chicago: U of Chicago P, 1976.

Hirsch, E. D. *Validity in Interpretation*. New Haven: Yale UP, 1967.

Howard, Roy J. *Three Faces of Hermeneutics*. Berkeley and Los Angeles: U of California P, 1982.

Mueller-Vollmer, Kurt. *The Hermeneutics Reader: Texts of the German Tradition from the Enlightenment to the Present*. New York: Continuum, 1985.

Ong, Walter J., S.J. *The Presence of the Word: Some Prolegomena for Cultural and Religious History*. New Haven: Yale UP, 1967.

Palmer, Richard E. *Hermeneutics: Interpretation Theory in Schleiermacher, Dilthey, Heidegger, and Gadamer*. Evanston: Northwestern UP, 1969.

Ricoeur, Paul. *The Conflict of Interpretations: Essays in Hermeneutics*. Ed. Don Idhe. Evanston: Northwestern UP, 1974.

Todorov, Tzvetan. *Symbolism and Interpretation*. Trans. Catherine Porter. Ithaca: Cornell UP, 1982.

Let us look more closely at some of these early hermeneuticists, and then move to a contemporary example. In the early 1800s the German theologian Schleiermacher defined hermeneutics as the art of understanding texts of all kinds. The philosopher Dilthey in the 1890s proposed a science of hermeneutics in the humanities and social sciences to interpret meaning in writing. He defined the hermeneutic circle or the idea that we cannot understand meaning in any text without a prior sense of an overall meaning, even though we cannot know the meaning of a whole except by knowing the meanings of its constituent parts. And in the 1960s the American critic Hirsch pursued the notion that there is a meaning to be interpreted by the reader—a meaning the author intends—and that this meaning is transmitted through shared linguistic conventions and norms.

Most significant has been the work of Gadamer, who, following the philosophical existentialism of Heidegger, found that the reader brings to the text a set of beliefs and predispositions that must be taken into account. Hirsch's notion of hermeneutics paralleled the close readings of the New Critics, who approached the text as an artifact to be understood, but Gadamer's idea has been more influential: that text is not an autonomous object to be dissected but rather something to be addressed in dialogue by the reader that can then readdress itself to the reader. Gadamer's notion of the reader as an "I" and the text as a "Thou" promotes openness; for understanding is thus a fusion of interests rather than a one-way process. Hermeneutics today is not a matter of establishing authoritative readings, but rather of studying how the historical, social, and psychological contexts of the text and those of the reader interact, and how meaning *changes* depending on the array of these elements. Meaning is described as co-determined, and there can never be one right interpretation. This repudiation of the intentional fallacy, or the idea that one can ascertain the author's intended meaning, underlies all contemporary criticism, whether it is called hermeneutics or not.

The most important hermeneuticist in American literary criticism is the contemporary American philosopher Richard Rorty. In his most significant work, *Philosophy and the Mirror of Nature* (Princeton: Princeton UP, 1979) Rorty addresses what philosophers call problems of knowledge by examining the shift in American philosophy away from the certainties of subject-object knowledge toward the community of hermeneutics. He defines epistemology, or the study of how we know, to mean the search for a single truth, and he characterizes epistemology as dependent upon the ocular metaphor for knowledge. To this he opposes hermeneutics as modeled by the conversational, or dialogic metaphor for knowledge. Rorty describes how American pragmatist philosophers such as John Dewey and William James replaced the search for a "foundational" philosophy, questioning motives for philosophizing at all. In imagining a culture without epistemology and metaphysics, they have, Rorty argues, brought us to a period of revolutionary philosophy similar to Thomas Kuhn's revolutionary notion of science as communally negotiated rather than pursued in private as a search for a single demonstrable proof (7–8).

Rorty attacks the traditional metaphor of the mind as a mirror of nature. The "original dominating metaphor" of having our beliefs determined by "being brought face-to-face with the object of the belief" suggests, he says, that we should think of knowledge only as an assemblage of accurate representations. But looking for a "special privileged class of representations so compelling that their accuracy cannot be doubted" means that philosophy as epistemology becomes "the search for the immutable structures within which knowledge, life, and culture must be contained," causing us to commit the moral error of substituting "*confrontation* for *conversation* as the determinant of our belief." Rorty feels that our certainty should be "a matter of conversation between persons, rather than a matter of interaction with nonhuman reality." "Personhood" is a matter of "decision rather than knowledge, an acceptance of another being into fellowship rather than a recognition of a common essence." We should "turn outward rather than inward, toward the social context of justification rather than to the relations between inner representations." What matters is not the commensurability of epistemology but "that there should be agreement about what would have to be done if a resolution *were* to be achieved. In the meantime, the interlocutors can agree to differ." Indeed, "coming to understand is more like getting acquainted with a person than like following a demonstration." We "play back and forth between guesses about how to characterize particular statements or other events, and guesses about the point of the whole situation, until gradually we feel at ease with what was hitherto strange" (35–46, 156, 162–63, 316–17).

Thus, Rorty concludes, philosophers as well as authors must "decry the very notion of having a view, while avoiding having a view about having views" if they fear the dehumanizing idea that "there will be objectively true or false answers to every question we ask, so that human worth will consist in knowing truths, and human virtue will be merely justified true belief. This is frightening because it cuts off the possibility of something new under the sun." With this in mind, a writer may be able to adopt "keeping a conversation going as a sufficient aim" and thus to address people "as generators of new descriptions rather than beings one hopes to be able to describe accurately" (378, 388–89).

What exactly happens when a critic takes a "hermeneutic"

approach to a text? For one thing, hermeneutics has helped
those who have reformed the canon of those texts that are
even to be considered in literary studies. New *voices* have
been the primary characteristic of canon reform, from contemporary Native American writers to long-neglected texts
by women. Furthermore, when a hermeneutic analysis is undertaken, diverse voices within a text are attended to rather
than defined by the single, overriding voice of the author or
narrator. In this sense, hermeneutics is closely related to dialogics. We learn to hear William Faulkner's black characters
rather than formulate them; to find a new kind of meaning in
Virginia Woolf's multiple perspectives in a work like *The
Waves;* to negotiate the difficult intersections of voice and
meaning in a work such as T. S. Eliot's *Waste Land.* Finality,
closure, and any attempt at determination of identity are the
sins to be avoided in hermeneutics, and the excitement of
this approach lies largely in the new ways it is able to open
up previously "solved" literary meanings in major works as
well as to create new interest in so-called minor ones. A
salient case in point is in Jack London's late South Seas fiction, in which the presumably authoritative voice of the
white newcomer to the South Seas islands is contrapuntally
challenged by the voices of the native peoples with whom
the narrator engages in dialogue.

In many areas today, hermeneutics offers a broad-based
approach to addressing and being addressed by literary texts.
Perhaps of all the current uses of hermeneutics, the rejection
by feminist critics of ocularity—of the paralyzing "male
gaze"—directed at women in various works of literature is
most obviously an attempt at reading within a hermeneutic
framework. Feminists hermeneutically seek to replace such
reductive visions with the voices of women characters and
authors, even as they also seek to redefine looking by reap-
propriating the power of the visual into new feminine definitions. Authors who have received this sort of reading include
Emily Dickinson, Zora Neale Hurston, Toni Morrison, and
Margaret Atwood. Finally, hermeneutics can certainly prove
useful in reading an especially complex author such as
Henry James, whose many (often competing) levels of meaning can best be addressed by an approach dedicated to the

polyphony of voices rather than the controlling word of definition or closure.

For additional insights into hermeneutics see the following:

Altieri, Charles. *Act and Quality: A Theory of Meaning and Humanistic Understanding.* Amherst: U of Massachusetts P, 1981.

Armstrong, Paul B. "The Conflict of Interpretations and the Limits of Pluralism." *PMLA* 98 (1983): 341–52.

Bleicher, Josef, ed. *Contemporary Hermeneutics: Hermeneutics as Method, Philosophy, and Critique.* London: Routledge, 1980.

Cain, William E. "Authors and Authority in Interpretation." *The Georgia Review* 34 (1980): 617–34.

Dilthey, Wilhelm. "The Rise of Hermeneutics." *New Literary History* 3 (1972): 229–44.

Gadamer, Hans-Georg. *Philosophical Hermeneutics.* Berkeley and Los Angeles: U of California P, 1976.

———. *Reason in the Age of Science.* Cambridge, MA: MIT P, 1981.

———. *Truth and Historicity.* The Hague: Nijhoff, 1972.

Habermas, Jurgen. *Knowledge and Human Interests.* Boston: Beacon, 1971.

Halliburton, David. *Poetic Thinking: An Approach to Heidegger.* Chicago: U of Chicago P, 1981.

Heidegger, Martin. *Heidegger: Basic Writings.* New York: Harper, 1977.

Kermode, Frank. *The Genesis of Secrecy: On the Interpretation of Narrative.* Cambridge, MA: Harvard UP, 1979.

Mailloux, Steven. "Rhetorical Hermeneutics." *Critical Inquiry* 11 (1985): 620–42.

Reesman, Jeanne Campbell. *American Designs: The Late Novels of James and Faulkner.* Philadelphia: U of Pennsylvania P, 1991.

Schleiermacher, Friedrich. "The Hermeneutics: Outline of the 1819 Lectures." *New Literary History* 10 (1978): 1–16.

Shapiro, Gary, and Alan Sica, eds. *Hermeneutics: Questions & Prospects.* Amherst: U of Massachusetts P, 1984.

Spanos, William V., et al. *Martin Heidegger and the Question of Literature.* Bloomington: Indiana UP, 1979.

Wolff, Janet. *Hermeneutic Philosophy and the Sociology of Art.* London: Routledge, 1975.

X. DIALOGICS

Dialogics is the key term used to describe the narrative theory of Mikhail Mikhailovich Bakhtin (1895–1975) and is specifically identified with his approach to questions of language in the novel. Dialogics refers to the inherent addressivity of all language; that is, all language is addressed, never uttered without consciousness of a relationship between the speaker and the addressee. In this humanistic emphasis, Bakhtin departed from linguistically based theories of literature (that later develop into structuralism) and from other Russian formalists. He also felt suspicious of what was to become the psychological approach to literature, for he saw such an approach as the materialization of a human soul and an attendant sacrifice of human freedom. It is safe to say that Bakhtin would have rejected *any* ism as an approach to the novel if it failed to recognize the essential indeterminacy of meaning outside the dialogic—and hence open—relationship between voices. Bakhtin would call such a closed view of meaning monological (single-voiced). For him, not only the interaction of characters but also the act of reading the novel in which they exist is a living event.

The writings of Bakhtin go back to the 1920s and 1930s, but he remained largely unknown outside of the Soviet Union until translations in the 1970s brought him to world attention. His thought emphasizes language as an area of social conflict, particularly in the ways the discourse of characters in a literary work may disrupt and subvert the authority of ideology as expressed in a single voice of a narrator. He contrasts the monologic novels of writers such as Leo Tolstoy with the dialogic works of Fyodor Dostoyevsky. Instead of subordinating the voices of all characters to an overriding authorial voice, a writer such as Dostoyevsky creates a polyphonic discourse in which the author's voice is only one among many, and the characters are allowed free speech. Indeed, Bakhtin seems to believe that a writer such as Dostoyevsky actually thought in *voices* rather than in *ideas* and wrote novels that were thus primarily dialogical exchanges. What is important in them is not the presentation of facts about a character, then, but the significance of facts voiced to the hero himself and to other characters. In a sense, the hero *is* a word and not a fact in himself. Bakhtin identifies such polyphony as a special property of the novel, and he traces it back to its carnivalistic sources in classical, medieval, Renaissance cultures (for "carnivalistic," see later discussion).

Bakhtin's constant focus is thus on the many voices in a novel, especially the way that some authors in particular, such as Dostoyevsky, allow characters' voices free play by actually placing them on the same plane as the voice of the author. In the last twenty years, as more of his works have been translated, Bakhtin has become very important to critics of many literatures and has been found to be especially appropriate to the many-voiced, open-ended American novel.

In a sense there are multiple Bakhtins. He is read differently by Marxist critics, for example, than by more traditional humanistic critics. He himself partook both of Christianity and revolutionary Marxism. Marxist critics respond more to his notions of chronotope, or how time is encoded in fiction, and to his notion of the hidden polemic in all speech, while humanistic or moral critics address themselves more to his notion of addressivity as promoting human connection and community. Since his emergence inside and outside the Soviet Union, his ideas have proved attractive to critics of all sorts of ideologies, most recently feminist critics. In any case, his idea that words are saturated with the various perceptions they contain from other words has continued to be provocative.

In the biographical introduction to Bakhtin's *The Dialogic Imagination: Four Essays by M. M. Bakhtin* (trans. Caryl Emerson and Michael Holquist [Austin: U of Texas P, 1981]), Michael Holquist tells us that Bakhtin was born into an old family of nobility in prerevolutionary Russia. After finishing his education, he joined the historical and philological faculty of the university in Odessa in 1913, but soon transferred to Saint Petersburg University. There he found an exhilarating intellectual and political climate, because Victor Shklovsky, a prominent critic of the time, and other formalists were engaged in exciting intellectual battles. Bakhtin studied prodigiously, concentrating at first on the classics and philosophy and then on literature. In 1918 he moved to a western city, Nevel, where he taught school for two years and first met with other intellectuals in what came to be known as the Bakhtin circle, discussing philosophical, literary, religious, and political issues of the day. In 1920 he moved again to Vitebsk, a refuge for avant-garde artists and thinkers, with lively journals, lectures, and discussion groups. Working at the Historical Institute, newly married, and beginning to be afflicted with the bone disease that would cause his leg to be amputated in 1938, Bakhtin decided to begin publishing his ideas.

Holquist notes that Bakhtin's publishing history is fraught with misfortune. His manuscripts were constantly being lost or suppressed, but he continued his writing and his discussion groups ceaselessly. His first major published work appeared in 1929, *Problems of Dostoevsky's Poetics* (ed. and trans. Caryl Emerson, introd. by Wayne C. Booth [Minneapolis: U of Minnesota P, 1984]), in which he advanced his concept of dialogism. During the purges of the 1930s, Bakhtin and his friends suffered tremendously. Unlike many of them, who simply disappeared forever, Bakhtin was only exiled for six years to Kazakhstan, where he spent his time as a bookkeeper. Some of his most important essays, such as "Discourse in the Novel," now included in *The Dialogic Imagination,* were written in these years. His friends in Leningrad sent him books, and he was finally allowed to teach again in 1936–1937 in the Mordovian Pedagogical Institute in Saransk and then in Kimry, a town closer to Moscow. Another book manuscript, on the German novel, disappeared, ironically enough, during the German invasion of the Soviet Union. Bakhtin, an inveterate smoker, used the only other copy of the manuscript to roll his cigarettes during the terrible days of the German invasion. From 1940 until the end of World War II Bakhtin lived in Moscow, where he prepared a dissertation on Rabelais that split the Moscow scholarly world when he defended it in 1946 and 1949. The originality and radicality of what was finally published as *Rabelais and Folk Culture of the Middle Ages and Renaissance* caused the government to step in to halt the dissertation proceedings, and the book was not published until 1965.

In the late 1930s Bakhtin returned to Saransk as chairman

of the literature department, where he remained until his retirement in 1961, seeing the school through its upgrading from teacher's college to university. He had a tremendous impact on generations of teachers in the Soviet Union. After he retired, he lived in Moscow in ill health until his death in 1975. In his last years he finally received some of the fulfillment he had long deserved, as colleagues and students in several institutions took up his cause. The Dostoyevsky book was republished in 1963, and his other works began to be published and translated around the world. He is now known as one of the greatest theorists of the novel who ever lived. (For more on Bakhtin's life, see Holquist, "Introduction" to *The Dialogic Imagination* xv–xxvi; and Katerina Clark and Michael Holquist, *Mikhail Bakhtin* [Cambridge, MA: Harvard UP, 1985].)

Bakhtin's definition of the modern polyphonic, dialogic novel made up of a plurality of voices that avoids reduction to a single perspective indicates a concern on his part about the dangers of knowledge, whether inside or outside a text. That is, he points toward a parallel between issues of knowledge and power among the characters and those between the author and the reader. In both cases, knowledge is best thought of as dialogic rather than monologic, as open to the other rather than closed, as *addressing* rather than *defining*. In this broad sense Bakhtin's thought is strongly similar to the general emphasis of hermeneutic and metafictive analyses upon openness of interpretation. Obviously Bakhtin's theory and criticism feature a powerful moral lesson concerning freedom.

Another of Bakhtin's key terms is carnivalization. Out of the primordial roots of the carnival tradition in folk culture, he argues, arises the many-voiced novel of the twentieth century. Dostoyevsky, for example, writes out of a rich tradition of seriocomic, dialogic, satiric literature that may be traced through Socratic dialogue and Menippian satire, Apuleius, Boethius, medieval mystery plays, Boccaccio, Rabelais, Shakespeare, Cervantes, Voltaire, Balzac, and Hugo. In the modern world this carnivalized antitradition appears most significantly in the novel. Just as the public ritual of carnival inverts values so as to question them, so may the novel call closed meanings into question. Of particular importance is the ritual crowning and decrowning of a mock king; in such actions, often through the medium of the grotesque, the people of a community express both their sense of being victims of power and their own power to subvert institutions. (One thinks of the Ugly King, El Rey Feo, of Hispanic tradition, as well as of the King of Comus in New Orleans' Mardi Gras.) As carnival concretizes the abstract in a culture, so Bakhtin claims that the novel carnivalizes through diversities of speech and voice reflected in its structure. Like carnival's presence in the public square, the novel takes place in the public sphere of the middle class, as Ian Watt and others have also convincingly shown. Carnival and the novel *relativize* power by *addressing* it. This makes the novel unique among other genres, many of which arose in the upper classes.

As Michael Holquist points out, rather than seeing the novel as a genre alongside others, such as epic, ode, or lyric, for Bakhtin it is a supergenre that has always been present in Western culture, always breaking traditional assumptions about form. Holquist explains that " 'novel' is the name Bakhtin gives to whatever force is at work within a given literary system to reveal the limits, the artificial constraints of that system. Literary systems are comprised of canons, and 'novelization' is fundamentally anticanonical." The novel, Bakhtin argues, is "the only developing genre" (Bakhtin, *The Dialogic Imagination* 261–62, 291; Holquist, "Introduction" to *The Dialogic Imagination* xxxi). One can easily see the importance of such a transforming or relativizing function for Bakhtin, living as he did through the oppressions of the czars, the gloomy years of Stalin's purges, and the institution of official Soviet bureaucracy. Through carnivalization in the novel, opposites may come to know and understand one another in a way not otherwise possible. The key is the unfettered but clearly addressed human voice.

In his insistence on the novel's dynamism, Bakhtin teaches us a great deal about its history and its future. As he observes, although the novel has existed since ancient times, its full potential was not developed until after the Renaissance. A major factor was the development of a sense of linear time, past, present, and especially future, moving away from the cyclical time of ancient epochs. Whereas the epic lives in cyclical time, the novel is oriented to contemporary reality. "From the very beginning, then," says Bakhtin, "the novel was structured . . . in the zone of direct contact with inconclusive present-day reality. At its core lay personal experience and free creative imagination." In its contemporaneity, the novel is "made of different clay [from] the other already completed genres," and "with it and in it is born the future of all literature." Bakhtin adds that the novel may absorb any other genre into itself and still remain a novel and that no other genre can do so. It is "every-questing, ever examining itself and subjecting its established forms to review" (*The Dialogic Imagination* 38–40).

Bakhtin's best-known idea is dialogicity, which moves past genre to describe language. The person is always the *"subject of an address"* because one "cannot talk about him; one can only address oneself to him." One cannot understand another person as an object of neutral analysis or "master him through a merging with him, through empathy with him." The solution, dialogue, "is not the threshold to action, it is the action itself." Indeed, "to be means to communicate dialogically. When dialogue ends, everything ends." Bakhtin's principles of dialogue of the hero are by no means limited to actual dialogue in novels; they refer to a novelist's entire undertaking. Yet in a polyphonic novel, dialogues are unusually powerful (*The Dialogic Imagination* 338–39, 342).

Bakhtin's major principles of the novel include the freedom of the hero, special placement of the idea in the polyphonic design, and the principles of linkage that shape the novel into a whole—including multiple voices, ambiguity, multiple genres, stylization, parody, the use of negatives, and the function of the double address of the word both to another word and to another speaker of words. An author may build indeterminacies into his or her polyphonic design, introduce multiple voices, render ideas intersubjective, and leave novels seemingly unfinished—all to leave characters free. And no reader may "objectify an entire event according to some ordinary monologic category." The novel does not recognize any overriding point of view outside the world of its dialogue, a "monologically all-encompassing consciousness—but on the contrary, everything in the novel is structured to make dialogic opposition interminable. Not a single element of the work is structured from the point of view of a non-participating 'third person'" (*Problems of Dostoevsky's Poetics* 17).

Bakhtin brilliantly describes how the novelist may voice a moral concern through narrative techinque, particularly the power of knowledge to enact a design on that which is known. As Bakhtin argues, to think about other people "means to *talk with them; otherwise they immediately turn to us their objectivized side:* they fall silent, close up and congeal into finished, objectivized images." For this reason, the author of the polyphonic novel does not renounce his or her own consciousness but "to an extraordinary extent broaden[s], deepen[s] and rearrange[s] this consciousness. . . in order to accommodate the consciousnesses of others," and he or she does not turn other consciousnesses, whether character or reader, into objects of a single vision, but instead 're-creates them in their authentic *unfinalizability"* (*Problems of Dostoevsky's Poetics* 6–7, 59, 68). As Bakhtin asks in "Author and Hero in Aesthetic Activity," an early essay, "What would I have to gain if another were to *fuse* with me? He would see and know only what I already see and know, he would only repeat in himself the inescapable closed circle of my own life; let him rather remain outside me" (quoted in Caryl Emerson, "The Tolstoy Connection in Bakhtin," *PMLA* 100 [1985]: 68–80).

By allowing characters their free speech, then, authors may thus ensure that they do not perpetrate a narrowing design using their knowledge of the characters, a design that would violate them by restricting their freedom. To do this the author must create a "design for discourse" that allows the reader to interpret the characters' actions and words without the direct intervention of the author. Such "dialogic opposition" means that the greatest challenge for an author, "to create out of heterogeneous and profoundly disparate materials of varying worth a unified and integral artistic creation," cannot be realized by using a single "philosophical design" as the basis of artistic unity, just as musical polyphony cannot be reduced to a single accent. Contrasting this polyphony with novels in which the hero is the "voiceless object" of the "ideologue" author's "deduction," Bakhtin decribes such intrusive narrators as those of many nineteenth-century British novelists. In the polyphonic novel, "there are only . . . voice-viewpoints." Through characterization, Dostoyevsky structurally dramatizes "internal contradictions and internal stages in the development of a single person," allowing his characters "to converse with their own doubles, with the devil, with their alter egos, with caricatures of themselves." Dialogicity in characterization thus leads to particular structures. A polyphonic novel seeks to "*juxtapose* and *counterpose* [forms] dramatically," to "*guess at their interrelationships in the cross-section of a single moment.*" Not "evolution" but "*coexistence* and *interaction*" characterize such structures. "It cannot be otherwise," Bakhtin insists, for "only a dialogic and participatory orientation takes another person's discourse seriously, and is capable of approaching it as both a semantic position and another point of view." It is only through such orientation that one can come into "intimate contact with someone else's discourse" and yet not "fuse with it, not swallow it up, not dissolve in itself the other's power to mean" (*Problems of Dostoevsky's Poetics* 7–8, 18–20, 28–30, 63–64, 82–85).

For additional discussion of Bakhtin's theories in practical criticism see "Forum on Mikhail Bakhtin," ed. Gary Saul Morson, *Critical Inquiry* 10 (December 1983): 225–320, particularly Morson's introduction, Caryl Emerson's "The Outer Word and Inner Speech: Bakhtin, Vygotsky, and the Internalization of Language," and Susan Stewart's "Shouts on the Street: Bakhtin's Anti-Linguistics." See also Morson, "The Heresiarch of *Meta*," *PTL: A Journal for Descriptive Poetics and Theory of Literature* 3 (1978): 407–27. Morson has been especially important in Bakhtin studies, particularly in his understanding of Bakhtin's broad notion of the novel. As he notes, "the novel is unlike other genres because they have a canon, but the novel, in its essential spirit, is anti-canonic. . . . It is aware of its own historicity, its immediate participation in the social flux. . . . In the novel, each truth is someone else's truth, reflects someone else's interests. . . . The epic knows, but the novel asks how we know." Further, he argues that "personality in the novel, like the genre itself, is always unfinished, incomplete, and incompletely integrated. It is always changing itself in the act of expressing itself" (Introduction 237–38). Elsewhere, he adds: "[The novelist creates] situations that will provoke the encounter of languages and maximize the dialogization of tongues" ("Heresiarch" 416).

Finally, as examples of specific applications, two recent books argue that Bakhtin can lead us to a new appreciation of heterogeneity in American literature. In *The Unusable Past* (New York: Methuen, 1986), Russell J. Reising is moved by Bakhtin toward a new appreciation of neglected "'social' or mimetic writers," particularly those minority writers who address social inequality and oppression. He uses Bakhtin as a strategy for reassessing works in their historical and economic settings. The "dialogization of inadequately examined divisions in the American canon" would yield, he says, "a new understanding of the interrelationships of now nearly mutually exclusive writers, themes, and world views" (2–6, 9–10, 235). Dale M. Bauer, in *Feminist Dialogics: A Theory of Failed Community* (Albany: SUNY P, 1988), stresses the necessity of tension between the central and the marginal in a dialogic community, specifically as she defines the other as the female voice. Bauer finds certain American heroines— such as Zenobia in Hawthorne's *Blithedale Romance* and Lily Bart in Wharton's *House of Mirth*—engaging in a "battle among voices" where there are "no interpretive communities willing to listen to women's alien and threatening discourse." Bauer's effort is to generate a feminist dialogics that will get these heroines "back into the dialogue in order to reconstruct the process by which [they were] read out in the first place" (3–4).

In addition to previously cited works, the reader may also find the following helpful:

Holquist, Michael. *Dialogism: Bakhtin and His World.* New York: Routledge, 1990.
Morson, Gary Saul, ed. *Literature and History: Theoretical Problems and Russian Case Studies.* Stanford: Stanford UP, 1986.

Morson, Gary Saul, and Caryl Emerson, eds. *Rethinking Bakhtin: Extensions and Challenges*. Evanston, IL: Northwestern UP, 1987.

Reesman, Jeanne Campbell. *American Designs: The Late Novels of James and Faulkner*. Philadelphia: U of Pennsylvania P, 1991.

XI. THE MARXIST APPROACH

Despite communism's political setbacks in the late 1980s and the early 1990s, Marxist literary theory still enjoys considerable academic critical support. It appeals to students of literature, as do some other theories, because it questions the authority and rationale for a fixed literary canon or tradition. It relates literature to social, economic, and historical processes, and it rejects any purely formalistic emphasis in literary interpretation. Most other critical approaches seem suspect to the Marxist critic because they allegedly support the interests of prevailing bourgeois societies. Instead, Marxism places literature within a wide context of social, political, economic and historical forces, and for a method it explicitly or implicitly offers dialectic—the constant opposition of forces in society and the resolution of that opposition as the inevitable course of the class struggle.

Idealistic forms of socialism, especially in America, appeared in such utopian communities as Robert Owens's New Harmony, Indiana, in 1825; at Brook Farm, the community of the New England Transcendentalists, in 1844; and among the Shakers, the Mormons, and various millennialists and revivalists. But the systematic doctrines that spawned socialism and communism came from the writings of Karl Marx (1818–1883) and Friedrich Engels (1820–1895), especially their *Communist Manifesto* (1848) and Marx's *Kapital* (1867). Their speculations cited historical patterns as evidence of successive class struggles. But they tended to concentrate on the nineteenth-century phenomenon of capitalism and predicted an inevitable revolution of the working class against their bourgeois capitalist masters, which would make possible a temporary dictatorship of the proletariat and ultimately an ideal, classless society—in effect, a new golden age. To Marx and Engels, the scientific scrutiny of history inevitably must reveal an undeviating evolutionary, dialectic process— the inevitable fusion of history thus becomes the succession of class struggles.

Before World War I, the appeal of Marxist socialism in the United States was its apparent adaptability to native populism and the idealism spawned by the New England Transcendentalists in the 1840s and the 1850s—Ralph Waldo Emerson and Henry David Thoreau—and by Walt Whitman. But after the Russian Revolution in 1917, which toppled the czarist government and installed the Union of Soviet Socialist Republics under Lenin and later Stalin, only those Americans disillusioned by the Great War and attracted to radical ideas could defend the notion of a finally successful Marxism entrenched in a new Russia committed to world revolution against the capitalist states and particularly against the United States, which had emerged from World War I as the most powerful Western capitalist society.

In the context of concerns about the negative social effects of the capitalist system the appeal of Marxist ideas is not difficult to understand. That appeal manifested itself in activities that ranged from an idealistic participation in the Spanish Civil War of the 1930s to literary compositions that reflect the class struggle to participation in the international Communist party. For some persons, Marxism was an attempt to help the poor and the weak, the laborers who labored for others. From the Marxist perspective any society can be understood only in terms of its base—the conditions of producing goods and making them available to satisfy the group's requirements for food, clothing, and shelter. Varieties of the base have been nomadic, agrarian, and industrial societies, each generating its typical ways to produce and distribute the material essentials of life. Beyond satisfying people's material desires, the Marxist critic would note, a society creates its own culture—the prevailing ideology, politics, myths, religion, morality, art, education, and literature, called the "superstructure." Scientific materialism, dialectical materialism, Marxism, or socialism (the terms tend to be synonymous) sets forth principles for studying ideological or cultural data "scientifically"—that is, in terms of the conflicts, contradictions, the dialectic in history.

During the 1930s, a number of writers expressed special interest in social reform, and there was considerable stress on the uses of literature in the proletarian revolt and on seeing literature as a projection of the movement of social history. Some believed that dialectical materialism could provide a sufficiently large frame of reference, a worldview, within which literature would have a practical, even a polemical role. Among American writers of some prominence who have written such criticism are Granville Hicks (*The Great Tradition* [New York: Macmillan, 1933; New York: Quadrangle, 1969] and *Figures of Transition* [New York: Macmillan, 1939]), Lionel Trilling (*The Liberal Imagination: Essays on Literature and Psychology* [1950; Garden City, NY: Doubleday (Anchor), 1957), and Edmund Wilson (*The Triple Thinkers* [New York: Harcourt, 1948]). But interest in this kind of criticism declined in light of the effects of World War II, the cold war with the Communist bloc, and the awakening to the crisis that faced the democracies. Stanley Edgar Hyman was able to write in the revised edition of *The Armed Vision* (New York: Random [Vintage], 1955) that "Marxism in practice has hardly made good its claim as an all-embracing integrative system" (389), and he omitted two chapters on sociologically oriented critics, Christopher Caudwell and Edmund Wilson, that he had included earlier.

But once again the pendulum swung. Partly because of the pressures brought on by the Vietnam War of the 1960s and early 1970s and partly because of a developing wave of critical effort, Marxist—or neo-Marxist—criticism has received renewed consideration. We can find, for example, Fredric Jameson admitting that the Marxist critics of the 1930s have been "relegated to the status of an intellectual and historical curiosity"; but then Jameson stresses that "In recent years . . . a different kind of Marxist criticism has begun to make its presence felt upon the English-language horizon. This is what may be called—as opposed to the Soviet tradition—a relatively Hegelian kind of Marxism . . ." (*Marxism and*

Form: Twentieth-Century Dialectical Theories of Literature [Princeton NJ: Princeton UP, 1971]: ix).

From a different perspective this renewed interest in Marxist criticism is the result of the opinion of many that the formalist approach, especially as practiced by the New Critics, has been inadequate in treating the literary work. The formalistic approach, a Marxist might say, is even elitist and deals too restrictedly with the made object, with the art work's internal or aesthetic form, and not enough with the social milieu in which it found its being or the social circumstances to which it ought to speak. Consequently, among the schools of criticism that have found formalistic criticism inadequate, none have been more direct in their attack than the new Marxists.

Though acknowledging, for example, that Marxist criticism, in its stress on content, has some difficulty in dealing with form, a Marxist-oriented writer can and will challenge the formalists: ". . . they need to explain how their own methodologies can come to grips with class, race, sex, with oppressions and liberation" according to Richard Wasson, in "New Marxist Criticism," an introduction to an edition of *College English* (34 [Nov. 1972]: 171) especially devoted to Marxist criticism. Similarly the coeditor of the issue, Ira Shor, noted in "Questions Marxists Ask about Literature" that "These questions are primarily theme-and-content oriented, though they can illuminate subtleties of consciousness and literature's paradigmatic relation to society" (179). The special issue, he said, should show that "Marxist criticism is a tense dialectic between society and literature, between an attitude toward history and an appreciation of art" (178). This is not to say that form is neglected by the Marxist critics—only that they themselves admit that content tends to be emphasized. For example, in the same special issue, James G. Kennedy, in "The Content and Form of *Native Son*" (269–83), consciously wants to preserve the "dialectic of content and form" of Richard Wright's novel. His first two sections, consequently, are "Content: Class Character" and "Content: World View," but his latter two are "Form: Plot" and "Form: Sub-Plot." Fredric Jameson's *Marxism and*

Form, already mentioned, is not as helpful to the beginning student in matters of form as its title might suggest. The book is quite theoretical, offering the reader an introduction to several major European figures. Jameson then provides (306–416) an essay entitled "Towards Dialectical Criticism," a synthesis that, in the words of M. L. Raina, "would account for the relationship of the achieved work to larger structures within which it draws its significance, as well as explain the uniqueness of the work, its intrinsic configurations of theme and technique."

The latter comment comes from the helpful, if brief, annotated bibliography of M. L. Raina, "Marxism and Literature— A Select Bibliography," in the November 1972 special issue of *College English* (308–14). Raina's bibliography, highly adequate for the beginning student, ranges from "Marxist Aesthetics" through "Marxism and Literary Criticism" to "Applications." The bibliography guides the user not only to major literary and philosophical figures (György Lukács, Herbert Marcuse, Christopher Caudwell), but also to major Marxist theoreticians (Marx himself, Lenin, Trotsky, Mao) and even to other figures who, though not Marxists, have written about realism or other socially oriented matters. For a comprehensive bibliography, Raina recommends Lee Baxandall's *Marxism and Aesthetics—An Annotated Bibliography* (New York: Humanities, 1968).

In any bibliography of new Marxist criticism, the preeminent figure, commanding respect regardless of one's political or philosophical leanings, is the Hungarian György Lukács (1885–1971), whose scholarly publishing spanned half a century. Besides his considerable work in aesthetics, other studies by him that are often cited are *The Theory of the Novel*, which first appeared in book form in 1920 (Cambridge, MA: MIT P, 1971); *Studies in European Realism* (New York: Grosset [Universal], 1971); *The Historical Novel* (London: Merlin, 1962; Boston: Beacon, 1962), and *The Meaning of Contemporary Realism* (London: Merlin, 1963), which appeared in paperback as *Realism in Our Time: Literature and the Class Struggle* (New York: Harper, 1964). His criticism ranges wide, and he has praise for figures as diverse as Walter Scott and Alexander Solzhenitsyn (in *Solzhenitsyn*

[Cambridge, MA: MIT P, 1971]). Lukács's works were first published in Hungarian and German; some are now available in English translations.

The name of György Lukács is associated with one particular direction of postwar Marxist criticism—the reflection theory. Lukács and his followers stress literature's reflection, conscious or unconscious, of the social reality surrounding it, not just a reflection of a flood of realistic detail but a reflection of the wholeness or essence of a society. Detriments to social wholeness reveal themselves in the literary work as aspects of capitalism. Like prewar Marxists, the reflectionists have praised nineteenth-century realists (including Sir Walter Scott, some of whose novels, like *Waverley*, reflect a time of social discord) because their complex pictures of society, no matter how unconsciously, show "truth" within social tensions, history in motion, and essence within change. Truly valuable writing reflects not only the myriad details of the objective world but also typical human nature and social relationships. Such enduring value and therefore useful models are identified in the fiction of the Russian realists—Tolstoy, Dostoyevsky, Gogol, and Chekhov—and of Western writers like Dickens, George Eliot, Conrad, Howells, and James. In truth every work reflects to some degree the age of its composition and thus the conditions of society; but the contention of the Marxist would be that fiction formed without benefit of Marxist principles can never show true social wholeness or meaningfulness.

It is not difficult to extend this perception of the reflectiveness of literature even to such older major works as Chaucer's *Canterbury Tales* or William Langland's *Piers Plowman*, for both works are commentaries on and reflections of fourteenth-century English society. Such an interaction between social milieu and literary work is clearly evident in *Huckleberry Finn*, as so much of the present book makes evident. A similar view might be taken of Twain's *Connecticut Yankee in King Arthur's Court*. Even more explicit in social commentary and therefore more clearly open to such a critical approach is William Dean Howells's *Traveler from Altruria* or any other utopian work that deals with the social fabric of a real or an imagined time (for example William

Morris's *News from Nowhere*). In one way or another, social themes are important in dealing with literature as diverse as Frank Norris's *Octopus*, Upton Sinclair's *Jungle*, John Steinbeck's *Grapes of Wrath*, William Faulkner's *Intruder in the Dust*, John Updike's *Couples*, and Kurt Vonnegut's *God Bless You, Mister Rosewater*. American naturalists like Stephen Crane (*Maggie, A Girl of the Streets* and Theodore Dreiser (*An American Tragedy*) can especially be studied from this point of view, for these authors themselves were conscious of the effect of society on individuals. Arthur Miller's *Death of a Salesman*, though much more than a sociological tract, is nevertheless a commentary on the economic system within which Willy Loman tries to find his place.

One theoretician who wrote in English and has been given some renewed attention because of the rise of the new Marxists is Christopher Caudwell (a pseudonym for Christopher St. John Sprigg). Though Caudwell is scarcely mentioned in Jameson's book, he is stressed in a useful overview of both his ideas and the 1930s done by Andrew Hawley, "Art for Man's Sake: Christopher Caudwell as Communist Aesthetician" (*College English* 30 [Oct. 1968]: 1–9). Perhaps this interest is partly the result of Caudwell's almost romantic death, a symbolic act. Like others who were drawn to the Spanish Civil War for ideological reasons, the Englishman Caudwell went to Spain in 1937, at the age of 29—and died on his first day of battle. Ultimately more significant, of course, is his work, some of which has been recently published or republished. In 1970, Samuel Hynes, editor of Caudwell's *Romance and Realism: A Study in English Bourgeois Literature* (Princeton, NJ: Princeton UP, 1970), went so far as to say that Caudwell's "essays on literature and literary figures, together with *Illusion and Reality*, [were] certainly the most important Marxist criticism in English" (14). Hynes, whose comment might need revision in light of work done since 1970, was alluding to *Studies in a Dying Culture* (New York: Dodd, 1938), *Further Studies in a Dying Culture* (London: Monthly Review, 1971), and *Illusion and Reality* (1937; New York: Russell, 1955). Noting that Caudwell's work is only partly oriented to literature as such, Hynes points out that Caudwell was interested in relationships, in offering a world view that he found lacking in others (Freud, for example):

> Caudwell did not choose analytic criticism because the task he had set himself was to synthesize. Caudwell acknowledged this point by making his only essay on literature a part of his synthetic study of bourgeois culture; *Romance and Realism* is far more concerned with literature than any other of Caudwell's writings, but like the other studies it sets its subject in relation to the general movement of society. (24–25)

Thus we note once again that the tendency of the Marxist critic is to deal with content, for in content is to be found literature's importance in the movement of history.

With that notion in mind we can understand better the words of George Levine when he expressed his concern about the breach between the practice of criticism and any concern for society: he hoped for a "step toward healing the terrible breach between the study of literature and the life that surrounds that study, between this hall, where we think about politics and those streets, which our thinking about politics and literature has so far helped keep as they are" ("Politics and the Form of Disenchantment," *College English* 36 [Dec. 1974]: 435).

If indeed our thinking about politics and literature has kept the streets as they are, the Marxist critic will wish to go beyond mere concern with literature's inevitable disclosure of tensions and contradictions within a society. He will espouse what has been called the production theory of Louis Althusser. According to this theory, through the ideology that capitalism has generated—the structures of thought, feelings, and behavior that maintain its control over society —capitalism exacts of its artists undeviating reproduction of those structures. Thus we get fictions that gloss over the contradictions in order to justify capitalism. For example, Emily Brontë's *Wuthering Heights* seems to effect the resolution of class and family ruptures through the union at last of the youthful lovers; but the resolution does not convince most readers because the explosive violence of Heathcliff and the passion linking him to Cathy will not be contained in it. On the other hand, a writer truly enlightened by Marxism would feel compelled to transform the modes of production so that his work would show the transformation of social relationships. These modes of production are whatever a culture places at hand—music, motion pictures, journalism, advertising, political speeches, painting, sports. The instinct to try for such transformation may be seen in Melville's *Moby-Dick,* in which disdain for the norms of nineteenth-century realism and sentimental romance drove Melville to claim epic, Elizabethan tragedy, cetology, whaling history and lore, and American humor for his purposes and in the process to transform them. Closer to the Marxist imperative was John Dos Passos, when he appropriated the news story and the newsreel for his *U.S.A.* trilogy in order to shock his reader into a perception of capitalism's failure and the dire necessity of socioeconomic change in the United States. There is further the perhaps more familiar example of John Steinbeck's *Grapes of Wrath,* in which the essay and the editorial are summoned as intervals between chapters of narrative to reinforce the massiveness of human despair and social disintegration in the wake of the Dust Bowl and the migration of the Okies to California.

If such devices as these can combine—within an excellent story or play—with Marxist theory and with a workable solution to socioeconomic ills, then perhaps the ideal Marxist piece of literature will have been written.

In addition to previously cited works the reader may also find the following helpful:

Baxandall, Lee, and Stefan Morawski, eds. *Marx & Engels on Literature and Art.* Saint Louis: Telos, 1973.

Bennett, Tony. *Formalism and Marxism.* London: Methuen, 1979.

Caudwell, Christopher. *Illusion and Reality: A Study of the Sources of Poetry.* New York: International, 1963.

Demetz, Peter. *Marx, Engels and the Poets: Origins of Marxist Literary Criticism.* Chicago: U of Chicago P, 1967.

Dowling, William. *Jameson, Althusser, Marx: An Introduction to The Political Unconscious.* Ithaca: Cornell UP, 1984.

Eagleton, Terry. *Criticism and Ideology: A Study in Marxist Literary Theory.* London: New Left, 1976.

———. *Marxism and Literary Criticism.* Berkeley: U of California P, 1976.

Foley, Barbara. *Telling the Truth: The Theory and Practice of Documentary Fiction.* Ithaca: Cornell UP, 1986.

Frow, John. *Marxism and Literary History.* Cambridge: Harvard UP, 1986.

Jameson, Fredric. *The Prison-House of Language: A Critical Account of Structuralism and Russian Formalism.* Princeton: Princeton UP, 1972.

————. *The Political Unconscious: Studies in the Ideology of Form.* Ithaca: Cornell UP, 1979.

Lentricchia, Frank. *Criticism and Social Change.* Chicago: U of Chicago P, 1983.

Marcuse, Herbert. *The Aesthetic Dimension: Toward a Critique of Marxist Aesthetics.* Boston: Beacon, 1978.

Orwell, George. *Critical Essays.* London: Secker, 1946.

Williams, Raymond. *Culture and Society, 1780–1950.* London: Chatto, 1958.

————. *The Country and the City.* New York: Oxford UP, 1973.

————. *Marxism and Literature.* New York: Oxford UP, 1977.

Wilson, Edmund. *Axel's Castle.* 1931; New York: Scribner's, 1961.

XII. THE NEW HISTORICISM

"Laputa"—"The whore." What did Jonathan Swift mean when he gave that name to the flying island in the third voyage of *Gulliver's Travels*? It is a question that has tantalized readers—and professional critics—since the eighteenth century. The science-fiction aspect can still captivate us in that island—but why "the whore"? There may be an answer, and the new historicism may be a way of finding it. More about that later in this chapter.

"If the 1970s could be called the Age of Deconstruction," Joseph Litvak writes, "some hypothetical survey of late twentieth-century criticism might well characterize the 1980s as marking the Return to History, or perhaps the Recovery of the Referent" ("Back to the Future: A Review-Article on the New Historicism, Deconstruction, and Nineteenth-Century Fiction," *Texas Studies in Literature and Language* 30 [1988]: 120). In fact, since the new historicism emerged on the critical scene in the 1980s, it has remained the most volatile of all current theoretical and critical approaches to literature. It is no accident that its practitioners were students in the 1960s, for the new historicism has held on to an interest in the sort of sociopolitical ferment characterized by that decade. The new historicism is concerned with reading, writing, and teaching as actions rather than as descriptions of actions; it has allied itself in varying degrees to such activist approaches as Marxism and feminism since its inception.

As Carolyn Porter has observed, the new historicism arises out of a diverse set of practices that are not in themselves new: "'the turn toward history' has been in evidence for some time." She brings into her discussion of the origins of new historicism such diverse figures as Louis Althusser, Michel Foucault, Frederic Jameson, Raymond Williams, Mikhail Bakhtin, Terry Eagleton, Hayden White, Myra Jehlen, Bruce Franklin, Annette Kolodny, Sacvan Bercovitch, and Eugene Genovese, all of whom began their practice before the advent of new historicism and all of whom have been important resources for those today calling themselves new historicists. She also credits the emergence of American Studies, Women's Studies, and Afro-American Studies programs on college and university campuses as an important sign of the changing nature of literary criticism ("Are We Being Historical Yet?" *South Atlantic Quarterly* 87 [Fall 1988]: 743–49).

The new historicism is notoriously difficult to define; as H. Aram Veeser notes, "It brackets together literature, ethnography, anthropology, art history, and other disciplines and sciences" in such a way that "its politics, its novelty, its historicality, [and] its relationship to other prevailing ideologies all remain open questions" (Introduction, *The New Historicism*, ed. Veeser [New York: Routledge, 1989]: xi). New historicists have consciously resisted identifying their approaches with a single methodology, for they believe that as history and culture must be described as constantly changing constructions made by variously interested men and women, so too must the new historicism maintain its multidimensionality if it is to remain true to its strengths, what Forrest G. Robinson identifies as "a principled flexibility, a sharp eye to the distortion in all perspectives, a cultivated pleasure in the discovery of doubleness and subversion in a world bound to have things both ways, and the genial refusal to claim an exemption from the general angling for position" ("The New Historicism and the Old West," *Western American Literature* 25 [Summer 1990]: 104).

Forrest Robinson and many others identify Stephen Greenblatt, a Renaissance scholar and a founding editor of the journal *Representations*, as the leader of the new historicism. Greenblatt's English Institute essay, "Improvisation and Power," along with a series of path-breaking articles and Modern Language Association sessions, found Greenblatt and other early new historicists such as Louis Montrose, Walter Benn Michaels, Catherine Gallagher, and others inventing a new approach. Greenblatt himself offered the term "new historicism" in his introduction to a special issue of the journal *Genre* (15.1–2 [1982]), though it seems to have been coined by another critic in reference to cultural semiotics in Renaissance studies (Michael McCanles, "The Authentic Discourse of the Renaissance," *Diacritics* 10.1 [Spring 1980]: 77–87). Greenblatt defines the approach as "a practice rather than a doctrine" and names the major influences on his own practice as Michel Foucault, Fredric Jameson, and Jean-François Lyotard, all French social and anthropological critics who raise the question of art and society as related institutionalized practices. Another important influence on this issue has been that of the French Marxist, Louis Althusser. If there is a controlling idea in new historicism, it is reflected in the shared interest of these thinkers in the conjunction of aesthetics with material reality. Greenblatt notes that as Jameson blames capitalism for perpetrating a false distinction between the public and the private, and as Lyotard argues that capitalism has forced a false integration of these worlds, new historicism exists between these two poles in the attempt to work with the "apparently contradictory historical effects of capitalism" without insisting upon an inflexible historical and economic theory (Greenblatt, "Towards a Poetics of Culture," in *The New Historicism* 1–6; Robinson 104).

The new historicism seeks to bring down the boundaries between separate disciplines, particularly politics and literature. Greenblatt's focus has been to rehistoricize Shakespeare's plays, but his Renaissance studies have now occasioned renewed interest in literature as a social act in the field of Romantic studies and, somewhat later, in Victorian studies, the American Renaissance, and Latin American literature. Jon Klancher has postulated that the new historicism first appeared in Renaissance studies because the Renaissance uniquely appeals to postmodern culture in that it did not separate politics and literature as subsequent eras did; similarly, he observes, British Romanticism has appealed to new historicists precisely because it so strongly attempted to keep literature and politics separate ("English Romanticism and Cultural Production," in *The New Historicism* 77). And the new historicism is moving into many other areas as well with its notion, as Forrest Robinson puts it, that "literature is at once socially produced and socially productive," always existing within a particular socioeconomic context and always transforming that context as well: "At the same time that it re-historicizes the text, the New Historicism re-textualizes history" (Robinson 105). Louis Montrose calls this "a shift from History to histories" and calls upon critics to think of what they do as a process rather than a system ("Professing the Renaissance: The Poetics and Politics of Culture," in *The New Historicism* 21). In H. Aram Veeser's view, "it encourages us to admire the sheer intricacy and unavoidability of exchanges between culture and power." Thus despite what Veeser calls its "portmanteau quality" (xi), Robinson emphasizes that the new historicism thus rejects any approach that "construe[s] the relations between the text and its context, its author, and its audience as more or less literal mirrorings or direct influences passing from one autonomous zone to another" (Robinson 105). In Veeser's words, its challenges have thus been to "the norm of disembodied objectivity" in literary criticism as well as to "the boundaries separating history, anthropology, art, politics, literature, and economics." It has "struck down the doctrine of noninterference that forbade humanists to intrude on questions of politics, power, indeed on all matters that deeply affect people's practical lives—matters best left, prevailing wisdom went, to experts who could be trusted to preserve order and stability in 'our' global and intellectual domains" (ix).

This crossing of boundaries has also threatened some critics: opponents of the new historicism warn that if "we"— critics—cross over into "their" territory, then "they"—historians, psychologists, economists—may very well threaten "our" territory and obscure what has been thought of previously as the special, protected practice of literary criticism. This attack has best been voiced by Edward Pechter, who seems mostly to be concerned about the Marxist influences on the new historicism, a certain "new politicization" of literary studies ("The New Historicism and its Discontents: Politicizing Renaissance Drama," *PMLA* 102 [May 1987]: 292–303). Yet, as Veeser—and others—argue, just as the new historicism is an attack on naive formalism, it is also "as much a reaction against Marxism as a continuation of it" (xi). Catherine Gallagher supports this view when she argues that good criticism should embody no particular political values but should always be open to the most stringent sort of debate on all topics ("Marxism and the New Historicism," in *The New Historicism* 37–48). Yet Carolyn Porter counters with the idea that when we ask "Why historicize?" however one responds, "certain political assumptions are operating." New historicists, she feels, need not to reject that question but to *address* it (781).

Yet despite its wish to dissociate itself from identification with Marxism, the new historicism frequently uses terminology from the marketplace: exchange, negotiation, circulation of ideas are often described. As Veeser notes, it is the "moment of exchange" that fascinates new historicists, as social values are said to be reflective of "symbolic capital" as may be presented in a literary text: "the critic's role is to dismantle the dichotomy of the economic and the non-economic, to show that the most purportedly disinterested and self-sacrificing practices, including art, aim to maximize material or symbolic profit" (xiv). As Greenblatt puts it, "contemporary theory must situate itself . . . not outside interpretation, but in the hidden places of negotiation and exchange" ("Towards a Poetics of Culture" 13).

Carolyn Porter has characterized the central tenet of the new historicism as the radical reproblematization of the relation between history and literature. As she argues, "We no longer regard 'history' as a given backdrop against which to see the literary text. It is neither something that springs forth from that backdrop, spotlighted by its transcendent expression of 'the human spirit,' nor is it the result of the fellow working the lights, exposing a fixed, historically 'set' scene. Neither lamps nor mirrors will do anymore" (770). Richard Lehan would agree: "[N]o theory of history—or of literary criticism—is neutral but carries within it an ideology that consciously or unconsciously fixes one's relationship to a culture and is reinforced by principles of power—academic, political, and legal forms of power" ("The Theoretical Limits of the New Historicism," *New Literary History* 21 [Spring 1990]: 535). Veeser offers a helpful list of the key assumptions of the new historicism:

1. that every expressive act is embedded in a network of material practices;
2. that every act of unmasking, critique, and opposition uses the tools it condemns and risks falling prey to the practice it exposes;
3. that literary and non-literary "texts" circulate inseparably;
4. that no discourse, imaginative or archival, gives access to unchanging truths nor expresses inalterable human nature;
5. . . . that a critical method and a language adequate to describe culture under capitalism participate in the economy they describe. (xi)

In practice, he observes, new historicists reject any controlling ideological grid in favor of discovering meaning in previously ignored events or statements, which they then reread in a manner similar to Clifford Geertz's method of "thick description" in order to reveal through analysis of particulars the codes and forces that control a culture. They eschew overarching hypothetical constructs in favor of "surprising coincidences" (an example of which we present at the end of this section) that may cross generic, historical, and cultural lines previously maintained, highlighting unsuspected lendings and borrowings of metaphor, ceremony, dance, dress, or popular culture. Retaining literary criticism's historical emphasis upon research and scholarship, the new historicism brings literature from the arm's length at which it has been held into broader contexts (xii–xiii). Veeser's list also helps us see the difference between the new historicism and the "old" historicism Greenblatt describes as "monological." Old historicism posited an "internally coherent and consistent" vision that "can serve as a stable point of reference, beyond contingency, to which literary interpretation can securely refer" (Introduction to *The Forms of Power and the Power of Forms in the Renaissance*, ed. Greenblatt [Norman: U of Oklahoma P, 1982]: 5). Porter has contrasted New Historicism to "old" historicism by describing the latter as wanting "to periodize [history] by reference to world views magisterially unfolding as a series of tableaux in a film called Progress," as though all Elizabethans, for example, held views in common. The new historicism, she says, "projects a vision of history as an endless skein of cloth smocked in a complex, overall pattern by the needle and thread of Power. You need only pull the thread at one place to find it connected to another" (765).

A key issue and methodology in the new historicism involves the simultaneity of two seemingly oppositional processes. On the one hand new historicists describe how texts and other agencies contest or subvert a dominant culture; on the other hand there are those who argue that in fact such subversion is created and contained by the dominant institutional culture—that such a culture produces subversion in order to contain such "threats" to itself. This issue has been addressed in depth by Jonathan Dollimore and Alan Sinfield, who identify themselves as cultural materialists and might be described as the British allies of the new historicists, though these two national approaches are quite distinct. (See *Political Shakespeare*, ed. Dollimore and Sinfield [Ithaca: Cornell UP], 1985; and Dollimore's "Shakespeare, Cultural Materialism, Feminism and Marxist Humanism," in *New Literary History* 21 [Spring 1990]: 471–93.) Raymond Williams has similarly addressed himself to such dynamic and dialogical structures in *Marxism and Literature*, noting that at any time one may address oneself to shifting conjunctures of oppositional values, creating conceptual sites within the ideological field from which the dominant can both be contested and redefined ([Oxford: Oxford UP, 1977]: 121). Williams was practicing new historicism long before it was called such, describing the essential need for contextualizing language in history and history in language.

Louis Montrose describes this "ceaseless jostling among dominant and subordinate positions, a ceaseless interplay of continuity and change . . . [as opening] both the *object* and the *practice* of cultural poetics to history." But as he reminds us, we must always be aware that as we attempt to understand the interplay of culture-specific discursive practices, we too are part of that interplay we seek to analyze (23). Alan Liu corroborates this: "Though *we* would understand the historical *them* in all their strangeness, the forms of our understanding are fated at last to reveal that *they* are a remembrance or prophecy of *us*." Liu believes the new historicism "imagines an existentially precarious power secured upon the incipient civil war between, on the one hand, cultural plurality, and, on the other . . . , the cultural dominant able to bind plurality within structure." As he aptly observes, "theatricality in particular is the paradigm that stresses the slender control of dominance *over* plurality" ("The Power of Formalism: The New Historicism," *ELH* 55 [1989]: 723, 733). Liu's observation causes one to wonder whether this similarity explains why Renaissance drama has been such an important focus for new historicists; as he notes, "the past is costume drama in which the interpreter's subject plays. Historical understanding . . . is an act" (733). In any case, this sort of constant attention to "interplay" rather than stasis, to dualism, dynamism, dialogism—to *relationship*—makes the new historicism indeed exciting.

Let us conclude with a brief example of a New Historicist reading of a well-known text. In "The Flying Island and Female Anatomy: Gynaecology and Power in *Gulliver's Travels*" (*Genders* 2 [July 1988]: 60–76), Susan Bruce offers a reading of book 3 of *Gulliver's Travels*, "A Voyage to Laputa," that makes sense of that mysterious book in a new and invigorating way. It has always been difficult to answer the inevitable question from students acquainted with Spanish: "Why does Swift name the flying island 'Laputa,' which

in Spanish means 'the whore'?" In her article Bruce ties to-
gether some events of the year 1727, soon after *Gulliver's
Travels* was published, including eighteenth-century gender
issues, the relationship between midwives and physicians,
and a famous scandal involving a "monstrous birth."
Bruce begins by describing a four-volume commentary on
Gulliver's Travels published in 1727 by one Corolini di
Marco, in which the author offers a dry account of his observations on Swift's work until he gets to the episode in Book
IV, "A Voyage to the Houyhnhnms," in which Gulliver
catches rabbits for food. At that point, di Marco loses control
of his rhetoric and launches into a tirade:

> But here I must observe to you, Mr. Dean, *en passant*, that Mr.
> Gulliver's Rabbits were wild Coneys, not tame Gutless ones,
> such as the consummate native effronterie of St. André has
> paulmed upon the publick to be generated in the Body of the
> Woman at Godalming in Surrey. St. André having, by I know
> not what kind of fatality, insinuated himself among the for-
> eigners, obtained the post of Anatomist-Royal.

Di Marco was referring to a notorious scandal involving the
then royal physician, St. André, and the so-called "rabbet-
woman" of Surrey, Mary Toft, who managed to convince
various members of the medical profession in 1727 that she
had given birth to a number of rabbits, which she had actually inserted into her vagina and then "labored" to produce.
Bruce asks the question as to why di Marco should have felt
it necessary to include this reference, and by researching
records of Mary Toft's famous trial and the ultimate ruin of
St. André, she connects this event with the literary depiction
of the female body in *Gulliver's Travels* and elsewhere.
Bruce describes the trend toward the education of midwives and the medical profession's desire to stamp them out.
Examining the phenomenon of books published for literate
midwives during the period and quoting testimony from
Mary Toft's trial allow Bruce to describe the hostility not
only toward the midwife that collaborated with Toft in the
hoax, but toward women in general. Bruce then connects this
issue of the male establishment's outrage at the female power
expressed in the hoax to Gulliver's observations on women

in general in *Gulliver's Travels*, particularly his nauseating descriptions of the female body, as in his description of the Queen of Brobdingnag at table, but more particularly of women's breasts. Perhaps the most memorable passage occurs when Gulliver sees a tumor on a Brobdingnagian woman's breast:

> One Day the Governess ordered our Coachman to stop at several shops; where the beggars watching their Opportunity, crouded to the Sides of the Coach, and gave me the most horrible Spectacles that ever an *European* Eye beheld. There was a Woman with a Cancer in her Breast, swelled to a monstrous Size, full of Holes, in two or three of which I could have easily crept, and covered my whole Body." (Swift, *Gulliver's Travels*, ed. R. A. Greenberg [New York: Norton, 1970]: 90).

The implication is that under the male gaze, the magnification of the female body leads not to enhanced appreciation of it, but rather to horror and disgust. But Bruce connects Gulliver's anxious fascination for what is *in* the body to the anxieties of his age involving the rise of science and to general implications for perceiving the body.

Finally, turning to book 3, Bruce brings her insights to bear on the floating island of Laputa, which she describes as a gigantic trope of the female body. A circular island with a round chasm in its center, through which the astronomers of the island descend to the domelike structure of the "Flandona Gagnole," or "astronomer's cave," Laputa has at its center a giant lodestone on which the movement of the island depends. Bruce compares the floating physical structure of Laputa to the uterus and vagina, noting that Gulliver and the Laputians are able to enter this cavity at will and by doing so control not only the movements of the lodestone but also of their entire society. As Bruce notes, "It is this which engenders the name of the island: in a paradigmatic instance of misogyny, the achievement of male control over the female body itself renders that body the whore: *la puta.*"

Yet, as Bruce goes on to observe, eventually the attempted control over the female body that drives Laputa becomes its undoing, for the more the men of the island try to restrict their women from traveling below to Balnibarbi (where the women are described as having sexual adventures with the Balnibarbian men), the more male impotence threatens Laputian society. One recalls that Gulliver notes the men's ineffectuality in several ways, from the descriptions of the "Flappers" that must slap them out of their scientific reveries in order to make them speak, to the way Gulliver tells us he is only able to communicate with the women of Laputa, who have

> Abundance of Vivacity; they contemn their Husbands, and are exceedingly fond of Strangers. . . . But the vexation is, that they act with too much Ease and Security, for the Husband is always so rapt in Speculation, that the Mistress and Lover may proceed to the greatest Familiarities before his Face, if he be but provided with Paper and implements, and without his *Flapper* by his side. (Swift 138)

Bruce concludes by connecting the "doomed attempt of various types of science to control the woman's body" to the debate about language in book 3, the failure in the Laputians' foolish experiments with language (such as the "Engine for Improving Speculative Knowledge" that produces only broken sentences). Swift thus locates "his critique of science . . . in the failure of science to bridge the gap between the empirical and the discursive, between praxis and language." Gulliver tells us that not only were the women the only "People from whom [he] could ever receive a reasonable answer," but that in Balnibarbi "the Women in Conjunction with the Vulgar and Illiterate . . . threatned to raise a Rebellion, unless they might be allowed the Liberty to speak with their Tongues, after the Manner of their Forefathers: Such constant irreconcileable Enemies to Science are the common People" (Swift 158).

Thus in "A Voyage to Laputa" the attempt to bring the female body under control clashes with the realization that such control would have to include control over the discourse produced by that body. The female body, Swift implies, cannot be so managed because no one has discovered a way of controlling female discourse or sexual desire. Thus di Marco's anxiety over the Mary Toft episode indicates that the most confusing and least understood section of the most canonized satire of the age "is directly connected to a crux of debate over female power taking place in this period, a power which revolved around a woman's control over her own body and her power over, or in relation to, discourse." One final historical note: a pamphlet published in 1727 was purportedly written by "Lemuel Gulliver, Surgeon and Anatomist to the Kings of Lilliput and Blefuscu, and Fellow of the Academy of Sciences in Balnibarbi." It is entitled: *The Anatomist Dissected: or the Man-Midwife finally brought to Bed.* Its subject is Mary Toft, the "rabbet-woman."

Such an innovative and illuminating approach as Bruce's makes all the more clear the power and potential of the new historicism. As Porter argues, once we are freed from the stasis of "world views" or controlling ideologies in criticism, we may take the opportunity to "approach literary texts as agents as well as effects of cultural change, as participating in a cultural conversation rather than merely re-presenting the conclusion reached in that conversation, as if it could have reached no other. . . . [F]urthermore, we ought to be able to produce a criticism that would make significant strides toward understanding 'language in history: that full field'" (782). Richard Lehan is similarly hopeful for his particular field, the American novel: "Once the idea of literary periods gives way to the idea of historical process, we can then see the connection between such literary and cultural movements as literary naturalism and the industrialization of Europe and America: the rise of the industrial city as an investment center, the creation of an industrial class, and the confrontation between and among nature, man, and the machine." Lehan is thus able to connect the novels of Émile Zola with the work of Frank Norris, Theodore Dreiser, and Upton Sinclair, and to make further connections: "[T]he world that Zola, Norris, and Dreiser depicted from within the boardroom, Henry James, Virginia Woolf, E. M. Forster, and T. S. Eliot depicted from the salon." And James Joyce and Ezra Pound, he continues, depicted the same world by superimposing upon the realist-naturalist realm a mythic realm; "modernism and naturalism are thus two different responses to the same historical moment" (552–53). As Liu urges, "the promise of New Historicism . . . is to develop the philosophy of allegory into a true speaking in the agora: a rhetorical notion of literature as text-cum-action performed by historical subjects upon other subjects" (Liu 756).

In additon to previously cited works the reader may also find the following helpful:

Greenblatt, Stephen. "The Power of Forms and the Forms of Power." Introduction to special issue. Genre 15.1–2 (1982): 1–4.

———. Renaissance Self-Fashioning: From More to Shakespeare. Chicago: U of Chicago P, 1980.

———. Shakespearean Negotiations: The Circulation of Social Energy in Renaissance England. Berkeley and Los Angeles: U of California P, 1988.

Liu, Alan. Wordsworth: The Sense of History. Stanford: Stanford UP, 1989.

Montrose, Louis. "Renaissance Literary Studies and the Subject of History." English Literary Renaissance 16 (1986): 5–12.

XIII. READER-RESPONSE CRITICISM

Reader-response theory, one of the important recent developments in literary analysis, arose in large measure as a reaction against the New Criticism, which dominated this field for roughly a half-century. The New Criticism, or formalist approach, numbered some of the most important names in American and British literature among its theorists, practitioners, and popularizers, figures such as I. A. Richards, T. S. Eliot, William Empson, John Crowe Ransom, Allen Tate, Cleanth Brooks, Robert Penn Warren, and R. P. Blackmur. (It should be added that these men were by no means of one mind on formalism; indeed, Richards and Blackmur both wrote essays that question basic New Critical trends and suggest more than a little sympathy with reader-oriented theory.)

Though formalism is treated in chapter 2 of this book, it will not be amiss to rehearse the main lines of it again. At the risk of oversimplifying and thereby misrepresenting it, it may be said that formalism regards a piece of literature as an art object with an existence of its own, independent of or not necessarily related to its author, its readers, the historical time it depicts, or the historical period in which it was written. Its meaning emerges when readers scrutinize it and it alone, without regard to any of the aforementioned considerations. Such scrutiny results in the perception of the literary work as an organic whole wherein all the parts fit together and are so perfectly related as to form an objective meaning. Formalism concentrates on the text as the sole source of interpretation. The poem or play or story has its meaning in itself and reveals that meaning to a critic-reader who examines it on its own terms. The aim of formalist criticism is to show how the work achieves its meaning.

This formalist critical perspective is the result of a view that sees literature itself not essentially as a means to an end (the way the Greeks looked at it), or as an expression of individualism and emotion and common connections of all human beings (salient romantic tendencies), or as the product of complex psychological impulses (a modern psychoanalytic point of view). Rather, formalism sees literature as a unique and peculiar kind of knowledge, which presents humans with the deepest truths related to them, truths that science is unable to disclose. It has its own language, a language different from and more intense than ordinary (scientific) language. This language is not, however, subjective or anarchic. It is comprehensible and discernible by a rigorous and systematic methodology: close reading and the application of the concepts and vocabulary of literary analysis. Paradoxically, though denigrating science as the sole means of knowing, formalist critics employ the techniques of science in interpreting literary art. Formalism, then, focuses on the *text*, finding all meaning and value in it and regarding everything else as extraneous, including readers, whom formalist critics regard as downright dangerous as sources of interpretation. To rely on readers as a source of meaning is to fall victim to subjectivism, relativism, and other types of critical madness.

Reader-response critics take a radically different approach. They feel that readers have been ignored in discussions of the reading process instead of being the central concern as they should have been. The argument goes something like this. A text does not even exist, in a sense, until it is read by some reader. Indeed, the reader has a part in creating or actually does create the text. It is somewhat like the old question posed in philosophy classes: if a tree falls in the forest and no one hears it, does it make a sound? Reader-response critics are saying that in effect, if a text does not have a reader, it does not exist—or at least, it has no meaning. It is readers, with whatever experience they bring to the text, who give it its meaning. Whatever meaning it may have inheres in the reader, and thus it is the reader who should say what a text means.

We should, perhaps, point out here that reader-response theory is by no means a monolithic critical position. Those who give an important place to *readers* and their *responses* in interpreting a work come from a number of different critical camps, not excluding formalism, which is the target of the heaviest reader-response attacks. Reader-response critics see formalist critics as narrow, dogmatic, elitist, and certainly wrong-headed in essentially refusing readers even a place in the reading-interpretive process. Conversely, reader-response critics see themselves, as Jane Tompkins has put it, "willing to share their critical authority with less tutored readers and at the same time to go into partnership with psychologists, linguists, philosophers, and other students of mental functioning" (*Reader-Response Criticism: From Formalism to Post-Structuralism* [Baltimore: Johns Hopkins UP, 1980]: 223).

Although reader-response ideas were present in critical writing as long ago as the 1920s, most notably in that of I. A. Richards, and in the 1930s in D. W. Harding's and Louise Rosenblatt's work, it is not until the mid-century that they begin to gain increasing currency. Walker Gibson, writing in *College English* in February 1950, talks about "mock readers," who enact roles which actual readers feel compelled to play because the author clearly expects them to by the way the text is presented ("Authors, Speakers, Readers, and Mock Readers," *College English* 11.5 [1950]: 265–69). By the 1960s and continuing into the present as a more or less concerted movement, reader-response criticism had gained enough advocates to mount a frontal assault on the bastions of formalism.

Because the ideas underlying reader-response criticism are complex, and because their proponents frequently present them in technical language, it will be well to enumerate the forms that have received most attention and to attempt as clear a definition of them as possible.

Let us review once again the basic premises of reader-oriented theory, realizing that individual reader-response theorists will differ on a given point but that the following tenets reflect the main perspectives in the position as a whole. First, in literary interpretation, the text is not the most important component; the reader is. In fact, there is no text unless there is a reader. And the reader is the only one who can say what the text is; in a sense, the reader creates the text as much as the author does. This being the case, to arrive at meaning, critics should reject the autonomy of the text and concentrate on the reader and the reading process, the interaction that takes place between the reader and the text.

This premise perplexes people trained in the traditional methods of literary analysis. It declares that reading-response theory is subjective and relative, whereas earlier theories sought for as much objectivity as possible in a field of study that has a high degree of subjectivity by definition. Paradoxically, the ultimate source of this subjectivity is modern science itself, which has become increasingly skeptical that any objective knowledge is possible. Einstein's theory of relativity stands as the best known expression of that doubt. Also, the philosopher Thomas S. Kuhn's demonstration that scientific fact is dependent on the observer's frame of reference reinforces the claims of subjectivity (*The Structure of Scientific Revolutions* [Chicago: U of Chicago P, 1962]).

Another special feature of reader-response theory is that it is based on rhetoric, the art of persuasion, which has a long tradition in literature dating back to the Greeks, who originally employed it in oratory. It now refers to the myriad devices or strategies used to get the reader to respond to the literary work in certain ways. Thus, by establishing the reader firmly in the literary equation, the ancients may be said to be precursors of modern reader-response theory. Admittedly, however, when Aristotle, Longinus, Horace, Cicero, and Quintilian applied rhetorical principles in judging a work, they concentrated on the presence of the formal elements within the work rather than on the effect they would produce on the reader.

In view, then, of the emphasis on the audience in reader-response criticism, its relationship to rhetoric is quite obvious. Wayne Booth in his *Rhetoric of Fiction* (Chicago: U of Chicago P, 1961) was among the earliest of modern critics to restore readers to consideration in the interpretive act. The New Criticism, which strongly influenced the study of literature, and still does, had actually proscribed readers, maintaining that it was a critical fallacy, the affective fallacy, to mention any effects that a piece of literature might have on them. And while Booth did not go as far as some critics in assigning readers the major role in interpretation, he certainly did give them prominence and called rhetoric "the author's means of controlling his reader" (Preface to *Rhetoric of Fiction*). For example, in a close reading of Jane Austen's *Emma*, Booth demonstrates the rhetorical strategies that Austen uses to ensure the reader's seeing things through the heroine's eyes.

In 1925 I. A. Richards, usually associated with the New Critics, published *Principles of Literary Criticism* (New York: Harcourt), in which he constructed an affective system of interpretation, that is, one based on emotional responses. Unlike the New Critics who were to follow in the next two decades, Richards conceded that the scientific conception of truth is the correct one and that poetry provides only pseudo-statements. These pseudo-statements, however, are crucial to the psychic health of humans because they have now replaced religion as fulfilling our desire—"appetency" is Richards's term—for truth, that is, for some vision of the world that will satisfy our deepest needs. Matthew Arnold had in the nineteenth century predicted that literature would fulfill this function. Richards tested his theory by asking Cambridge students to write their responses to and assessments of a number of short unidentified poems of varying quality. He then analyzed and classified the responses and published them along with his own interpretations in *Practical Criticism* (New York: Harcourt, 1929). Richards's methodology is decidedly reader-response, but the use he made of his data is new critical. He arranged the responses he had received into categories according to the degrees in which they differed from the "right" or "more adequate" interpretation, which he demonstrated by referring to "the poem itself."

Louise Rosenblatt, Walker Gibson, and Gerald Prince are critics who, like Richards, affirm the importance of the reader but are not willing to relegate the text to a secondary role. Rosenblatt feels that irrelevant responses finally have to be excluded in favor of relevant ones and that a text can exist independently of readers. However, she advances a transactional theory: a poem comes into being only when it receives a proper ("aesthetic") reading, that is, when readers "compenetrate" a given text (*The Reader, the Text, the Poem* [Carbondale: Southern Illinois UP, 1978]). Gibson, essentially a formalist, proposes a mock reader, a role which the real reader plays because the text asks him or her to play it "for the sake of the experience." Gibson posits a dialogue between a speaker (the author?) and the mock reader. The critic, overhearing this dialogue, paraphrases it, thereby revealing the author's strategies for getting readers to accept or reject whatever the author wishes them to. Gibson by no means abandons the text, but he injects the reader further into the interpretive operation as a way of gaining fresh critical insights. Using a different terminology, Prince adopts a perspective similar to Gibson's. Wondering why critics have paid such close attention to narrators (omniscient, first person, unreliable, etc.) and have virtually ignored readers, Prince too posits a reader, whom he calls the narratee, one of a number of hypothetical readers to whom the story is directed. These readers, actually produced by the narrative, include the real reader, with book in hand; the virtual reader, for whom the author thinks he is writing; and the ideal reader of perfect understanding and sympathy; yet none of these is necessarily the narratee. Prince demonstrates the strategies by which the narrative creates the readers ("Introduction to the Study of the Narratee," in *Reader-Response Criticism*, ed. Jane Tompkins, 7–25).

The critics mentioned so far—except Prince—are the pioneers, or perhaps more accurately, the advance guard of the reader-response movement. While continuing to insist on the importance of the text in the interpretive act, they equally insist that the reader be taken into account; not to do so will, they maintain, either impoverish the interpretation or render it defective. As the advance guard, they have cleared the way for those who have become the principal theorists of reader-response criticism. Though there will be disagreement on who belongs in this latter group, most scholars would recognize Wolfgang Iser, Hans Robert Jauss, Norman Holland, and Stanley Fish as having major significance in the movement.

Wolfgang Iser is a German critic who applies the philosophy of phenomenology to the interpretation of literature. Phenomenology stresses the perceiver's (in this case, the reader's) role in any perception (in this case, reading experience) and asserts the difficulty, if not the impossibility, of separating anything known from the mind that knows it. According to Iser, the critic should not explain the text *as an object* but *its effect on the reader.* Iser's espousal of this position, however, has not taken him away from the text as a central part of interpretation. He also has posited an implied reader, one with "roots firmly planted in the structure of the text" (*The Act of Reading* [Baltimore: Johns Hopkins UP, 1978]: 34). Still, his phenomenological beliefs keep him from the formalist notion that there is one essential meaning of a text that all interpretations must try to agree on. Readers' experiences will govern the effects the text produces on them. Moreover, Iser says, a text does not tell readers everything; there are "gaps" or "blanks," which he refers to as the indeterminacy of the text. Readers must fill these in and thereby assemble the meaning(s), thus becoming coauthors in a sense. Such meanings may go far beyond the single "best" meaning of the formalist because they are the products of such varied reader backgrounds. To be sure, Iser's implied readers are fairly sophisticated: they bring to the contemplation of the text a conversance with the conventions that enables them to decode the text. But the text can transcend any set of literary or critical conventions, and readers with widely different backgrounds may fill in those blanks and gaps with new and unconventional meanings. Iser's stance, then, is phenomenological: at the center of interpretation lies the reader's experience. Nor does this creation of text by the reader mean that the resultant text is subjective and no longer the author's. It is rather, says Iser, proof of the text's inexhaustibility.

Yet another kind of reader-oriented criticism, also rhetorically grounded, is reception theory, which documents reader responses to authors and/or their works in any given period. Such criticism depends heavily on reviews in newspapers, magazines, and journals and on personal letters for evidence of public reception. There are varieties of reception theory, one of the most important recent types promulgated by Hans Robert Jauss, another German scholar, in his *Toward an Aesthetic of Reception* (trans. Timothy Bahti; Minneapolis: U of Minnesota P, 1982). Jauss seeks to bring about a compromise between that interpretation which ignores history and that which ignores the text in favor of social theories. To describe the criteria he would employ, Jauss has proposed the term horizons of expectations of a reading public. These result from what the public already understands about a genre and its conventions. For example, Pope's poetry was judged highly by his contemporaries, who valued clarity, decorum, and wit. The next century had different horizons of expectations and thus actually called into question Pope's claim to being considered a poet at all. Similarly, Flaubert's *Madame Bovary* was not well received by its mid-nineteenth-century readers, who objected to the impersonal, clinical, naturalistic style. Their horizons of expectations had conditioned them to appreciate an impassioned, lyrical, sentimental, and florid narrative method. Delayed hostile reader response to firmly established classics surfaced in the latter half of the twentieth century. *Huckleberry Finn* became the target of harsh and misguided criticism on the grounds that it contained racial slurs in the form of epithets like "nigger" and demeaning portraits of Negroes. Schools were in some instances required to remove the book from curriculums or reading lists of approved books and in extreme cases from library shelves. In like manner, feminists have resented what they considered male-chauvinist philosophy and attitudes in Marvell's "To His Coy Mistress." The horizon of expectations of these readers incorporated hot partisanship on contemporary issues into their literary analyses of earlier works.

Horizons of expectations do not establish the final meaning of a work. Thus, according to Jauss, we cannot say that a work is universal, that it will make the same appeal to or impact on readers of all eras. Is it possible, then, ever to reach a critical verdict about a piece of literature? Jauss thinks it is possible only to the extent that we regard our interpretations as stemming from a dialogue between past and present and thereby representing a fusion of horizons.

The importance of psychology in literary interpretation has long been recognized. Plato and Aristotle, for example, attributed strong psychological influence to literature. Plato saw this influence as essentially baneful: literature aroused people's emotions, especially those that ought to be stringently controlled. Conversely, Aristotle argued that literature exerted a good psychological influence; in particular, tragedy did, by effecting in audiences a catharsis or cleansing or purging of emotions. Spectators were thus calmed and satisfied, not excited or frenzied, after their emotional encounter.

As we noted in our earlier chapter on the psychological approach, one of the world's preeminent depth psychologists, Sigmund Freud, has had an incalculable influence on literary analysis with his theories about the unconscious and about the importance of sex in explaining much human behavior. Critics, then, have looked to Plato and Aristotle in examining the psychological relations between a literary work and its audience and to Freud in seeking to understand the unconscious psychological motivations of the characters in the literary work and in the author.

If, however, followers of Freud have been more concerned with the unconscious of literary characters and their creators, more recent psychological critics have focused on the unconscious of readers. Norman Holland, one such critic, argues that all people inherit from their mother an identity theme or fixed understanding of the kind of person they are. Whatever they read is processed to make it fit their identity theme ("The Miller's Wife and the Professors: Questions about the Transactive Theory of Reading," *New Literary History* 17 [1986]: 423–47). In other words, readers interpret texts as expressions of their own personalities or psyches and thereby use their interpretations as a means of coping with life. Holland illustrates this thesis in an essay entitled "Hamlet—My Greatest Creation" (*Journal of the American Academy of Psychoanalysis* 3 [1975]: 419–27). This highly personal response to literature appears in another Holland article, "Recovering 'The Purloined Letter': Reading as a Personal Transaction" (in Suleiman and Crosman, eds., *The Reader in the Text* [Princeton, NJ: Princeton UP, 1980]: 350–70). Here, Holland relates the story to his own attempt to hide an adolescent masturbatory experience.

Holland's theory, for all of its emphasis on readers and their psychology, does not deny or destroy the independence of the text. It exists as an object and as the expression of another mind, something different from readers themselves, something they can project onto. But David Bleich, who calls his variety of reader-response subjectivism, does deny that the text exists independent of readers (*Subjective Criticism* [Baltimore: Johns Hopkins UP, 1978]). Bleich accepts the arguments of such contemporary philosophers of science as Thomas S. Kuhn who deny that objective facts exist. Such a position asserts that even what passes for scientific observation of something—of anything—is still merely individual and subjective perception occurring in a special context. Bleich claims that individuals everywhere classify things into three essential groups: objects, symbols, and people. Literature, a mental creation (as opposed to a concrete one), would thus be considered a symbol. A text may be an object in that it is paper (or other matter) and print, but its meaning depends on the symbolization in the minds of readers. Meaning is not found; it is developed. Better human relations will result from readers with widely differing views sharing and comparing their responses and thereby discovering more about motives and strategies for reading. The honesty and tolerance required in such operations is bound to help in self-knowledge, which, according to Bleich, is the most important goal for everyone.

The last of the theorists to be treated in this discussion is Stanley Fish, who calls his technique of interpretation affective stylistics. Like other reader-oriented critics, Fish rebels against the so-called rigidity and dogmatism of the New Critics and especially against the tenet that a poem is a single, static object, a whole that has to be understood in its entirety at once. Fish's pronouncements on reader-response theory have come in stages. In an early stage, he argued that meaning in a literary work is not something to be extracted, as a dentist might pull a tooth; meaning must be negotiated by readers, a line at a time. Moreover, they will be surprised by rhetorical strategies as they proceed. Meaning is *what happens to readers during this negotiation.* A text, in Fish's view, could lead readers on, even set them up, to make certain interpretations, only to undercut them later and force readers into new and different readings. So, the focus is on the reader; the process of reading is dynamic and sequential. Fish does insist, however, on a high degree of sophistication in readers: they must be familiar with literary conventions and must be capable of changing when they perceive they have been tricked by the strategies of the text. His term for such readers is "informed" (*Surprised by Sin: The Reader in "Paradise Lost"* [Berkeley: U of California P, 1967]).

Fish later modified the method described above by attributing more initiative to the reader and less control by the text in the interpretive act. Fish's altered position holds that readers actually create a piece of literature as they read it. Fish concludes that every reading results in a new interpretation that comes about because of the strategies that readers use. The text as an independent director of interpretation has in effect disappeared. For Fish, interpretation is a communal affair. The readers just mentioned are informed; they possess linguistic competence, they form interpretive communities that have common assumptions; and, to repeat, they create texts when they pool their common reading techniques. These characteristics mean that such readers are employing the same or similar interpretive strategies and are thus members of the same interpretive community (*Is There a Text in This Class?* [Cambridge, MA: Harvard UP, 1980]).

It seems reasonable to say that there may be more than one response or interpretation of a work of literature and that this is true because responders and interpreters see things differently. It seems equally accurate to observe that to claim the meaning of literature rests exclusively with individual readers, whose opinions are equally valid, is to make literary analysis ultimately altogether relative. Somewhere within these two points of view most critics and interpreters will fall.

The procedure that we have followed in the other chapters, defining a critical approach and then applying it to four major pieces of literature representing the principal genres, will not work in as definite a way in reader-response criticism. Here, however, to illustrate, we shall cite, arbitrarily, two reader-responses to well-known works. Steven Mailloux's reading of Hawthorne's "Rappaccini's Daughter" is an engaging and convincing analysis of this complex narrative based on the thesis that snares and entanglements laid by an unreliable narrator serve to confuse readers until they learn to avoid such traps and arrive at some understanding based on their own interpretation of the characters' actions and not the omniscient author's (*Interpretive Conventions: The Reader in the Study of American Fiction* [Itaca and London: Cornell UP, 1984]: 73–92). Mailloux's reading of *Huckleberry Finn* combines a rhetorical approach, decidedly reader-oriented, and the new historicism, which stresses contemporary newspapers, magazine articles, public sentiment, prevailing ideologies, and so on ("Reading *Huckleberry Finn*: The Rhetoric of Performed Ideology," *New Essays on Huckleberry Finn*, ed. Louis Budd [Cambridge: Cambridge UP, 1985]: 107–33). A highly personalized, psychologized reader-response to *Hamlet* appears in Norman Holland's "*Hamlet*—My Greatest Creation," pages 171–76. Here, the focus is on the connection between words—prolixity—and parental violence or neglect. Though ingenious, this interpretation is less idiosyncratic than his reading of Poe's "The Purloined Letter," mentioned earlier.

To summarize, two distinguishing features characterize reader-response criticism. One is the effect of the literary work on the reader, hence the moral-philosophical-psychological-rhetorical emphases in reader-response analysis. (How does the work *affect* the reader, and what strategies or devices have come into play in the production of those effects?) The second feature is the relegation of the text to secondary importance. (The reader is of primary importance.) Thus, reader-response criticism attacks the authority of the text. This is where subjectivism comes in. If a text cannot have any existence except in the mind of the reader, then the text loses its authority. There is a shift from objective to subjective perspective. Texts mean what individual readers say they mean or what interpretive communities of readers say they mean. Since this is the case, the application of the reader-response approach—to *Huckleberry Finn*, for example—could result, at least theoretically, in as many readings of a work as there are readers. It would involve positing a hypothetical reader, whose response, while possibly interesting, would be random and arbitrary. Indeed, *which* reader response should we employ, since there are a number? If we have made the main reader-responses clear in principle to the readers of this book, we shall have accomplished our purpose. They may then apply them as they will. Thus, interpretation becomes the key to meaning—as it always is—but without the ultimate authority of the text or the author. The important element in reader-response criticism is the reader, and the effect (or affect) of the text on the reader.

When reader-response critics begin to analyze the effect of the text on the reader, the analysis often resembles formalist criticism or rhetorical criticism or psychological criticism. The major distinction is the emphasis on the reader's response in the analysis. Meaning inheres in the reader and not in the text. This is where reception theory fits in. The same text can be interpreted by different readers or communities of readers in very different ways. A text's interpretive history may vary considerably, as with Freudian interpretations of *Hamlet* versus earlier interpretations. Readers bring their own cultural heritage along with them in their responses to literary texts, a fact which allows for the principle that texts speak to other texts only through the intervention of particular readers. Thus, reader-response criticism can appropriate other theories—as all theories attempt to do.

Reader-response theory is likely to strike many people as both esoteric and too subjective. Unquestionably, readers had been little considered in the New Criticism; but they may have been over-emphasized by the theorists who seek to give them the final word in interpreting literature. Communication as a whole is predicated on the demonstrable claim that there are common, agreed-upon meanings in language, however rich, metaphorical, or symbolic. To contend that there are, even in theory, as many meanings in a poem as there are readers strongly calls into question the possibility of intelligible discourse. That some of the theorists themselves are not altogether comfortable with the logical implications of their position is evidenced by their positing of mock readers, informed readers, real readers, and implied readers—by which they mean readers of education, sensitivity, and sophistication.

Despite the potential dangers of subjectivism, reader-response criticism has been a corrective to literary dogmatism and a reminder of the richness, complexity, and diversity of viable literary interpretations, and it seems safe to predict that readers will never again be completely ignored in arriving at verbal meaning.

Epilogue

"How do you learn to read this way? Where do you learn this? Do you take a course in symbols or something?"

With a rising, plaintive pitch to his voice, with puzzled eyes and shaking head, a college student once asked those questions after his class had participated in a lively discussion of the multiple levels in Henry James's Turn of the Screw. As with most students when they are first introduced to a serious study of literature, members of this group were delighted, amazed—and dismayed—as they themselves helped to unfold the rich layers of the work, to see it from perspectives of form and of psychology, to correlate it with the author's biography and its cultural and historical context.

But that particular student, who was both fascinated and dismayed by the "symbols," had not yet taken a crucial step in the learning process: he had not perceived that the practice of close reading, the bringing to bear of all kinds of knowledge, and the use of several approaches are in themselves the "course in symbols or something." What we have traced in this volume is not a course in the occult or something only for those who have access to the inner sanctum. For, after all, as Wordsworth wrote in 1800 in his preface to Lyrical Ballads, the poet "is a man speaking to men." A poet or a dramatist embodies an experience in a poem or a play, embodies it—usually—for us, the readers; and we respond simply by reliving that experience as fully as possible.

345

To be sure, not all of us may want to respond to that extent. There was, for example, the secondary school teacher who listened to a fairly long and detailed explication of "The Death of the Ball-Turret Gunner," a five-line poem by Randall Jarrell. Later she took the lecturer aside and said, with something more than asperity, "I'd *never* make my class try to see all of *that* in a poem." Perhaps not. But is the class better or worse because of that attitude?

Clearly, the authors of this book believe that we readers are the losers when we fail to see in a work of literature all that may be legitimately seen there. We have presented a number of critical approaches to literature, aware that many have been only briefly treated and that much has been generalized. But we have suggested here some of the tools and some of the approaches that enable a reader to criticize—that is, to judge and to discern so that he or she may see better the literary work, to relate it to the range of human experience, to appreciate its form and style.

Having offered these tools and approaches, we would also urge caution against undue or misdirected enthusiasm in their use, for judgment and discernment imply reason and caution. Too often even seasoned critics, forgetting the etymology of the word *critic*, become personal and subjective or preoccupied with tangential concerns. Not-so-seasoned students, their minds suddenly open to psychoanalytic criticism, run gaily through a pastoral poem and joyously find Freudian symbols in every rounded hill and stately conifer. Some read a simple poem like William Carlos Williams's "The Red Wheelbarrow" and, unwilling to see simplicity and compression as virtues enough (and as much more than mere simpleness), stray from the poem into their individual mazes. They forget that any interpretation must be supported logically and fully from the evidence within the literary work and that the ultimate test of the validity of an interpretation must be its self-consistency. Conversely, sometimes they do establish a fairly legitimate pattern of interpretation for a work, only to find something that seems to be at odds with it; then, fascinated with or startled by what they assume to be a new element, they forget that their reading is not valid unless it permits a unified picture of both the original pattern and the new insight. In short, an object of literary art has its unique aesthetic experience; the reader is no more at liberty to mar it with careless extensions than the author would have been free to damage its organic unity with infelicitous inclusions.

Having said this, we acknowledge that individual readers bring their own unique experiences to the perception of a literary work of art; since these experiences may and will be vastly different, they will color the readers' perceptions. As we have shown elsewhere in this handbook, important recent critical theories have acknowledged and furthered this potential for more subjective interpretations.

We must therefore remember to be flexible and eclectic in our choices of critical approaches to a given literary work. Our choices are determined by the same discretion that controls what we exclude, by our concern for the unique experience and nature of a piece of literature. Not all approaches are useful in all cases. Perhaps we would not be too far wrong to suggest that there are as many approaches to literary works as there are literary works. All we can do is to draw from the many approaches the combination that best fits a particular literary creation. As David Daiches said at the end of *Critical Approaches to Literature* (Englewood Cliffs, NJ: Prentice, 1956), "Every effective literary critic sees some facet of literary art and develops an awareness with respect to it; but the total vision, or something approximating it, comes only to those who learn how to blend the insights yielded by many critical approaches" (393).

That is why we have chosen to present a variety of approaches and why some of the chapters in this book have even blended several methods. This blending is as it should be. It is not easy—and it would be unwise to try—to keep the work always separate from the life of the author and a view of his or her times; to divide the study of form from the study of basic imageries; to segregate basic imageries from archetypes or from other components of the experience of the work. And it would be unwise to ignore how, for example, a work long known and interpreted by conventional methods might yield fresh insights if examined from the perspectives of neo-Marxist criticism, feminist criticism, or phenomenological criticism.

Our final word, then, is this: we admit that literary criticism can be difficult and sometimes esoteric, but it is first of all an attempt of readers to understand fully what they are reading. To understand in that manner, they do well to bring to bear whatever is in the human province that justifiably helps them to achieve that understanding. For literature is a part of the richness of human experience: it at once thrives on it, feeds it, and constitutes a significant part of it. When we realize this, we never again can be satisfied with the simple notions that a story is something only for the idler or the impractical dreamer, that a poem is merely a pretty combination of sounds and sights, that a significant drama is equivalent to an escapist motion picture or a television melodrama. Browning's Fra Lippo Lippi says:

> This world's no blot for us,
> Nor blank; it means intensely, and means good:
> To find its meaning is my meat and drink.

So, too, must be our attitude toward any worthy piece of literature in that world.

Appendixes

Andrew Marvell
TO HIS COY MISTRESS

Had we but world enough, and time,
This coyness, Lady, were no crime.
We would sit down and think which way
To walk and pass our long love's day.
Thou by the Indian Ganges' side   5
Shouldst rubies find; I by the tide
Of Humber would complain. I would
Love you ten years before the Flood,
And you should, if you please, refuse
Till the conversion of the Jews.   10
My vegetable love should grow
Vaster than empires, and more slow;
An hundred years should go to praise
Thine eyes and on thy forehead gaze;
Two hundred to adore each breast,   15
But thirty thousand to the rest;
An age at least to every part,
And the last age should show your heart.
For, Lady, you deserve this state,
Nor would I love at lower rate.   20

But at my back I always hear
Time's wingèd chariot hurrying near;
And yonder all before us lie
Deserts of vast eternity.

Handwritten annotation: — Dramatic monolouge the speaker is the Character & is adressing another person but the listenener doesn't talk.

349

Thy beauty shall no more be found,   25
Nor, in thy marble vault, shall sound
My echoing song; then worms shall try
That long preserved virginity,
And your quaint honor turn to dust,
And into ashes all my lust: 30
The grave's a fine and private place,
But none, I think, do there enbrace.

Now therefore, while the youthful hue
Sits on thy skin like morning dew,
And while thy willing soul transpires   35
At every pore with instant fires,
Now let us sport us while we may,
And now, like amorous birds of prey,
Rather at once our time devour
Than languish in his slow-chapped power.   40
Let us roll all our strength and all
Our sweetness up into one ball,
And tear our pleasures with rough strife
Thorough* the iron gates of life:
Thus, though we cannot make our sun   45
Stand still, yet we will make him run.

*thorough: through

Nathaniel Hawthorne
YOUNG GOODMAN BROWN

Young Goodman Brown came forth at sunset into the street at Salem Village; but put his head back, after crossing the threshold, to exchange a parting kiss with his young wife. And Faith, as the wife was aptly named, thrust her own pretty head into the street, letting the wind play with the pink ribbons of her cap while she called to Goodman Brown.

"Dearest heart," whispered she, softly and rather sadly, when her lips were close to his ear, "prithee put off your journey until sunrise and sleep in your own bed to-night. A lone woman is troubled with such dreams and such thoughts that she's afeard of herself sometimes. Pray tarry with me this night, dear husband, of all nights in the year."

"My love and my Faith," replied young Goodman Brown, "of all nights in the year, this one night must I tarry away from thee. My journey, as thou callest it, forth and back again, must needs be done 'twixt now and sunrise. What, my sweet, pretty wife, dost thou doubt me already, and we but three months married?"

"Then God bless you!" said Faith, with the pink ribbons; "and may you find all well when you come back."

"Amen!" cried Goodman Brown. "Say thy prayers, dear Faith, and go to bed at dusk, and no harm will come to thee."

So they parted; and the young man pursued his way until, being about to turn the corner by the meeting-house, he looked back and saw the head of Faith still peeping after him with a melancholy air, in spite of her pink ribbons.

"Poor little Faith!" thought he, for his heart smote him. "What a wretch am I to leave her on such an errand! She talks of dreams, too. Methought as she spoke there was trouble in her face, as if a dream had warned her what work is to be done to-night. But no, no; 'twould kill her to think it. Well, she's a blessed angel on earth; and after this one night I'll cling to her skirts and follow her to heaven."

With this excellent resolve for the future, Goodman Brown felt himself justified in making more haste on his present evil purpose. He had taken a dreary road, darkened by all the gloomiest trees of the forest, which barely stood aside to let the narrow path creep through, and closed immediately behind. It was all as lonely as could be; and there is this peculiarity in such a solitude, that the traveller knows not who may be concealed by the innumerable trunks and the thick boughs overhead; so that with lonely footsteps he may yet be passing through an unseen multitude.

"There may be a devilish Indian behind every tree," said Goodman Brown to himself; and he glanced fearfully behind him as he added, "What if the devil himself should be at my very elbow!"

His head being turned back, he passed a crook of the road, and, looking forward again, beheld the figure of a man, in grave and decent attire, seated at the foot of an old tree. He arose at Goodman Brown's approach and walked onward side by side with him.

"You are late, Goodman Brown," said he. "The clock of the Old South was striking as I came through Boston, and that is full fifteen minutes agone."

"Faith kept me back a while," replied the young man, with a tremor in his voice, caused by the sudden appearance of his companion, though not wholly unexpected.

It was now deep dusk in the forest, and deepest in that part of it where these two were journeying. As nearly as could be discerned, the second traveller was about fifty years old, apparently in the same rank of life as Goodman Brown, and bearing a considerable resemblance to him, though perhaps more in expression than features. Still they might have been taken for father and son. And yet, though the elder person was as simply clad as the younger, and as simple in manner too, he had an indescribable air of one who knew the world, and who would not have felt abashed at the governor's dinner table or in King William's court, were it possible that his affairs should call him thither. But the only thing about him that could be fixed upon as remarkable was his staff, which bore the likeness of a great black snake, so curiously wrought that it might almost be seen to twist and wriggle itself like a living serpent. This, of course, must have been an ocular deception, assisted by the uncertain light.

"Come, Goodman Brown," cried his fellow-traveller, "this

is a dull pace for the beginning of a journey. Take my staff, if you are so soon weary."

"Friend," said the other, exchanging his slow pace for a full stop, "having kept covenant by meeting thee here, it is my purpose now to return whence I came. I have scruples touching the matter thou wot'st of."

"Sayest thou so?" replied he of the serpent, smiling apart. "Let us walk on, nevertheless, reasoning as we go; and if I convince thee not thou shalt turn back. We are but a little way in the forest yet."

"Too far! too far!" exclaimed the goodman, unconsciously resuming his walk. "My father never went into the woods on such an errand, nor his father before him. We have been a race of honest men and good Christians since the days of the martyrs; and shall I be the first of the name of Brown that ever took this path and kept—"

"Such company, thou wouldst say," observed the elder person, interpreting his pause. "Well said, Goodman Brown! I have been as well acquainted with your family as with ever a one among the Puritans; and that's no trifle to say. I helped your grandfather, the constable, when he lashed the Quaker woman so smartly through the streets of Salem; and it was I that brought your father a pitch-pine knot, kindled at my own hearth, to set fire to an Indian village, in King Philip's war. They were my good friends, both; and many a pleasant walk have we had along this path, and returned merrily after midnight. I would fain be friends with you for their sake."

"If it be as thou sayest," replied Goodman Brown, "I marvel they never spoke of these matters; or, verily, I marvel not, seeing that the least rumor of the sort would have driven them from New England. We are a people of prayer, and good works to boot, and abide no such wickedness."

"Wickedness or not," said the traveller with the twisted staff, "I have a very general acquaintance here in New England. The deacons of many a church have drunk the communion wine with me; the selectmen of divers towns make me their chairman; and a majority of the Great and General

Court are firm supporters of my interest. The governor and I, too— But these are state secrets."

"Can this be so?" cried Goodman Brown, with a stare of amazement at his undisturbed companion. "Howbeit, I have nothing to do with the governor and council; they have their own ways, and are no rule for a simple husbandman like me. But, were I to go on with thee, how should I meet the eye of that good old man, our minister, at Salem village? Oh, his voice would make me tremble both Sabbath day and lecture day."

Thus far the elder traveller had listened with due gravity; but now burst into a fit of irrepressible mirth, shaking himself so violently that his snake-like staff actually seemed to wriggle in sympathy.

"Ha! ha! ha!" shouted he again and again; then composing himself, "Well, go on, Goodman Brown, go on; but, prithee, don't kill me with laughing."

"Well, then, to end the matter at once," said Goodman Brown, considerably nettled, "there is my wife, Faith. It would break her dear little heart; and I'd rather break my own."

"Nay, if that be the case," answered the other, "e'en go thy ways, Goodman Brown. I would not for twenty old women like the one hobbling before us that Faith should come to any harm."

As he spoke he pointed his staff at a female figure on the path, in whom Goodman Brown recognized a very pious and exemplary dame, who had taught him his catechism in youth, and was still his moral and spiritual adviser, jointly with the minister and Deacon Gookin.

"A marvel, truly, that Goody Cloyse should be so far in the wilderness at nightfall," said he. "But with your leave, friend, I shall take a cut through the woods until we have left this Christian woman behind. Being a stranger to you, she might ask whom I was consorting with and whither I was going."

"Be it so," said his fellow-traveller. "Betake you to the woods, and let me keep the path."

Accordingly the young man turned aside, but took care to watch his companion, who advanced softly along the road until he had come within a staff's length of the old dame. She, meanwhile, was making the best of her way, with singular speed for so aged a woman, and mumbling some indistinct words—a prayer, doubtless—as she went. The traveller put forth his staff and touched her withered neck with what seemed the serpent's tail.

"The devil!" screamed the pious old lady.

"Then Goody Cloyse knows her old friend?" observed the traveller, confronting her and leaning on his writhing stick.

"Ah, forsooth, and is it your worship indeed?" cried the good dame. "Yea, truly it is, and in the very image of my old gossip, Goodman Brown, the grandfather of the silly fellow that now is. But—would your worship believe it?—my broomstick hath strangely disappeared, stolen, as I suspect, by that unhanged witch, Goody Cory, and that, too, when I was anointed with the juice of smallage, and cinquefoil, and wolf's bane—"

"Mingled with fine wheat and the fat of a new-born babe," said the shape of old Goodman Brown.

"Ah, your worship knows the recipe," cried the old lady, cackling aloud. "So, as I was saying, being all ready for the meeting, and no horse to ride on, I made up my mind to foot it; for they tell me there is a nice young man to be taken into communion to-night. But now your good worship will lend me your arm, and we shall be there in a twinkling."

"That can hardly be," answered her friend. "I may not spare you my arm, Goody Cloyse; but here is my staff, if you will."

So saying, he threw it down at her feet, where, perhaps, it assumed life, being one of the rods which its owner had formerly lent to the Egyptian magi. Of this fact, however, Goodman Brown could not take cognizance. He had cast up his eyes in astonshiment, and, looking down again, beheld neither Goody Cloyse nor the serpentine staff, but his fellow-traveller alone, who waited for him as calmly as if nothing happened.

"That old woman taught me my catechism," said the young man; and there was a world of meaning in this simple comment.

They continued to walk onward, while the elder traveller exhorted his companion to make good speed and persevere in the path, discoursing so aptly that his arguments seemed rather to spring up in the bosom of his auditor than to be suggested by himself. As they went, he plucked a branch of maple to serve for a walking stick, and began to strip it of the twigs and little boughs, which were wet with evening dew. The moment his fingers touched them they became strangely withered and dried up as with a week's sunshine. Thus the pair proceeded, at a good free pace, until suddenly, in a gloomy hollow of the road, Goodman Brown sat himself down on the stump of a tree and refused to go any farther.

"Friend," said he, stubbornly, "my mind is made up. Not another step will I budge on this errand. What if a wretched old woman do choose to go to the devil when I thought she was going to heaven: is that any reason why I should quit my dear Faith and go after her?"

"You will think better of this by and by," said his acquaintance, composedly. "Sit here and rest yourself a while; and when you feel like moving again, there is my staff to help you along."

Without more words, he threw his companion the maple stick, and was as speedily out of sight as if he had vanished into the deepening gloom. The young man sat a few moments by the roadside, applauding himself greatly, and thinking with how clear a conscience he should meet the minister in his morning walk, nor shrink from the eye of good old Deacon Gookin. And what calm sleep would be his that very night, which was to have been spent so wickedly, but so purely and sweetly now, in the arms of Faith! Amidst these pleasant and praiseworthy meditations, Goodman Brown heard the tramp of horses along the road, and deemed it advisable to conceal himself within the verge of the forest, conscious of the guilty purpose that had brought him thither, though now so happily turned from it.

On came the hoof tramps and the voices of the riders, two grave old voices, conversing soberly as they drew near. These mingled sounds appeared to pass along the road, within a few yards of the young man's hiding-place; but, owing doubtless to the depth of the gloom at that particular spot, neither the travellers nor their steeds were visible. Though their figures brushed the small boughs by the wayside, it could not be seen that they intercepted, even for a moment, the faint gleam from the strip of bright sky athwart which they must have passed. Goodman Brown alternately crouched and stood on tiptoe, pulling aside the branches and thrusting forth his head as far as he durst without discerning so much as a shadow. It vexed him the more, because he could have sworn, were such a thing possible, that he recognized the voices of the minister and Deacon Gookin, jogging along quietly, as they were wont to do, when bound to some ordination or ecclesiastical council. While yet within hearing, one of the riders stopped to pluck a switch.

"Of the two, reverend sir," said the voice like the deacon's, "I had rather miss an ordination dinner than to-night's meeting. They tell me that some of our community are to be here from Falmouth and beyond, and others from Connecticut and Rhode Island, besides several of the Indian powwows, who, after their fashion, know almost as much deviltry as the best of us. Moreover, there is a goodly young woman to be taken into communion."

"Mighty well, Deacon Gookin!" replied the solemn old tones of the minister. "Spur up, or we shall be late. Nothing can be done, you know, until I get on the ground."

The hoofs clattered again; and the voices, talking so strangely in the empty air, passed on through the forest, where no church had ever been gathered or solitary Christian prayed. Whither, then, could these holy men be journeying so deep into the heathen wilderness? Young Goodman Brown caught hold of a tree for support, being ready to sink down on the ground, faint and overburdened with the heavy sickness of his heart. He looked up to the sky, doubting whether there really was a heaven above him. Yet there was the blue arch, and the stars brightening in it.

"With heaven above and Faith below, I will yet stand firm against the devil!" cried Goodman Brown.

While he still gazed upward into the deep arch of the firmament and had lifted his hands to pray, a cloud, though no wind was stirring, hurried across the zenith and hid the brightening stars. The blue sky was still visible, except directly overhead, where this black mass of cloud was sweeping swiftly northward. Aloft in the air, as if from the depths of the cloud, came a confused and doubtful sound of voices. Once the listener fancied that he could distinguish the accents of townspeople of his own, men and women, both pious and ungodly, many of whom he had met at the communion table, and had seen others rioting at the tavern. The next moment, so indistinct were the sounds, he doubted whether he had heard aught but the murmur of the old forest, whispering without a wind. Then came a stronger swell of those familiar tones, heard daily in the sunshine at Salem village, but never until now from a cloud of night. There was one voice, of a young woman, uttering lamentations, yet with an uncertain sorrow, and entreating for some favor, which, perhaps, it would grieve her to obtain; and all the unseen multitude, both saints and sinners, seemed to encourage her onward.

"Faith!" shouted Goodman Brown, in a voice of agony and desperation; and the echoes of the forest mocked him, crying, "Faith! Faith!" as if bewildered wretches were seeking her all through the wilderness.

The cry of grief, rage, and terror was yet piercing the night, when the unhappy husband held his breath for a response. There was a scream, drowned immediately in a louder murmur of voices, fading into far-off laughter, as the dark cloud swept away, leaving the clear and silent sky above Goodman Brown. But something fluttered lightly down through the air and caught on the branch of a tree. The young man seized it, and beheld a pink ribbon.

"My Faith is gone!" cried he, after one stupefied moment. "There is no good on earth; and sin is but a name. Come, devil; for to thee is this world given."

And, maddened with despair, so that he laughed loud and long, did Goodman Brown grasp his staff and set forth again, at such a rate that he seemed to fly along the forest path rather than to walk or run. The road grew wilder and drearier and more faintly traced, and vanished at length, leaving him in the heart of the dark wilderness, still rushing onward with the instinct that guides mortal man to evil. The whole forest was peopled with frightful sounds—the creaking of the trees, the howling of wild beasts, and the yell of Indians; while sometimes the wind tolled like a distant church bell, and sometimes gave a broad roar around the traveller, as if all Nature were laughing him to scorn. But he was himself the chief horror of the scene, and shrank not from its other horrors.

"Ha! ha! ha!" roared Goodman Brown when the wind laughed at him. "Let us hear which will laugh loudest. Think not to frighten me with your deviltry. Come witch, come wizard, come Indian powwow, come devil himself, and here comes Goodman Brown. You may as well fear him as he fear you."

In truth, all through the haunted forest there could be nothing more frightful than the figure of Goodman Brown. On he flew among the black pines, brandishing his staff with frenzied gestures, now giving vent to an inspiration of horrid blasphemy, and now shouting forth such laughter as set all the echoes of the forest laughing like demons around him. The fiend in his own shape is less hideous than when he rages in the breast of man. Thus sped the demoniac on his course, until, quivering among the trees, he saw a red light before him, as when the felled trunks and branches of a clearing have been set on fire, and throw up their lurid blaze against the sky, at the hour of midnight. He paused, in a lull of the tempest that had driven him onward, and heard the swell of what seemed a hymn, rolling solemnly from a distance with the weight of many voices. He knew the tune; it was a familiar one in the choir of the village meeting-house. The verse died heavily away, and was lengthened by a chorus, not of human voices, but of all the sounds of the benighted wilderness pealing in awful harmony together. Goodman Brown cried out, and his cry was lost to his own ear by its unison with the cry of the desert.

In the interval of silence he stole forward until the light glared full upon his eyes. At one extremity of an open space, hemmed in by the dark wall of the forest, arose a rock, bearing some rude, natural resemblance either to an altar or a pulpit, and surrounded by four blazing pines, their tops aflame, their stems untouched, like candles at an evening meeting. The mass of foliage that had overgrown the summit of the rock was all on fire, blazing high into the night and fitfully illuminating the whole field. Each pendent twig and leafy festoon was in a blaze. As the red light arose and fell, a numerous congregation alternately shone forth, then disappeared in shadow, and again grew, as it were, out of the darkness, peopling the heart of the solitary woods at once.

"A grave and dark-clad company," quoth Goodman Brown.

In truth they were such. Among them, quivering to and fro between gloom and splendor, appeared faces that would be seen next day at the council board of the province, and others which, Sabbath after Sabbath, looked devoutly heavenward, and benignantly over the crowded pews, from the holiest pulpits in the land. Some affirm that the lady of the governor was there. At least there were high dames well known to her, and wives of honored husbands, and widows, a great multitude, and ancient maidens, all of excellent repute, and fair young girls, who trembled lest their mothers should espy them. Either the sudden gleams of light flashing over the obscure field bedazzled Goodman Brown, or he recognized a score of the church members of Salem village famous for their special sanctity. Good old Deacon Gookin had arrived, and waited at the skirts of that venerable saint, his revered pastor. But, irreverently consorting with these grave, reputable, and pious people, these elders of the church, these chaste dames and dewy virgins, there were men of dissolute lives and women of spotted fame, wretches given over to all mean and filthy vice, and suspected even of horrid crimes. It was strange to see that the good shrank not from the wicked, nor were the sinners abashed by the saints. Scattered also among their pale-faced enemies were the Indian priests, or powwows, who had often scared their native forest with more hideous incantations than any known to English witchcraft.

"But where is Faith!" thought Goodman Brown; and, as hope came into his heart, he trembled.

Another verse of the hymn arose, a slow and mournful strain, such as the pious love, but joined to the words which expressed all that our nature can conceive of sin, and darkly hinted at far more. Unfathomable to mere mortals is the lore of fiends. Verse after verse was sung; and still the chorus of the desert swelled between like the deepest tone of a mighty organ; and with the final peal of that dreadful anthem there came a sound, as if the roaring wind, the rushing streams, the howling beasts, and every other voice of the unconverted wilderness were mingling and according with the voice of guilty man in homage to the prince of all. The four blazing pines threw up a loftier flame, and obscurely discovered shapes and visages of horror on the smoke wreaths above the impious assembly. At the same moment the fire on the rock shot redly forth and formed a glowing arch above its base, where now appeared a figure. With reverence be it spoken, the apparition bore no slight similitude, both in garb and manner, to some grave divine of the New England churches.

"Bring forth the converts!" cried a voice that echoed through the field and rolled into the forest.

At the word, Goodman Brown stepped forth from the shadow of the trees and approached the congregation, with whom he felt a loathful brotherhood by the sympathy of all that was wicked in his heart. He could have well-nigh sworn that the shape of his own dead father beckoned him to advance, looking downward from a smoke wreath, while a woman, with dim features of despair, threw out her hand to warn him back. Was it his mother? But he had no power to retreat one step, nor to resist, even in thought, when the minister and good old Deacon Gookin seized his arms and led him to the blazing rock. Thither came also the slender form of a veiled female, led between Goody Cloyse, that pious teacher of the catechism, and Martha Carrier, who had received the devil's promise to be queen of hell. A rampant hag was she. And there stood the proselytes beneath the canopy of fire.

"Welcome, my children," said the dark figure, "to the communion of your race. Ye have found thus young your nature and your destiny. My children, look behind you!"

They turned; and flashing forth, as it were, in a sheet of flame, the fiend worshippers were seen; the smile of welcome gleamed darkly on every visage.

"There," resumed the sable form, "are all whom ye have reverenced from youth. Ye deemed them holier than yourselves, and shrank from your own sin, contrasting it with their lives of righteousness and prayerful aspirations heavenward. Yet here are they all in my worshipping assembly. This night it shall be granted you to know their secret deeds: how hoary-bearded elders of the church have whispered wanton words to the young maids of their households; how many a woman, eager for widows' weeds, has given her husband a drink at bedtime and let him sleep his last sleep in her bosom; how beardless youths have made haste to inherit their fathers' wealth; and how fair damsels—blush not, sweet ones—have dug little graves in the garden, and bidden me, the sole guest, to an infant's funeral. By the sympathy of your human hearts for sin ye shall scent out all the places— whether in church, bed-chamber, street, field, or forest where crime has been committed, and shall exult to behold the whole earth one stain of guilt, one mighty blood spot. Far more than this. It shall be yours to penetrate, in every bosom, the deep mystery of sin, the fountain of all wicked arts, and which inexhaustibly supplies more evil impulses than human power—than my power at its utmost—can make manifest in deeds. And now, my children, look upon each other."

They did so; and, by the blaze of the hell-kindled torches, the wretched man beheld his Faith, and the wife her husband, trembling before that unhallowed altar.

"Lo, there ye stand, my children," said the figure, in a deep and solemn tone, almost sad with its despairing awfulness, as if his once angelic nature could yet mourn for our miserable race. "Depending upon one another's hearts, ye had still hoped that virtue were not all a dream. Now are ye undeceived. Evil is the nature of mankind. Evil must be your only happiness. Welcome again, my children, to the communion of your race."

"Welcome," repeated the fiend worshippers in one cry of despair and triumph.

And there they stood, the only pair, as it seemed, who were yet hesitating on the verge of wickedness in this dark world. A basin was hollowed, naturally, in the rock. Did it contain water, reddened by the lurid light? or was it blood? or, perchance, a liquid flame? Herein did the shape of evil dip his hand and prepare to lay the mark of baptism upon their foreheads, that they might be partakers of the mystery of sin, more conscious of the secret guilt of others, both in deed and thought, than they could now be of their own. The husband cast one look at his pale wife, and Faith at him. What polluted wretches would the next glance show them to each other, shuddering alike at what they disclosed and what they saw!

"Faith! Faith!" cried the husband, "look up to heaven, and resist the wicked one."

Whether Faith obeyed he knew not. Hardly had he spoken when he found himself amid calm night and solitude, listening to a roar of the wind which died heavily away through the forest. He staggered against the rock, and felt it chill and damp; while a hanging twig, that had been all on fire, besprinkled his cheek with the coldest dew.

The next morning young Goodman Brown came slowly in to the street of Salem village, staring around him like a bewildered man. The good old minister was taking a walk along the graveyard to get an appetite for breakfast and meditate his sermon, and bestowed a blessing, as he passed, on Goodman Brown. He shrank from the venerable saint as if to avoid an anathema. Old Deacon Gookin was at domestic worship, and the holy words of his prayer were heard through the open window. "What God doth the wizard pray to?" quoth Goodman Brown. Goody Cloyse, that excellent old Christian, stood in the early sunshine at her own lattice, catechizing a little girl who had brought her a pint of morning's milk. Goodman Brown snatched away the child as from the grasp of the fiend himself. Turning the corner by the meetinghouse, he spied the head of Faith, with the pink ribbons, gazing anxiously forth, and bursting into such joy at sight of him that she skipped along the street and almost kissed her husband before the whole village. But Goodman Brown looked sternly and sadly into her face, and passed on without a greeting.

Had Goodman Brown fallen asleep in the forest and only dreamed a wild dream of a witch-meeting?

Be it so if you will; but, alas! it was a dream of evil omen for young Goodman Brown. A stern, a sad, a darkly meditative, a distrustful, if not a desperate man did he become from the night of that fearful dream. On the Sabbath day, when the congregation were singing a holy psalm, he could not listen because an anthem of sin rushed loudly upon his ear and drowned all the blessed strain. When the minister spoke from the pulpit with power and fervid eloquence, and, with his hand on the open Bible, of the sacred truths of our religion, and of saint-like lives and triumphant deaths, and of future bliss or misery unutterable, then did Goodman Brown turn pale, dreading lest the roof should thunder down upon the gray blasphemer and his hearers. Often, awaking suddenly at midnight, he shrank from the bosom of Faith; and at morning or eventide, when the family knelt down at prayer, he scowled and muttered to himself, and gazed sternly at his wife, and turned away. And when he had lived long, and was borne to his grave a hoary corpse, followed by Faith, an aged woman, and children and grandchildren, a goodly procession, besides neighbors not a few, they carved no hopeful verse upon his tombstone, for his dying hour was gloom.

Bibliography

We have found the following works to be particularly useful in our own study and teaching of the various critical approaches discussed in the foregoing pages. As with this book, our list is intended to be suggestive rather than exhaustive. Generally speaking, the numerous important works cited in individual chapters are not repeated in this bibliography.

Works of General Interest and Surveys of Critical History

Atkins, G. Douglass, and Laura Morrow, eds. *Contemporary Literary Theory*. Amherst: U of Massachusetts, 1989.

Baldrick, Chris. *Concise Oxford Dictiomary of Literary Terms*. New York: Oxford UP, 1990.

Daiches, David. *Critical Approaches to Literature*. Englewood Cliffs: Prentice, 1956.

Holman, C. Hugh, and William Harmon, eds. *A Handbook to Literature*. 5th ed. New York: Macmillan, 1986.

Hyman, Stanley Edgar. *The Armed Vision: A Study of the Methods of Literary Criticism*. Rev. ed. New York: Random, 1955.

Leitch, Vincent B. *American Literary Criticism from the Thirties to the Eighties*. New York: Columbia UP, 1988.

Lentricchia, Frank. *After the New Criticism*. Chicago: U of Chicago P, 1980.

Lipking, Lawrence I., and A. Walton Litz, eds. *Modern Literary Criticism, 1900–1970*. New York: Atheneum, 1972.

Nelson, Cary, ed. *Theory in the Classroom*. Urbana: U of Illinois P, 1986.

Polletta, Gregory T., ed. *Issues in Contemporary Literary Criticism.* Boston: Little, 1973.

Renza, Louis A. *"A White Heron" and the Question of Minor Literature.* Madison: U of Wisconsin P, 1984.

Scott, Wilbur, ed. *Five Approaches to Literary Criticism.* New York: Macmillan, 1962.

Selden, Raman. *A Reader's Guide to Contemporary Literary Theory.* Lexington: UP of Kentucky, 1985.

Thorpe, James, ed. *Relations of Literary Study: Essays on Interdisciplinary Study.* New York: MLA, 1967.

Wimsatt, William K., Jr., and Cleanth Brooks. *Literary Criticism: A Short History.* New York: Knopf, 1957.

Works Helpful for Individual Approaches

Abel, Elizabeth, ed. Special Issue on Writing and Sexual Difference. *Critical Inquiry,* 8 (1981). Reprinted as *Writing and Sexual Difference.* Chicago: U of Chicago P, 1983.

Adams, Hazard. *Philosophy of the Literary Symbolic.* Tallahassee: U of Florida P, 1983.

Altick, Richard D. *The Art of Literary Research.* Rev. ed. New York: Norton, 1975.

Aristotle. *Poetics* (available in several paperback editions; for example, U of North Carolina P, U of Michigan P, Dutton [Everyman]).

Auden, W. H. *The Enchafed Flood, or The Romantic Iconography of the Sea.* London: Faber, 1951.

Auerbach, Nina. *Woman and the Demon: The Life of a Victorian Myth.* Cambridge: Harvard UP, 1982.

Baird, James. *Ishmael: A Study of the Symbolic Mode in Primitivism.* Baltimore: Johns Hopkins UP, 1956.

Barchilon, Jose, and Joel S. Kovel. Huckleberry Finn: A Psychoanalytic Study." *Journal of the American Psychoanalytic Association* 14 (Oct. 1966): 775–814.

Basler, Roy P. *Sex, Symbolism, and Psychology in Literature.* 1948; New York: Octagon, 1967.

Baym, Nina. *Woman's Fiction: A Guide to Novels By and About Women in America, 1820–1870.* Ithaca: Cornell UP, 1978.

Bell, Roseann, P., et al., eds. *Sturdy Black Bridges: Visions of Black Women in Literature.* New York: Anchor-Doubleday, 1979.

Bercovitch, Sacvan, ed. *Reconstructing American Literary History.* Cambridge: Harvard UP, 1986.

Bodkin, Maud. *Archetypal Patterns in Poetry: Psychological Studies of Imagination.* 1934; New York: Random, 1958.

Bonaparte, Marie. *The Life and Works of Edgar Allan Poe: A Psychoanalytic Interpretation.* Trans. John Rodker. London: Imago, 1949.

Booth, Wayne C. "Freedom of Interpretation: Bakhtin and the Challenge of Feminist Criticism." *Critical Inquiry* 9 (1982): 45–76.

Bradbury, John M. *The Fugitives: A Critical Account.* Chapel Hill: U of North Carolina P, 1958.

Bradley, A. C. *Shakespearean Tragedy.* London: Macmillan, 1941.

Brooks, Cleanth. *The Well Wrought Urn: Studies in the Structure of Poetry.* Rev. ed. London: Dobson, 1968.

Brooks, Cleanth, and Robert B. Heilman, eds. *Understanding Drama: Twelve Plays.* New York: Holt, 1948.

Brooks, Cleanth, and Robert Penn Warren, eds. *Understanding Fiction.* 2nd ed. Englewood Cliffs: Prentice, 1959.

———. *Understanding Poetry.* 3rd ed. New York: Holt, 1960.

Brooks, Van Wyck. *The Ordeal of Mark Twain.* New York: Dutton, 1920.

Burke, Kenneth. *Counter-Statement.* Rev. ed. Los Altos, CA: Hermes, 1953.

Cain, William E. *The Crisis in Criticism: Theory, Literature, and Reform in English Studies.* Baltimore: Johns Hopkins UP, 1984.

———. "Theory and Practice in Contemporary Criticism." *The CEA Critic* 49 (Winter 1986–Summer 1987): 3–17.

Campbell, Joseph. *The Flight of the Wild Gander: Explorations in the Mythological Dimension.* New York: Viking, 1969.

———. *The Hero with a Thousand Faces.* Rev. ed. Princeton: Princeton UP, 1968.

Carpenter, Frederic I. *American Literature and the Dream.* New York: Philosophical, 1955.

Christian, Barbara. *Black Feminist Criticism: Perspectives on Black Women Writers.* New York: Pergamon, 1985.

Cirlot, J. E. *A Dictionary of Symbols.* Trans. Jack Sage. New York: Philosophical, 1962.

Cixous, Hélène, and Catherine Clement. *The Newly Born Woman.* Trans. Betsy Wing. Intro. by Sandra M. Gilbert. Minneapolis: U of Minnesota P, 1986.

Coleridge, Samuel Taylor. *Biographia Literaria,* 1817 (available in collections of Coleridge's work and in anthologies of English literature, especially English Romanticism).

Cowan, Louise. *The Fugitive Group: A Literary History.* Baton Rouge: Louisiana State UP, 1950.

Crews, Frederick C. *The Sins of the Fathers: Hawthorne's Psychological Themes.* New York: Oxford UP, 1966.

———, ed. *Psychoanalysis and Literary Process.* Cambridge: Winthrop, 1970.

———. *Out of My System: Psychoanalysis, Ideology, and Critical Method.* New York: Oxford UP, 1975.

———. "The Future of an Illusion." *The New Republic* Jan. 21, 1985: 28–33.

Culler, Jonathan. *Structuralist Poetics: Structuralism, Linguistics, and the Study of Literature.* Ithaca: Cornell UP, 1975.

———. *On Deconstruction: Theory and Criticism after Structuralism.* Ithaca: Cornell UP, 1982.

De George, Richard T. and Fernande M., eds. *The Structuralists: From Marx to Lévi-Strauss.* Garden City: Doubleday, 1972.

De Man, Paul. *Allegories of Reading: Figural Language in Rousseau, Nietzsche, Rilke, and Proust.* New Haven: Yale UP, 1979.

———. *Blindness and Insight: Essays in the Rhetoric of Contemporary Criticism.* New York: Oxford UP, 1971.

Drew, Elizabeth. *Discovering Poetry.* New York: Norton, 1933.

Eliot, T. S. *The Sacred Wood: Essays on Poetry and Criticism.* London: Methuen, 1920.

Faderman, Lillian. *Surpassing the Love of Men: Romantic Friendship and Love Between Women From the Renaissance to the Present.* New York: Morrow, 1981.

Felman, Shoshana. *Jacques Lacan and the Adventure of Insight: Psychoanalysis in Contemporary Culture.* Cambridge: Harvard UP, 1987.

———, ed. *Literature and Psychoanalysis—The Question of Reading: Otherwise.* Baltimore: Johns Hopkins UP, 1982.

Fiedler, Leslie. *An End to Innocence: Essays on Culture and Politics.* Boston: Beacon, 1955.

———. *What Was Literature? Class Culture and Mass Society.* New York: Simon, 1982.

Fogle, Richard Harter. *Hawthorne's Fiction: The Light and the Dark.* Rev. ed. Norman: U of Oklahoma P, 1964.

Frazer, Sir James George. *The Golden Bough: A Study in Magic and Religion.* One vol. abridged ed. New York: Macmillan, 1958.

Freud, Sigmund. *New Introductory Lectures on Psychoanalysis.* Trans. and ed. James Strachey. New York: Norton, 1933.

———. *On Dreams.* Trans. and ed. James Strachey. New York: Norton, 1949.

Friedman, Albert B. *Myth, Symbolic Modes and Ideology: A Discursive Bibliography*. Claremont, CA: Claremont Graduate School, 1976.

Fromm, Erich. *The Forgotten Language: An Introduction to the Understanding of Dreams, Fairy Tales and Myths*. New York: Grove,1957.

Frye, Northrop. *Anatomy of Criticism: Four Essays*. Princeton: Princeton UP, 1957.

Gallop, Jane. *Feminism and Psychoanalysis: The Daughter's Seduction*. London: Macmillan, 1982.

———. "The Ladies' Man." *Diacritics* 6 (1976): 28–34.

Garner, Shirley Nelson, Claire Kahane, and Madelon Sprengnether, eds. *The (M)other Tongue: Essays in Feminist Psychoanalytic Interpretation*. Ithaca: Cornell UP, 1985.

Gilbert, Sandra, and Susan Gubar. *The Madwoman in the Attic*. New Haven: Yale UP, 1979.

Gottesman, Ronald, and Scott Bennett, eds. *Art and Error: Modern Textual Editing*. Bloomington: Indiana UP, 1970.

Graff, Gerald, and Reginald Gibbons, eds. *Criticism in the University*. Evanston: Northwestern UP, 1985.

Handy, William J. *Kant and the Southern New Critics*. Austin: U of Texas P, 1963.

Hartman, Geoffrey H. *Criticism in the Wilderness: The Study of Literature Today*. New Haven: Yale UP, 1980.

Heilbrun, Carolyn. *Toward a Recognition of Androgyny*. New York: Harper Colophon, 1973.

Hoffman, Daniel. *Form and Fable in American Fiction*. New York: Oxford UP, 1965.

Hoffman, Frederick J. *Freudianism and the Literary Mind*. 2nd ed. Baton Rouge: Louisiana State UP, 1957.

Holland, Norman N. *The Dynamics of Literary Response*. New York: Oxford UP, 1968.

Howe, Florence. *Myths of Coeducation: Selected Essays, 1964–1983*. Bloomington: Indiana UP, 1984.

Humm, Maggie. *Feminist Criticism: Women as Contemporary Critics*. Brighton, Eng.: Harvester, 1986.

Jacobi, Jolande. *The Psychology of C. G. Jung*. Trans. Ralph Manheim. Rev. ed. New Haven: Yale UP, 1962.

James, Henry. *Theory of Fiction*. Ed. James E. Miller, Jr. Lincoln: U of Nebraska P, 1972.

Jameson, Fredric. *The Prison House of Language: A Critical Account of Structuralism and Russian Formalism*. Princeton: Princeton UP, 1972.

Jardine, Alice. "Genesis." *Diacritics* 12 (1982): 56–61.

Johnson, Samuel. *Lives of the Poets.* New York: Dutton, 1925, 2 vols.

Jones, Ann Rosalind. "Writing the Body: Toward an Understanding of *l'écriture féminine.*" *Feminist Studies* 7, 2 (1981): 247–63.

Jung, C. G. *The Archetypes and the Collective Unconscious.* Vol. 9, part 1 of the *Collected Works.* Trans. R. F. C. Hull. 2d ed. Princeton: Princeton UP, 1968.

———. *Modern Man in Search of a Soul.* Trans. W. S. Dell and Cary F. Baynes. 1933; New York: Harcourt, n. d.

———, ed. *Man and His Symbols.* Garden City: Doubleday, 1964.

Jung, C. G., and C. Kerenyi. *Essays on a Science of Mythology: The Myths of the Divine Child and the Divine Maiden.* New York: Pantheon, 1949.

Karanikas, Alexander. *Tillers of a Myth: Southern Agrarians as Social and Literary Critics.* Madison: U of Wisconsin P, 1966.

Kenner, Hugh. *The Art of Poetry.* New York: Holt, 1959.

Kirk, G. S. *Myth: Its Meaning and Functions in Ancient and Other Cultures.* Berkeley: U of California P, 1970.

———. *The Nature of Greek Myths.* Woodstock, NY: Overlook, 1975.

Kolodny, Annette. "A Map of Misreading: Or, Gender and the Interpretation of Literary Texts." *New Literary History* 11 (1980): 451–67.

———. "Some Notes on Defining a 'Feminist Literary Criticism.'" *Critical Inquiry* 2 (1975): 75–92.

Kris, Ernst. *Psychoanalytic Explorations in Art.* New York: Schocken, 1975.

Kristeva, Julia. *Desire in Language: A Semiotic Approach to Literature and Art.* Oxford: Blackwell, 1980; New York: Columbia UP, 1980.

———. "Women's Time." *Signs* 7, 1 (1981): 13–35.

Latimer, Dan, ed. *Contemporary Critical Theory.* San Diego: Harcourt, 1989.

Lauter, Estella, and Carol Schreiner, eds. *Feminist Archetypal Theory.* Knoxville: U of Tennessee P, 1985.

Lentricchia, Frank. *Criticism and Social Change.* Chicago: U of Chicago P, 1983.

Lesser, Simon O. *Fiction and the Unconscious.* Boston: Beacon, 1957.

Lewis, R. W. B. *The American Adam: Innocence, Tragedy, and Tradition in the Nineteenth Century.* Chicago: U of Chicago P, 1955.

Lowes, John Livingston. *The Road to Xanadu: A Study in the Ways of the Imagination*. Boston: Houghton, 1927.

Marcuse, Herbert. "Marxism and Feminism." *Women's Studies* 2 (1974): 279–88.

Marks, Elaine, and Isabelle de Courtivron, eds. *New French Feminisms: An Anthology*. Brighton, Eng.: Harvester; Amherst: U of Massachusetts P; New York: Schocken, 1981.

Miller, J. Hillis. *Fiction and Repetition: Seven English Novels*. Cambridge: Harvard UP, 1982.

———. *The Linguistic Moment: From Wordsworth to Stevens*. Princeton: Princeton UP, 1985.

Morris, Wesley. *Toward a New Historicism*. Princeton: Princeton UP, 1972.

Morrison, Claudia S. *Freud and the Critic: The Early Use of Depth Psychology in Literary Criticism*. Chapel Hill: U of North Carolina P, 1968.

Nethercot, Arthur H. *The Road to Tryermaine: A Study of the History, Background, and Purposes of Coleridge's "Cristabel."* 1939; New York: Russel, 1962.

Neumann, Erich. *The Great Mother: An Analysis of the Archetype*. Trans. Ralph Manheim. Princeton: Princeton UP, 1963.

———. *The Origins and History of Consciousness*. Trans. R. F. C. Hull. Princeton: Princeton UP, 1970.

Ohmann, Richard M., ed. *The Making of Myth*. New York: Putnam, 1962.

Paris, Bernard J. *A Psychological Approach to Fiction: Studies in Thackeray, Stendhal, George Eliot, Dostoevsky, and Conrad*. Bloomington: Indiana UP, 1974.

Pearce, Roy Harvey. *Historicism Once More: Problems and Occasions for the American Scholar*. Princeton: Princeton UP, 1969.

Perrine, Laurence. *Sound and Sense: An Introduction to Poetry*. 7th ed. New York: Harcourt, 1987.

Radway, Janice A. *Reading the Romance: Women, Patriarchy, and Popular Literature*. Chapel Hill: U of North Carolina P, 1984.

Raine, Kathleen. *On the Mythological*. Fullerton, CA: CEA, 1969.

Ransom, John Crowe. *The New Criticism*. New York: New Directions, 1941.

Ray, William. *Literary Meaning: From Phenomenology to Deconstruction*. Oxford: Blackwell. 1984.

Rich, Adrienne. "Compulsory Heterosexuality and Lesbian Existence." In *The SIGNS Reader: Women, Gender, and Scholarship.* Ed. Elizabeth Abel and Emily Abel. Chicago: U of Chicago P, 1983.

Richards, I. A. *Practical Criticism: A Study of Literary Judgment.* New York: Harcourt, 1929.

Richter, David H. *Fable's End: Completeness and Closure in Rhetorical Fiction.* Chicago: U of Chicago P, 1974.

Righter, William. *Myth and Literature.* London: Routledge, 1975.

Robinson, Lillian S. "Dwelling in Decencies: Radical Criticism and the Feminist Perspective." *Critical Exchange* 32 (1971): 879–89.

Ruitenbeek, Hendrik M., ed. *Psychoanalysis and Literature.* New York: Dutton, 1964.

Ruthven, K. K. *Feminist Literary Studies: An Introduction.* Cambridge: Cambridge UP, 1984.

———. "Male Critics and Feminist Criticism." *Essays in Criticism* 33, 4 (1983): 263–72.

Saussure, Ferdinand de. *Course in General Linguistics.* New York: McGraw, 1966.

Sebeok, Thomas A., ed. *Myth: A Symposium.* Bloomington: Indiana UP,1965.

Showalter, Elaine, ed. *The New Feminist Criticism: Essays on Women, Literature and Theory.* New York: Pantheon, 1985.

———. *A Literature of Their Own: British Women Novelists From Brontë to Lessing.* Princeton: Princeton UP, 1977.

———. "Towards a Feminist Poetics." In *Women Writing and Writing About Women.* Ed. by Mary Jacobus. London: Croom Helm, 1979. 22–41.

Shumaker, Wayne. *Literature and the Irrational: A Study in Anthropological Backgrounds.* Englewood Cliffs: Prentice, 1960.

Sidney, Sir Philip. *The Defense of Poesy,* 1580 (available in collections of Sidney's work and in anthologies).

Simon, John K., ed. *Modern French Criticism: From Proust and Valéry to Structuralism.* Chicago: U of Chicago P, 1972.

Skura, Meredith. *The Literary Use of Psychoanalytic Process.* New Haven: Yale UP, 1981.

Slochower, Harry. *Mythopoesis: Mythic Patterns in the Literary Classics.* Detroit: Wayne State UP, 1970.

Smith, Barbara. *Toward a Black Feminist Literary Criticism.* New York: Out and Out Books, 1977.

Smith, Henry Nash. *Virgin Land: The American West as Symbol and Myth.* 1950; New York: Random, 1957; reissued with new preface, Cambridge: Harvard UP, 1970.

Spacks, Patricia Meyer. *The Female Imagination: A Literary and Psychological Investigation of Women's Writing*. New York: Knopf, 1975.

Spitzer, Leo. *Linguistics and Literary History: Essays in Stylistics.* 1948; New York: Russell, 1962.

Spivak, Gayatri Chakravorty. "Displacement and the Discourse of Woman." In *Displacement: Derrida and After.* Ed. Mark Krupnick. Bloomington: Indiana UP, 1983. 169–95.

Spurgeon, Caroline. *Shakespeare's Inagery and What It Tells Us.* 1935; Boston: Beacon, 1958.

Stallman, Robert Wooster, ed. *Critiques and Essays in Criticism, 1920–1948, Representing the Achievement of Modern British and American Critics*. New York: Ronald, 1949.

Stevick, Phillip. *The Chapter in Fiction: Theories of Narrative Division*. Syracuse, NY: Syracuse UP, 1970.

Stimpson, Catherine. "The New Feminism and Women's Studies." *Change* 5 (1973): 43–48.

Storr, Anthony, ed. *The Essential Jung*. Princeton: Princeton UP, 1983.

Taine, H. A. *Histoire de la littérature anglaise*, 1863.

Thorpe, James. *Principles of Textual Criticism*. San Marino, CA: Huntington, 1972.

Ulanov, Ann Belford. *The Feminine in Jungian Psychology and in Christian Theology*. Evanston: Northwestern UP, 1971.

Van Ghent, Dorothy. *The English Novel: Form and Function*. New York: Harper, 1961.

Vendler, Helen. "Feminism and Literature." *The New York Review of Books* 31 May 1990: 19–25.

Vendler, Helen, et al., "Feminism and Literature: An Exchange." *The New York Review of Books* 16 August 1990: 58–59.

Vickery, John B. *The Literary Impact of The Golden Bough*. Princeton: Princeton UP, 1973.

————, ed. *Myth and Literature: Contemporary Theory and Practice*. Lincoln: U of Nebraska P (Bison), 1969.

Walcutt, Charles Child, and J. Edwin Whitesell, eds. *The Explicator Cyclopedia*. 3 vols. New York: Quadrangle, 1968.

Wetherill, P. M. *The Literary Text: An Examination of Critical Methods*. Berkeley: U of California P, 1974.

Wilson, J. Dover. *What Happens in Hamlet*. London: Cambridge UP, 1935.

Wimsatt, W. K., Jr., ed. *The Verbal Icon: Studies in the Meaning of Poetry*. Lexington: U of Kentucky P, 1954.

Wimsatt, W. K., Jr., ed. *Explication as Criticism: Selected Papers* from the English Institute, 1941–1952. New York: Columbia UP, 1963.

Wright, Elizabeth. *Psychoanalytic Criticism: Theory in Practice.* London: Methuen, 1984.

Young, Philip. *Ernest Hemingway: A Reconsideration.* University Park: Pennsylvania State UP, 1966. First published as *Ernest Hemingway.* New York: Holt, 1952.

Index

Abel, Elizabeth, 192
Abrams, M. H., 258
Action code, 249, 250, 252
Adler, Alfred, 119
Aeschylus. See *Agamemnon*
Affective fallacy, 78–79, 335
Agamemnon (Aeschylus), 160
Allegory, 92–93
Allusion
    formalistic approach and, 66
    in "To His Coy Mistress,"
        84–85
    in "Young Goodman
        Brown," 91–92
Alter, Robert, 258
Althusser, Louis, 317, 321
Altick, Richard D., 25
Ambiguity
    Hamlet and, 253–54
    Moby Dick and, 175
    "Young Goodman Brown"
        and, 89–90, 97–98
American Adam concept, 176–
    78. See *also* Hero arche-
    types
American Dream, 175–78
American novel, 330

Anima archetype, 169–71
Antagonist, 9
Anthropology. See *also* Lévi-
    Strauss, Claude
    Jungian psychology and,
        173–75
    mythological criticism and,
        156
Arch, principle of, 81–82
Archetypal approaches. See
    Archetypes; Mythological
    approaches
Archetypal woman, 152–53,
    179
Archetypes, 149–55
    defined, 149–50
    as genres, 155
    images and, 150–54
    Jungian psychology and,
        166–75
    motifs and patterns and, 154,
        159, 178–80
    in Oedipus myth, 159
    in "To His Coy Mistress,"
        164–66
Aristocracy and Twain, 48–49
Aristotelian criticism, 261–64